Multivariate Approximation and Applications

Edited by

N. DYN
Tel Aviv University

D. LEVIATAN
Tel Aviv University

D. LEVIN
Tel Aviv University

A. PINKUS
Technion–Israel Institute of Technology

PUBLISHED BY THE PRESS SYNDICATE OF THE UNIVERSITY OF CAMBRIDGE
The Pitt Building, Trumpington Street, Cambridge, United Kingdom

CAMBRIDGE UNIVERSITY PRESS
The Edinburgh Building, Cambridge, CB2 2RU, UK
40 West 20th Street, New York, NY 10011–4211, USA
10 Stamford Road, Oakleigh, VIC 3166, Australia
Ruiz de Alarcón 13, 28014 Madrid, Spain
Dock House, The Waterfront, Cape Town 8001, South Africa

http://www.cambridge.org

© Cambridge University Press 2001

This book is in copyright. Subject to statutory exception
and to the provisions of relevant collective licensing agreements,
no reproduction of any part may take place without
the written permission of Cambridge University Press.

First published 2001

Printed in the United Kingdom at the University Press, Cambridge

Typeface Computer Modern 10/12pt   *System* LATEX   [UPH]

A catalogue record for this book is available from the British Library

Library of Congress Cataloguing in Publication data
ISBN 0 521 80023 4 hardback

Contents

List of contributors		*page* v
Preface		vii
1	Characterization and construction of radial basis functions	
	R. Schaback and H. Wendland	1
2	Approximation and interpolation with radial functions	
	M.D. Buhmann	25
3	Representing and analyzing scattered data on spheres	
	H.N. Mhaskar, F.J. Narcowich and J.D. Ward	44
4	A survey on L_2-approximation orders from shift-invariant spaces K. Jetter and G. Plonka	73
5	Introduction to shift-invariant spaces. Linear independence	
	A. Ron	112
6	Theory and algorithms for nonuniform spline wavelets	
	T. Lyche, K. Mørken and E. Quak	152
7	Applied and computational aspects of nonlinear wavelet approximation A. Cohen	188
8	Subdivision, multiresolution and the construction of scalable algorithms in computer graphics P. Schröder	213
9	Mathematical methods in reverse engineering J. Hoschek	252
Index		285

Contributors

M.D. Buhmann
Mathematical Institute, Justus-Liebig University, 35392 Gießen, Germany
email: `Martin.Buhmann@@math.uni-giessen.de`

A. Cohen
Laboratoire d'Analyse Numérique, Université Pierre et Marie Curie, Paris, France
email: `cohen@@ann.jussieu.fr`

J. Hoschek
Department of Mathematics, Darmstadt University of Technology, 64289 Darmstadt, Germany
email: `hoschek@@mathematik.tu-darmstadt.de`

K. Jetter
Institut für Angewandte Mathematik und Statistik, Universität Hohenheim, 70593 Stuttgart, Germany
email: `kjetter@@uni-hohenheim.de`

T. Lyche
Department of Informatics, University of Oslo, P.O. Box 1080 Blindern, 0316 Oslo, Norway
email: `tom@@ifi.uio.no`

H.N. Mhaskar
Department of Mathematics, California State University, Los Angeles, CA 90032, USA
email: `hmhaska@@calstatela.edu`

K. Mørken
Department of Informatics, University of Oslo, P.O. Box 1080 Blindern, 0316 Oslo, Norway
email: `knutm@@ifi.uio.no`

F.J. Narcowich
Department of Mathematics, Texas A&M University, College Station, TX 77843, USA
email: `fnarc@@math.tamu.edu`

G. Plonka
Fachbereich Mathematik, Universität Duisburg, 47048 Duisburg, Germany
email: `plonka@@math.uni-duisburg.de`

E. Quak
SINTEF Applied Mathematics, P.O. Box 124 Blindern, 0314 Oslo, Norway
email: `Ewald.Quak@@math.sintef.no`

A. Ron
Computer Sciences Department, 1210 West Dayton, University of Wisconsin-Madison, Madison, WI 57311, USA
email: `amos@@cs.wisc.edu`

R. Schaback
Institut für Numerische und Angewandte Mathematik, Universität Göttingen, Lotzestraße 16-18, 37083 Göttingen, Germany
email: `schaback@@math.uni-goettingen.de`

P. Schröder
Department of Computer Science, California Institute of Technology, Pasadena, CA 91125, USA
email: `ps@@cs.caltech.ed`

J.D. Ward
Department of Mathematics, Texas A&M University, College Station, TX 77843, USA
email: `jward@@math.tamu.edu`

H. Wendland
Institut für Numerische und Angewandte Mathematik, Universität Göttingen, Lotzestraße 16-18, 37083 Göttingen, Germany
email: `wendland@@math.uni-goettingen.de`

Preface

Multivariate approximation theory is today an increasingly active research area. It deals with a multitude of problems in areas such as wavelets, multi-dimensional splines, and radial-basis functions, and applies them, for example, to problems in computer aided geometric design, geometric modeling, geodesic applications and image analysis. The field is both fascinating and intellectually stimulating since much of the mathematics of the classical univariate theory does not straightforwardly generalize to the multivariate setting which models many real-world problems; so new tools have had to be, and must continue to be, developed.

This advanced introduction to multivariate approximation and related topics consists of nine chapters written by leading experts that survey many of the new ideas and tools and their applications. Each chapter introduces a particular topic, takes the reader to the forefront of research and ends with a comprehensive list of references.

This book will serve as an ideal introduction for researchers and graduate students who wish to learn about the subject and see how it may be applied.

A more detailed description of each chapter follows:

Chapter 1: Characterization and construction of radial basis functions, by R. Schaback (Göttingen) and H. Wendland (Göttingen)

This chapter introduces characterizations of (conditional) positive definiteness and shows how they apply to the theory of radial basis functions. Complete proofs of the (conditional) positive definiteness of practically all relevant basis functions are provided. Furthermore, it is shown how some of these characterizations may lead to construction tools for positive definite functions. Finally, a new construction technique is given which is based on discrete methods which leads to non-radial, even non-translation invariant, local basis functions.

Chapter 2: Approximation and interpolation with radial functions, by M.D. Buhmann (Giessen)

This chapter provides a short, up-to-date survey of some of the recent developments in the research of radial basis functions. Among these new developments are results on convergence rates of interpolation with radial basis functions, recent contributions concerning approximation on spheres, and computations of interpolants with Krylov space methods.

Chapter 3: Representing and analyzing scattered data on spheres, by H.N. Mhaskar (California State Univ. at Los Angeles), F.J. Narcowich (Texas A & M) and J.D. Ward (Texas A & M)

Geophysical or meteorological data collected over the surface of the earth via satellites or ground stations will invariably come from scattered sites. There are two extremes in the problems one faces when handling such data. The first is representing sparse data by fitting a surface to it. The second is analyzing dense data to extract features of interest. In this chapter various aspects of fitting surfaces to scattered data are reviewed. Analyzing data is a more recent problem that is currently being addressed via various spherical wavelet schemes, which are discussed along with multilevel schemes. Finally quadrature methods, which arise in many of the wavelet schemes as well as some interpolation methods, are touched upon.

Chapter 4: A survey on L_2-approximation orders from shift-invariant spaces, by K. Jetter (Hohenheim) and G. Plonka (Duisburg)

The aim of this chapter is to provide a self-contained introduction to notions and results connected with the L_2-approximation order of finitely generated shift-invariant spaces. Special attention is given to the principal shift-invariant case, where the shift-invariant space is generated from the multi-integer translates of a single generator. This case is of special interest because of its possible applications in wavelet methods. The general finitely generated shift-invariant space case is considered subject to a stability condition being satisfied, and the recent results on so-called superfunctions are developed. For the case of a refinable system of generators, the sum rules for the matrix mask and the zero condition for the mask symbol, as well as invariance properties of the associated subdivision and transfer operators, are discussed.

Chapter 5: Introduction to shift-invariant spaces. Linear independence, by A. Ron (Madison)

Shift-invariant spaces play an increasingly important role in various areas of mathematical analysis and its applications. They appear either implicitly or explicitly in studies of wavelets, splines, radial basis function approximation, regular sampling, Gabor systems, uniform subdivision schemes, and perhaps in some other areas. One must keep in mind, however, that the shift-invariant system explored in one of the above-mentioned areas might be very different from those investigated in others. The theory of shift-invariant spaces attempts to provide a uniform platform for all these different investigations. The two main pillars of that theory are the study of the *approximation properties* of shift-invariant spaces, and the study of *generating sets* for such spaces. Chapter 4 had already provided an excellent up-to-date account of the first topic. The present chapter is devoted to the second topic, and its goal is to provide the reader with an easy and friendly introduction to the basic principles of that topic. The core of the presentation is devoted to the study of local principal shift-invariant spaces, while the more general cases are treated as extensions of that basic setup.

Chapter 6: Theory and algorithms for nonuniform spline wavelets, by T. Lyche (Oslo), K. Mørken (Oslo), E. Quak (Oslo)

This chapter discusses mutually orthogonal spline wavelet spaces on nonuniform partitions of a bounded interval, addressing the existence, uniqueness and construction of bases of minimally supported spline wavelets. The relevant algorithms for decomposition and reconstruction are considered as well as some questions related to stability. In addition, a brief review is given of the bivariate case for tensor products and arbitrary triangulations. The chapter concludes with a discussion of some special cases.

Chapter 7: Applied and computational aspects of nonlinear wavelet approximation, by A. Cohen (Paris)

Nonlinear approximation is recently being applied to computational applications such as data compression, statistical estimation and adaptive schemes for partial differential and integral equations, especially through the development of wavelet-based methods. The goal of this chapter is to provide a short survey of nonlinear wavelet approximation from the perspective of these applications, as well as to highlight some remaining open questions.

Chapter 8: Subdivision, multiresolution and the construction of scalable algorithms in computer graphics, by P. Schröder (Caltech)

Multiresolution representations are a critical tool in addressing complexity issues (time and memory) for the large scenes typically found in computer graphics applications. Many of these techniques are based on classical subdivision techniques and their generalizations. In this chapter we review two exemplary applications from this area: multiresolution surface editing and semi-regular remeshing. The former is directed towards building algorithms which are fast enough for interactive manipulation of complex surfaces of arbitrary topology. The latter is concerned with constructing smooth parameterizations for arbitrary topology surfaces as they typically arise from 3D scanning techniques. Remeshing such surfaces then allows the use of classical subdivision ideas. The particular focus here is on the practical aspects of making the well-understood mathematical machinery applicable and accessible to the very general settings encountered in practice.

Chapter 9: Mathematical methods in reverse engineering, by J. Hoschek (Darmstadt)

In many areas of industrial applications it is desirable to create a computer model of existing objects for which no such model is available. This process is called *reverse engineering*. Reverse engineering typically starts with digitizing an existing object. These discrete data must then be converted into smooth surface models. This chapter provides a survey of the practical algorithms including triangulation, segmentation, feature lines detection, B-spline approximation and trimming.

No book of this sort can be compiled without the help of many people. First and foremost there are the authors. We thank each of them for their efforts, and for their willingness to humor the whims of the editors. Secondly we thank Diana Yellin, our TEX expert, for bringing the various manuscripts into a uniform format, and for her typing, retyping, retyping and patience.

N. Dyn
D. Leviatan
D. Levin
A. Pinkus

1
Characterization and construction of radial basis functions

R. SCHABACK and H. WENDLAND

Abstract

We review characterizations of (conditional) positive definiteness and show how they apply to the theory of radial basis functions. We then give complete proofs for the (conditional) positive definiteness of all practically relevant basis functions. Furthermore, we show how some of these characterizations may lead to construction tools for positive definite functions. Finally, we give new construction techniques based on discrete methods which lead to non-radial, even non-translation invariant, local basis functions.

1.1 Introduction

Radial basis functions are an efficient tool for solving multivariate scattered data interpolation problems. To interpolate an unknown function $f \in C(\Omega)$ whose values on a set $X = \{x_1, \ldots, x_N\} \subset \Omega \subset \mathbb{R}^d$ are known, a function of the form

$$s_{f,X}(x) = \sum_{j=1}^{N} \alpha_j \Phi(x, x_j) + p(x) \qquad (1.1)$$

is chosen, where p is a low degree polynomial and $\Phi : \Omega \times \Omega \to \mathbb{R}$ is a fixed function. The numerical treatment can be simplified in the special situations

(i) $\Phi(x, y) = \phi(x - y)$ with $\phi : \mathbb{R}^d \to \mathbb{R}$ (*translation invariance*),
(ii) $\Phi(x, y) = \phi(\|x - y\|_2)$ with $\phi : [0, \infty) \to \mathbb{R}$ (*radiality*),

and this is how the notion of *radial basis functions* arose. The most prominent examples of radial basis functions are:

$$\phi(r) = r^\beta, \quad \beta > 0, \quad \beta \notin 2\mathbb{N};$$
$$\phi(r) = r^{2k} \log(r), \quad k \in \mathbb{N} \quad \text{(thin-plate splines)};$$

$$\phi(r) = (c^2 + r^2)^\beta, \quad \beta < 0, \quad \text{(inverse multiquadrics)};$$
$$\phi(r) = (c^2 + r^2)^\beta, \quad \beta > 0, \quad \beta \notin \mathbb{N} \quad \text{(multiquadrics)};$$
$$\phi(r) = e^{-\alpha r^2}, \quad \alpha > 0 \quad \text{(Gaussians)};$$
$$\phi(r) = (1-r)_+^4 (1 + 4r).$$

All of these basis functions can be uniformly classified using the concept of (conditionally) positive definite functions:

Definition 1.1.1 A continuous function $\Phi : \Omega \times \Omega \to \mathbb{C}$ is said to be *conditionally positive (semi-)definite of order m* on Ω if for all $N \in \mathbb{N}$, all distinct $x_1, \ldots, x_N \in \Omega$, and all $\alpha \in \mathbb{C}^N \setminus \{0\}$ satisfying

$$\sum_{j=1}^{N} \alpha_j p(x_j) = 0 \tag{1.2}$$

for all polynomials p of degree less than m, the quadratic form

$$\sum_{j=1}^{N} \sum_{k=1}^{N} \alpha_j \overline{\alpha_k} \Phi(x_j, x_k) \tag{1.3}$$

is positive (nonnegative). The function Φ is *positive definite* if it is conditionally positive definite of order $m = 0$.

Note that in case of a positive definite function the conditions (1.2) are empty and hence (1.3) has to be positive for all $\alpha \in \mathbb{C}^N \setminus \{0\}$. Finally, if Φ is a symmetric real-valued function, it is easy to see that it suffices to test only real α.

The use of this concept in the context of multivariate interpolation problems is explained in the next theorem, which also shows the connection between the degree of the polynomial p in (1.1) and the order m of conditional positive definiteness of the basis function Φ. We will denote the space of d-variate polynomials of degree at most m by $\pi_m(\mathbb{R}^d)$.

Theorem 1.1.2 Suppose Φ is conditionally positive definite of order m on $\Omega \subseteq \mathbb{R}^d$. Suppose further that the set of centers $X = \{x_1, \ldots, x_N\} \subseteq \Omega$ is $\pi_{m-1}(\mathbb{R}^d)$ unisolvent, i.e. the zero polynomial is the only polynomial from $\pi_{m-1}(\mathbb{R}^d)$ that vanishes on X. Then, for given $f_1, \ldots, f_N$, there is exactly one function $s_{f,X}$ of the form (1.1) with a polynomial $p \in \pi_{m-1}(\mathbb{R}^d)$ such that $s_{f,X}(x_j) = f_j$, $1 \leq j \leq N$ and $\sum_{j=1}^{N} \alpha_j q(x_j) = 0$ for all $q \in \pi_{m-1}(\mathbb{R}^d)$.

It is the goal of this chapter to give full proofs for the conditional positive definiteness of all the aforementioned radial basis functions and to use the ideas behind these proofs to construct new ones. We only rely on certain analytical tools that are not directly related to radial basis functions.

1.2 The Schoenberg–Micchelli characterization

Given a continuous univariate function $\phi : [0,\infty) \to \mathbb{R}$ we can form the function $\Phi(x,y) := \phi(\|x-y\|_2)$ on $\mathbb{R}^d \times \mathbb{R}^d$ for arbitrary space dimension d. Then we can say that ϕ is conditionally positive definite of order m on $\mathbb{R}^d$, iff Φ is conditionally positive definite of order m on $\mathbb{R}^d$ in the sense of Definition 1.1.1.

Taking this point of view, we are immediately led to the question of whether a univariate function ϕ is conditionally positive definite of some order m on $\mathbb{R}^d$ *for all* $d \geq 1$. This question was fully answered in the positive definite case by Schoenberg (1938) in terms of completely monotone functions. In the case of conditionally positive definite functions, Micchelli (1986) generalized the sufficiency part of Schoenberg's result, suspecting that it was also necessary. This was finally proved by Guo et al. (1993).

Definition 1.2.1 A function $\phi : (0,\infty) \to \mathbb{R}$ is said to be *completely monotone on* $(0,\infty)$ if $\phi \in C^\infty(0,\infty)$ and

$$(-1)^\ell \phi^{(\ell)}(r) \geq 0, \qquad \ell \in \mathbb{N}_0, \quad r > 0. \tag{1.4}$$

A function $\phi : [0,\infty) \to \mathbb{R}$ is said to be *completely monotone on* $[0,\infty)$ if it is completely monotone on $(0,\infty)$ and continuous at zero.

Theorem 1.2.2 (Schoenberg) *Suppose $\phi : [0,\infty) \to \mathbb{R}$ is not the constant function. Then ϕ is positive definite on every $\mathbb{R}^d$ if and only if the function $t \mapsto \phi(\sqrt{t})$, $t \in [0,\infty)$ is completely monotone on $[0,\infty)$.*

Schoenberg's characterisation of positive definite functions allows us to prove the positive definiteness of Gaussians and inverse multiquadrics without difficulty:

Theorem 1.2.3 *The Gaussians $\phi(r) = e^{-\alpha r^2}$, $\alpha > 0$, and the inverse multiquadrics $\phi(r) = (c^2 + r^2)^\beta$, $c > 0$, $\beta < 0$, are positive definite on $\mathbb{R}^d$ for all $d \geq 1$.*

Proof For the Gaussians note that

$$f(r) := \phi(\sqrt{r}) = e^{-\alpha r}$$

satisfies $(-1)^\ell f^{(\ell)}(r) = \alpha^\ell e^{-\alpha r} > 0$ for all $\ell \in \mathbb{N}_0$ and $\alpha, r > 0$. Similarly, for the inverse multiquadrics we find, with $f(r) := \phi(\sqrt{r}) = (c^2+r)^{-|\beta|}$, that
$$(-1)^\ell f^{(\ell)}(r) = (-1)^{2\ell}|\beta|(|\beta|+1)\cdots(|\beta|+\ell-1)(r+c^2)^{-|\beta|-\ell} > 0.$$

Since in both cases ϕ is not the constant function, the Gaussians and inverse multiquadrics are positive definite. $\square$

There are several other characterizations of completely monotone functions (see Widder (1946)), which by Schoenberg's theorem also apply to positive definite functions. The most important is the following one by Bernstein (see Widder (1946)). It implies that the proper tool for handling positive definite functions on $\mathbb{R}^d$ for all $d \geq 1$, is the Laplace transform.

Theorem 1.2.4 (Bernstein) *A function ϕ is positive definite on $\mathbb{R}^d$ for all $d \geq 1$, if and only if there exists a nonzero, finite, nonnegative Borel measure μ, not supported in zero, such that ϕ is of the form*
$$\phi(r) = \int_0^\infty e^{-r^2 t} d\mu(t). \tag{1.5}$$

Note that the sufficient part of Bernstein's theorem is easy to prove if we know that the Gaussians are positive definite. For every $\alpha \in \mathbb{R}^N \setminus \{0\}$ and every distinct $x_1, \ldots, x_N \in \mathbb{R}^d$ the quadratic form is given by
$$\sum_{j,k=1}^N \alpha_j \alpha_k \phi(\|x_j - x_k\|_2) = \int_0^\infty \left|\sum_{j=1}^N \alpha_j e^{-t\|x_j-x_k\|_2^2}\right|^2 d\mu(t).$$

Another consequence of this theory is the following.

Theorem 1.2.5 *Suppose $\phi : [0,\infty) \to \mathbb{R}$ is positive definite on $\mathbb{R}^d$ for all $d \geq 1$. Then ϕ has no zero. In particular, there exists no compactly supported univariate function that is positive definite on $\mathbb{R}^d$ for all $d \geq 1$.*

Proof Since ϕ is positive definite on $\mathbb{R}^d$ for all $d \geq 1$, there exists a finite, nonzero, nonnegative Borel measure μ on $[0,\infty)$ such that (1.5) holds. If r_0 is a zero of ϕ this gives
$$0 = \int_0^\infty e^{-r_0^2 t} d\mu(t).$$
Since the measure is nonnegative and the weight function $e^{-r_0^2 t}$ is positive we find that the measure must be the zero measure. $\square$

Thus the compactly supported function $\phi(r) = (1-r)_+^4(1+4r)$ given in the introduction cannot be positive definite on $\mathbb{R}^d$ for all $d \geq 1$, and it is actually only positive definite on $\mathbb{R}^d$, $d \leq 3$. If one is interested in constructing basis functions with compact support, one has to take into account the above negative result. We shall see in the next section that the Fourier transform is the right tool to handle positive definite translation-invariant functions on $\mathbb{R}^d$ with a prescribed d. But before that, let us have a look at *conditionally positive definite* functions. We will state only the sufficient part as provided by Micchelli (1986).

Theorem 1.2.6 (Micchelli) *Given a function $\phi \in C[0, \infty)$, define $f = \phi(\sqrt{\cdot})$. If there exists an $m \in \mathbb{N}_0$ such that $(-1)^m f^{(m)}$ is well-defined and completely monotone on $(0, \infty)$, then ϕ is conditionally positive semi-definite of order m on $\mathbb{R}^d$ for all $d \geq 1$. Furthermore, if f is not a polynomial of degree at most m, then ϕ is conditionally positive definite.*

This theorem allows us to classify all the functions in the introduction, with the sole exception of the compactly supported one. However, to comply with the notion of conditional positive definiteness, we shall have to adjust the signs properly. To do this we denote the smallest integer greater than or equal to x by $\lceil x \rceil$.

Theorem 1.2.7 *The multiquadrics $\phi(r) = (-1)^{\lceil \beta \rceil}(c^2 + r^2)^\beta$, $c, \beta > 0$, $\beta \notin \mathbb{N}$, are conditionally positive definite of order $m \geq \lceil \beta \rceil$ on $\mathbb{R}^d$ for all $d \geq 1$.*

Proof If we define $f_\beta(r) = (-1)^{\lceil \beta \rceil}(c^2 + r)^\beta$, we find
$$f_\beta^{(k)}(r) = (-1)^{\lceil \beta \rceil}\beta(\beta - 1) \cdots (\beta - k + 1)(c^2 + r)^{\beta - k},$$
which shows that $(-1)^{\lceil \beta \rceil} f_\beta^{(\lceil \beta \rceil)}(r) = \beta(\beta - 1) \cdots (\beta - \lceil \beta \rceil + 1)(c^2 + r)^{\beta - \lceil \beta \rceil}$ is completely monotone, and that $m = \lceil \beta \rceil$ is the smallest possible choice of m to make $(-1)^m f^{(m)}$ completely monotone. □

Theorem 1.2.8 *The functions $\phi(r) = (-1)^{\lceil \beta/2 \rceil} r^\beta$, $\beta > 0$, $\beta \notin 2\mathbb{N}$, are conditionally positive definite of order $m \geq \lceil \beta/2 \rceil$ on $\mathbb{R}^d$ for all $d \geq 1$.*

Proof Define $f_\beta(r) = (-1)^{\lceil \frac{\beta}{2} \rceil} r^{\frac{\beta}{2}}$ to get
$$f_\beta^{(k)}(r) = (-1)^{\lceil \frac{\beta}{2} \rceil} \frac{\beta}{2}\left(\frac{\beta}{2} - 1\right) \cdots \left(\frac{\beta}{2} - k + 1\right) r^{\frac{\beta}{2} - k}.$$
This shows that $(-1)^{\lceil \frac{\beta}{2} \rceil} f_\beta^{(\lceil \frac{\beta}{2} \rceil)}(r)$ is completely monotone and $m = \lceil \frac{\beta}{2} \rceil$ is the smallest possible choice. □

Theorem 1.2.9 *The thin-plate or surface splines $\phi(r) = (-1)^{k+1} r^{2k} \log(r)$ are conditionally positive definite of order $m = k+1$ on every $\mathbb{R}^d$.*

Proof Since $2\phi(r) = (-1)^{k+1} r^{2k} \log(r^2)$ we set $f_k(r) = (-1)^{k+1} r^k \log(r)$. Then it is easy to see that

$$f_k^{(\ell)}(r) = (-1)^{k+1} k(k-1) \cdots (k-\ell+1) r^{k-\ell} \log(r) + p_\ell(r), \qquad 1 \leq \ell \leq k,$$

where p_ℓ is a polynomial of degree $k - \ell$. This means in particular

$$f_k^{(k)}(r) = (-1)^{k+1} k! \log(r) + c$$

and finally $(-1)^{k+1} f_k^{(k+1)}(r) = k! r^{-1}$ which is obviously completely monotone on $(0, \infty)$. □

1.3 Bochner's characterization

We saw in the last section that the Laplace transform is the right tool for analyzing positive definiteness of radial functions for all space dimensions d. However, we did not prove Schoenberg's and Micchelli's theorems. We also saw that the approach via Laplace transforms excludes functions with compact support, which are desirable from a numerical point of view. To overcome this problem and to work around these theorems, we shall now look at *translation-invariant* positive definite functions on $\mathbb{R}^d$ for some *fixed* d. We shall give the famous result of Bochner (1932,1933), which characterizes translation-invariant positive definite functions via Fourier transforms. In the next section we generalize this result to enable us to also handle translation-invariant *conditionally* positive definite functions, following an approach of Madych and Nelson (1983). Of course, we define a continuous function $\Phi : \mathbb{R}^d \to \mathbb{C}$ to be a translation-invariant conditionally positive (semi-)definite function of order m on $\mathbb{R}^d$ iff $\Phi_0(x,y) := \Phi(x-y)$ is conditionally positive (semi-)definite of order m on $\mathbb{R}^d$.

Theorem 1.3.1 (Bochner) *A continuous function $\Phi : \mathbb{R}^d \to \mathbb{C}$ is a translation-invariant positive semi-definite function if and only if it is the inverse Fourier transform of a finite nonnegative Borel measure μ on $\mathbb{R}^d$, i.e.,*

$$\Phi(x) = \mu^\vee(x) = (2\pi)^{-d/2} \int_{\mathbb{R}^d} e^{ix^T \omega} d\mu(\omega), \qquad x \in \mathbb{R}^d. \tag{1.6}$$

Again, the sufficient part is easy to prove since

$$\sum_{j,k=1}^N \alpha_j \overline{\alpha_k} \Phi(x_j - x_k) = \int_{\mathbb{R}^d} \left| \sum_{j=1}^N \alpha_j e^{ix_j^T \omega} \right|^2 d\mu(\omega), \tag{1.7}$$

and later we shall use this argument repeatedly to prove positive definiteness of certain functions without referring to Bochner's theorem. In the Fourier transform setting it is not straightforward to separate positive definite from positive semi-definite functions as it was in Schoenberg's characterization. But since the exponentials are linearly independent on every open supset of $\mathbb{R}^d$, we have

Corollary 1.3.2 *Suppose that the carrier of the measure μ of Theorem 1.3.1 contains an open subset of $\mathbb{R}^d$. Then Φ is a translation-invariant positive definite function.*

For a complete classification of positive definite functions via Bochner's theorem, see Chang (1996a,1996b). Here, we want to cite a weaker formulation, which we shall not use for proving positive definiteness of special functions. A proof can be found in Wendland (1999).

Theorem 1.3.3 *Suppose $\Phi \in L_1(\mathbb{R}^d)$ is a continuous function. Then Φ is a translation-invariant positive definite function if and only if Φ is bounded and its Fourier transform is nonnegative and not identically zero.*

Since a non-identically zero function cannot have an identically zero Fourier transform, we see that an integrable, bounded function, Φ, that is not identically zero, is translation-invariant and positive definite if its Fourier transform is nonnegative. This can be used to prove the positive definiteness of the Gaussian along the lines of the sufficiency argument for Theorem 1.3.1. Since this is easily done via (1.7), we skip over the details and only remark that

$$\Phi(x) = e^{-\alpha \|x\|_2^2}$$

has the Fourier transform

$$\widehat{\Phi}(\omega) = (2\pi)^{-d/2} \int_{\mathbb{R}^d} \Phi(x) e^{-ix^T \omega} dx = (2\alpha)^{-d/2} e^{-\|\omega\|_2^2/(4\alpha)}. \quad (1.8)$$

This allows us to circumvent Schoenberg's and Bochner's theorem for a direct proof of the positive definiteness of the Gaussians (see also Powell (1987)).

Now let us have a closer look at the Fourier transform of the inverse multiquadrics. To do this let us recall the definition of the modified Bessel functions. For $z \in \mathbb{C}$ with $|\arg(z)| < \pi/2$ they are given by

$$K_\nu(z) := \int_0^\infty e^{-z \cosh t} \cosh \nu t \, dt.$$

Theorem 1.3.4 *The function $\Phi(x) = (c^2 + \|x\|_2^2)^\beta$, $x \in \mathbb{R}^d$, with $c > 0$ and $\beta < -d/2$ is a translation-invariant positive definite function with Fourier transform*

$$\widehat{\Phi}(\omega) = \frac{2^{1+\beta}}{\Gamma(-\beta)} \left(\frac{\|\omega\|_2}{c}\right)^{-\beta-\frac{d}{2}} K_{\frac{d}{2}+\beta}(c\|\omega\|_2).$$

Proof Since $\beta < -d/2$ the function Φ is in $L_1(\mathbb{R}^d)$. From the representation of the Gamma function for $-\beta > 0$ we see that

$$\begin{aligned}\Gamma(-\beta) &= \int_0^\infty t^{-\beta-1} e^{-t} dt \\ &= s^{-\beta} \int_0^\infty u^{-\beta-1} e^{-su} du\end{aligned}$$

by substituting $t = su$ with $s > 0$. Setting $s = c^2 + \|x\|_2^2$ this implies

$$\Phi(x) = \frac{1}{\Gamma(-\beta)} \int_0^\infty u^{-\beta-1} e^{-c^2 u} e^{-\|x\|_2^2 u} du. \qquad (1.9)$$

Inserting this into the Fourier transform and changing the order of integration, which can be easily justified, leads to

$$\begin{aligned}\widehat{\Phi}(x) &= (2\pi)^{-d/2} \int_{\mathbb{R}^d} \Phi(\omega) e^{-ix^T\omega} d\omega \\ &= (2\pi)^{-d/2} \frac{1}{\Gamma(-\beta)} \int_{\mathbb{R}^d} \int_0^\infty u^{-\beta-1} e^{-c^2 u} e^{-\|\omega\|_2^2 u} e^{-ix^T\omega} du\, d\omega \\ &= (2\pi)^{-d/2} \frac{1}{\Gamma(-\beta)} \int_0^\infty u^{-\beta-1} e^{-c^2 u} \int_{\mathbb{R}^d} e^{-\|\omega\|_2^2 u} e^{-ix^T\omega} d\omega\, du \\ &= \frac{1}{\Gamma(-\beta)} \int_0^\infty u^{-\beta-1} e^{-c^2 u} (2u)^{-d/2} e^{-\frac{\|x\|_2^2}{4u}} du \\ &= \frac{1}{2^{d/2} \Gamma(-\beta)} \int_0^\infty u^{-\beta-\frac{d}{2}-1} e^{-c^2 u} e^{-\frac{\|x\|_2^2}{4u}} du,\end{aligned}$$

where we have used (1.8). On the other hand we can conclude from the definition of the modified Bessel function that for every $a > 0$

$$\begin{aligned}K_\nu(r) &= \frac{1}{2} \int_{-\infty}^\infty e^{-r \cosh t} e^{\nu t} dt \\ &= \frac{1}{2} \int_{-\infty}^\infty e^{-\frac{r}{2}(e^t + e^{-t})} e^{\nu t} dt \\ &= a^{-\nu} \frac{1}{2} \int_0^\infty e^{-\frac{r}{2}(\frac{s}{a} + \frac{a}{s})} s^{\nu-1} ds\end{aligned}$$

by substituting $s = ae^t$. If we now set $r = c\|x\|_2$, $a = \|x\|_2/(2c)$, and $\nu = -\beta - d/2$ we obtain

$$K_{-\beta-\frac{d}{2}}(c\|x\|_2) = \frac{1}{2}\left(\frac{\|x\|_2}{2c}\right)^{\frac{d}{2}+\beta} \int_0^\infty e^{-sc^2} e^{-\frac{\|x\|_2^2}{4s}} s^{-\beta-\frac{d}{2}-1} ds$$

$$= 2^{-\beta-1}\Gamma(-\beta)\left(\frac{\|x\|_2}{c}\right)^{\frac{d}{2}+\beta} \widehat{\Phi}(x),$$

which leads to the stated Fourier transform using $K_{-\nu} = K_\nu$. Since the modified Bessel function is nonnegative and non-vanishing, the proof is complete. □

Note that this result is somewhat weaker than the one given in Theorem 1.2.3, since we require $\beta < -d/2$ for integrability reasons. Furthermore, we can read from (1.9) the representing measure for Φ in the sense of Theorem 1.3.1.

1.4 The Madych–Nelson approach

So far we have seen that the Schoenberg–Micchelli approach is an elegant way of proving conditional positive definiteness of basis functions for all space dimensions. But these characterization theorems are rather abstract, hard to prove, and restricted to globally supported and radial basis functions.

On the other hand, Bochner's characterization provides direct proofs for translation-invariant and possibly nonradial functions, but is not applicable to *conditionally* positive definite functions.

Thus in this section we follow Madych and Nelson (1983) to generalize the approach of Bochner to the case of conditionally positive definite translation-invariant functions. It will turn out that the proof of the basic result is quite easy, but it will be technically difficult to apply the general result to specific basis functions. But our efforts will pay off by yielding explicit representations of generalized Fourier transforms of the classical radial basis functions, and these are important for further study of interpolation errors and stability results.

Recall that the Schwartz space $\mathcal{S}$ consists of all $C^\infty(\mathbb{R}^d)$-functions that, together with all their derivatives, decay faster than any polynomial.

Definition 1.4.1 For $m \in \mathbb{N}_0$ the set of all functions $\gamma \in \mathcal{S}$ which satisfy $\gamma(\omega) = \mathcal{O}(\|\omega\|_2^{2m})$ for $\|\omega\|_2 \to 0$ will be denoted by $\mathcal{S}_m$.

Recall that a function Φ is called slowly increasing if there exists an integer $\ell \in \mathbb{N}_0$ such that $|\Phi(\omega)| = \mathcal{O}(\|\omega\|_2^\ell)$ for $\|\omega\|_2 \to \infty$.

Definition 1.4.2 Suppose $\Phi : \mathbb{R}^d \to \mathbb{C}$ is continuous and slowly increasing. A continuous function $\widehat{\Phi} : \mathbb{R}^d \setminus \{0\} \to \mathbb{C}$ is said to be the *generalized Fourier transform* of Φ if there exists an integer $m \in \mathbb{N}_0$ such that

$$\int_{\mathbb{R}^d} \Phi(x)\widehat{\gamma}(x)dx = \int_{\mathbb{R}^d} \widehat{\Phi}(\omega)\gamma(\omega)d\omega$$

is satisfied for all $\gamma \in \mathcal{S}_m$. The smallest of such m is called the *order* of $\widehat{\Phi}$.

We omit the proof that the generalized Fourier transform is uniquely defined, but rather give a nontrivial example:

Proposition 1.4.3 *Suppose $\Phi = p$ is a polynomial of degree less than $2m$. Then for every test function $\gamma \in \mathcal{S}_m$ we have*

$$\int_{\mathbb{R}^d} \Phi(x)\widehat{\gamma}(x)dx = 0.$$

Proof Suppose Φ has the representation $\Phi(x) = \sum_{|\beta|<2m} c_\beta x^\beta$. Then

$$\begin{aligned}
\int_{\mathbb{R}^d} \Phi(x)\widehat{\gamma}(x)dx &= \sum_{|\beta|<2m} c_\beta i^{-|\beta|} \int_{\mathbb{R}^d} (ix)^\beta \widehat{\gamma}(x)dx \\
&= \sum_{|\beta|<2m} c_\beta i^{-|\beta|} \int_{\mathbb{R}^d} \left(\frac{\partial^{|\beta|}\gamma}{\partial x^\beta}\right)^\wedge dx \\
&= (2\pi)^{d/2} \sum_{|\beta|<2m} c_\beta i^{-|\beta|} \frac{\partial^{|\beta|}\gamma}{\partial x^\beta}(0) \\
&= 0
\end{aligned}$$

since $\gamma \in \mathcal{S}_m$. □

Note that the above result implies that the "inverse" generalized Fourier transform is not unique, because one can add a polynomial of degree less than $2m$ to a function Φ without changing its generalized Fourier transform. Note further that there are other definitions of generalized Fourier transforms, e.g. in the context of tempered distributions.

The next theorem shows that the order of the generalized Fourier transform, which is nothing but the order of the singularity of the generalized Fourier transform at the origin, determines the minimal order of a conditionally positive definite function, provided that the function has a *nonnegative* and *nonzero* generalized Fourier transform. We will state and prove only the sufficient part, but point out that the reverse direction also holds. We need the following auxiliary result:

Lemma 1.4.4 *Suppose that distinct $x_1, \ldots, x_N \in \mathbb{R}^d$ and $\alpha \in \mathbb{C}^N \setminus \{0\}$ are given such that (1.2) is satisfied for all $p \in \pi_{m-1}(\mathbb{R}^d)$. Then*

$$\sum_{j=1}^{N} \alpha_j e^{ix_j^T \omega} = \mathcal{O}(\|\omega\|_2^m)$$

holds for $\|\omega\|_2 \to 0$.

Proof The expansion of the exponential function leads to

$$\sum_{j=1}^{N} \alpha_j e^{ix_j^T \omega} = \sum_{k=0}^{\infty} \frac{i^k}{k!} \sum_{j=1}^{N} \alpha_j (x_j^T \omega)^k.$$

For fixed $\omega \in \mathbb{R}^d$ we have $p_k(x) := (x^T \omega)^k \in \pi_k(\mathbb{R}^d)$. Thus (1.2) ensures that the first $m-1$ terms vanish:

$$\sum_{j=1}^{N} \alpha_j e^{ix_j^T \omega} = \sum_{k=m}^{\infty} \frac{i^k}{k!} \sum_{j=1}^{N} \alpha_j (x_j^T \omega)^k,$$

which yields the stated behavior. □

Theorem 1.4.5 *Suppose $\Phi : \mathbb{R}^d \to \mathbb{C}$ is continuous, slowly increasing, and possesses a generalized Fourier transform $\widehat{\Phi}$ of order m which is nonnegative and non-vanishing. Then Φ is a translation-invariant conditionally positive definite function of order m.*

Proof Suppose that distinct $x_1, \ldots, x_N \in \mathbb{R}^d$ and $\alpha \in \mathbb{C}^N \setminus \{0\}$ satisfy (1.2) for all $p \in \pi_{m-1}(\mathbb{R}^d)$. Define

$$f(x) := \sum_{j,k=1}^{N} \alpha_j \overline{\alpha_k} \Phi(x + (x_j - x_k))$$

and

$$\gamma_\ell(x) = \left| \sum_{j=1}^{N} \alpha_j e^{ix^T x_j} \right|^2 \widehat{g_\ell}(x) = \sum_{j,k=1}^{N} \alpha_j \overline{\alpha_k} e^{ix^T (x_j - x_k)} \widehat{g_\ell}(x),$$

where $g_\ell(x) = (\ell/\pi)^{d/2} e^{-\ell \|x\|_2^2}$. On account of $\gamma_\ell \in \mathcal{S}$ and Lemma 1.4.4 we have $\gamma_\ell \in \mathcal{S}_m$. Furthermore,

$$\begin{aligned}
\widehat{\gamma_\ell}(x) &= (2\pi)^{-d/2} \int_{\mathbb{R}^d} \sum_{j,k=1}^N \alpha_j \overline{\alpha_k} e^{i\omega^T(x_j-x_k)} \widehat{g_\ell}(\omega) e^{-ix^T\omega} d\omega \\
&= \sum_{j,k=1}^N \alpha_j \overline{\alpha_k} (2\pi)^{-d/2} \int_{\mathbb{R}^d} \widehat{g_\ell}(\omega) e^{-i\omega^T(x-(x_j-x_k))} d\omega \\
&= \sum_{j,k=1}^N \alpha_j \overline{\alpha_k} g_\ell(x - (x_j - x_k)),
\end{aligned}$$

since $\widehat{\widehat{g_\ell}} = g_\ell$. Collecting these facts together with Definition 1.4.2 gives

$$\begin{aligned}
\int_{\mathbb{R}^d} f(x) g_\ell(x) dx &= \int_{\mathbb{R}^d} \Phi(x) \sum_{j,k=1}^N \alpha_j \overline{\alpha_k} g_\ell(x-(x_j-x_k)) dx \\
&= \int_{\mathbb{R}^d} \Phi(x) \widehat{\gamma_\ell}(x) dx \\
&= \int_{\mathbb{R}^d} \widehat{\Phi}(\omega) \gamma_\ell(\omega) d\omega \\
&= \int_{\mathbb{R}^d} \left| \sum_{j=1}^N \alpha_j e^{i\omega^T x_j} \right|^2 \widehat{g_\ell}(\omega) \widehat{\Phi}(\omega) d\omega \\
&\geq 0.
\end{aligned}$$

Since Φ is only slowly increasing, we have

$$\sum_{j,k=1}^N \alpha_j \overline{\alpha_k} \Phi(x_j - x_k) = \lim_{\ell \to \infty} \int_{\mathbb{R}^d} f(x) g_\ell(x) dx \geq 0$$

by means of approximation by convolution. Furthermore, the quantity

$$\left| \sum_{j=1}^N \alpha_j e^{i\omega^T x_j} \right|^2 \widehat{g_\ell}(\omega) \widehat{\Phi}(\omega)$$

is non-decreasing in ℓ and we already know that the limit

$$\lim_{\ell \to \infty} \int_{\mathbb{R}^d} \left| \sum_{j=1}^N \alpha_j e^{i\omega^T x_j} \right|^2 \widehat{g_\ell}(\omega) \widehat{\Phi}(\omega) d\omega$$

exists. Hence, the limit function $(2\pi)^{-d/2} \left| \sum_{j=1}^N \alpha_j e^{i\omega^T x_j} \right|^2 \widehat{\Phi}(\omega)$ is integrable due to the monotone convergence theorem. Thus we have established the equality

$$\sum_{j,k=1}^{N} \alpha_j \overline{\alpha_k} \Phi(x_j - x_k) = (2\pi)^{-d/2} \int_{\mathbb{R}^d} \left| \sum_{j=1}^{N} \alpha_j e^{i\omega^T x_j} \right|^2 \widehat{\Phi}(\omega) d\omega.$$

This quadratic form cannot vanish if $\widehat{\Phi}$ is non-vanishing, since the exponentials are linearly independent. $\square$

1.5 Classical radial basis functions

Now we use this generalization of the Bochner approach to compute the generalized Fourier transforms of the most popular translation-invariant or radial basis functions. Since it will turn out that these generalized Fourier transforms are nonnegative and non-vanishing, we can read off the order of conditional positive definiteness of the functions from the order of the singularity of their generalized Fourier transforms at the origin.

We start with the positive definite inverse multiquadrics treated in Theorem 1.3.4 and use analytic continuation to deal with the case of the conditionally positive definite (non-inverse) multiquadrics. To do this we need two results on the modified Bessel functions.

Lemma 1.5.1 *The modified Bessel function K_ν, $\nu \in \mathbb{C}$, has the uniform bound*

$$|K_\nu(r)| \leq \sqrt{\frac{2\pi}{r}} e^{-r} e^{\frac{|\Re(\nu)|^2}{2r}}, \qquad r > 0 \tag{1.10}$$

describing its behavior for large r.

Proof With $b = |\Re(\nu)|$ we have

$$\begin{aligned}
|K_\nu(r)| &\leq \frac{1}{2} \int_0^\infty e^{-r \cosh t} |e^{\nu t} + e^{-\nu t}| dt \\
&\leq \frac{1}{2} \int_0^\infty e^{-r \cosh t} (e^{bt} + e^{-bt}) dt \\
&= K_b(r).
\end{aligned}$$

Furthermore, from $e^t \geq \cosh t \geq 1 + \frac{t^2}{2}$, $t \geq 0$, we can conclude

$$\begin{aligned}
K_b(r) &\leq \int_0^\infty e^{-r(1+\frac{t^2}{2})} e^{bt} dt \\
&\leq e^{-r} e^{\frac{b^2}{2r}} \frac{1}{\sqrt{r}} \int_{\frac{-b}{\sqrt{r}}}^\infty e^{-s^2/2} ds \\
&\leq \sqrt{2\pi} e^{-r} e^{\frac{b^2}{2r}} \sqrt{\frac{1}{r}}.
\end{aligned}$$

$\square$

Lemma 1.5.2 *For $\nu \in \mathbb{C}$ the modified Bessel function K_ν satisfies*

$$|K_\nu(r)| \leq \begin{cases} 2^{|\Re(\nu)|-1}\Gamma(|\Re(\nu)|)r^{-|\Re(\nu)|}, & \Re(\nu) \neq 0, \\ \frac{1}{e} - \log \frac{r}{2}, & r < 2, \Re(\nu) = 0 \end{cases} \quad (1.11)$$

for $r > 0$, describing its behavior for small r.

Proof Let us first consider the case $\Re(\nu) \neq 0$. We set again $b = |\Re(\nu)|$ and already know that $|K_\nu(r)| \leq K_b(r)$ from the proof of the preceding lemma. Furthermore, from the proof of Theorem 1.3.4 we get

$$K_b(r) = \frac{1}{2} \int_0^\infty e^{-\frac{r}{2}(\frac{s}{a}+\frac{a}{s})} \left(\frac{s}{a}\right)^b \frac{ds}{s}$$

for every $a > 0$. By setting $a = r/2$ we see that

$$K_b(r) = 2^{b-1} r^{-b} \int_0^\infty e^{-s} e^{-\frac{r^2}{4s}} s^{b-1} ds \leq 2^{b-1}\Gamma(b)r^{-b}.$$

For $\Re(\nu) = 0$ we use $\cosh t \geq e^t/2$ to derive

$$\begin{aligned} K_0(r) &= \int_0^\infty e^{-r\cosh t} dt \\ &\leq \int_0^\infty e^{-\frac{r}{2}e^t} dt \\ &= \int_{\frac{r}{2}}^\infty e^{-u} \frac{1}{u} du \\ &\leq \int_1^\infty e^{-u} du + \int_{\frac{r}{2}}^1 \frac{1}{u} du \\ &= \frac{1}{e} - \log \frac{r}{2}. \quad \square \end{aligned}$$

We are now able to compute the generalized Fourier transform of the general multiquadrics. The basic idea of the proof goes back to Madych and Nelson (1983). It starts with the classical Fourier transform of the inverse multiquadrics given in Theorem 1.3.4, and then uses analytic continuation.

Theorem 1.5.3 *The function $\Phi(x) = (c^2 + \|x\|_2^2)^\beta$, $x \in \mathbb{R}^d$, with $c > 0$ and $\beta \in \mathbb{R} \setminus \mathbb{N}_0$ possesses the (generalized) Fourier transform*

$$\widehat{\Phi}(\omega) = \frac{2^{1+\beta}}{\Gamma(-\beta)} \left(\frac{\|\omega\|_2}{c}\right)^{-\beta-\frac{d}{2}} K_{\frac{d}{2}+\beta}(c\|\omega\|_2), \quad \omega \neq 0, \quad (1.12)$$

of order $m = \max(0, \lceil \beta \rceil)$.

Proof Define $G = \{\lambda \in \mathbb{C} : \Re(\lambda) < m\}$ and denote the right-hand side of (1.12) by $\varphi_\beta(\omega)$. We are going to show by analytic continuation that

$$\int_{\mathbb{R}^d} \Phi_\lambda(\omega)\widehat{\gamma}(\omega)d\omega = \int_{\mathbb{R}^d} \varphi_\lambda(\omega)\gamma(\omega)d\omega, \qquad \gamma \in \mathcal{S}_m, \quad (1.13)$$

is valid for all $\lambda \in G$, where $\Phi_\lambda(\omega) = (c^2 + \|\omega\|_2^2)^\lambda$. First, note that (1.13) is valid for $\lambda \in G$ with $\lambda < -d/2$ by Theorem 1.3.4, and, when $m > 0$, also for $\lambda = 0, 1, \ldots, m-1$, by Proposition 1.4.3 and the fact that $1/\Gamma(-\lambda)$ is zero in these cases. Analytic continuation will lead us to our stated result if we can show that both sides of (1.13) exist and are analytic functions in λ. We will do this only for the right-hand side, since the left-hand side can be handled more easily. Thus we define

$$f(\lambda) = \int_{\mathbb{R}^d} \varphi_\lambda(\omega)\gamma(\omega)d\omega$$

and study this function of λ. Suppose $\mathcal{C}$ is a closed curve in G. Since φ_λ is an analytic function in $\lambda \in G$ it has the representation

$$\varphi_\lambda(\omega) = \frac{1}{2\pi i} \int_\mathcal{C} \frac{\varphi_z(\omega)}{z - \lambda} dz$$

for $\lambda \in \text{Int } \mathcal{C}$. Now suppose that we have already shown that the integrand in the definition of $f(\lambda)$ can be bounded uniformly on $\mathcal{C}$ by an integrable function. This ensures that $f(\lambda)$ is well-defined in G and by Fubini's theorem we can conclude

$$\begin{aligned}
f(\lambda) &= \int_{\mathbb{R}^d} \varphi_\lambda(\omega)\gamma(\omega)d\omega \\
&= \frac{1}{2\pi i} \int_{\mathbb{R}^d} \int_\mathcal{C} \frac{\varphi_z(\omega)}{z - \lambda} dz \gamma(\omega)d\omega \\
&= \frac{1}{2\pi i} \int_\mathcal{C} \frac{1}{z - \lambda} \int_{\mathbb{R}^d} \varphi_z(\omega)\gamma(\omega)d\omega\, dz \\
&= \frac{1}{2\pi i} \int_\mathcal{C} \frac{f(z)}{z - \lambda} dz
\end{aligned}$$

for $\lambda \in \text{Int } \mathcal{C}$, which means that f is analytic in G. Thus it remains to bound the integrand uniformly.

Let us first consider the asymptotic behavior in a neighborhood of the origin, say for $\|\omega\|_2 < 1/c$. If we set $b = \Re(\lambda)$ we can use Lemma 1.5.2 and $\gamma \in \mathcal{S}_m$ to get in the case $b \neq -d/2$:

$$|\varphi_\lambda(\omega)\gamma(\omega)| \leq C_\gamma \frac{2^{b+|b+d/2|}\Gamma(|b+d/2|)}{|\Gamma(-\lambda)|} c^{b+d/2-|b+d/2|} \|\omega\|_2^{-b-d/2-|b+d/2|+2m},$$

and in the case $b = -d/2$:
$$|\varphi_\lambda(\omega)\gamma(\omega)| \leq C_\gamma \frac{2^{1-d/2}}{|\Gamma(-\lambda)|} \left(\frac{1}{e} - \log\frac{c\|\omega\|_2}{2}\right)\|\omega\|_2^{2m}.$$

Since $\mathcal{C}$ is compact and $1/\Gamma$ is analytic, this gives for all $\lambda \in \mathcal{C}$
$$|\varphi_\lambda(\omega)\gamma(\omega)| \leq C_{\gamma,m,c,\mathcal{C}}\left(1 + \|\omega\|_2^{-d+2\epsilon} - \log\frac{c\|\omega\|_2}{2}\right), \qquad \|\omega\|_2 \leq 1/c$$

with $\epsilon = m - b > 0$. For large arguments, the integrand in the definition of $f(\lambda)$ can be estimated via Lemma 1.5.1 by
$$|\varphi_\lambda(\omega)\gamma(\omega)| \leq C_\gamma \frac{2^{1+b}\sqrt{2\pi}}{|\Gamma(-\lambda)|}c^{b+\frac{d-1}{2}}\|\omega\|_2^{-b-\frac{d+1}{2}}e^{-c\|\omega\|_2}e^{\frac{|b+\frac{d}{2}|^2}{2c\|\omega\|_2}}$$

using that $\gamma \in \mathcal{S}$ is bounded. Since $\mathcal{C}$ is compact, this can be bounded independently of $\lambda \in \mathcal{C}$ by
$$|\varphi_\lambda(\omega)\gamma(\omega)| \leq C_{\gamma,\mathcal{C},m,c}e^{-c\|\omega\|_2/2},$$

completing the proof. $\square$

Theorem 1.5.4 *The function $\Phi(x) = \|x\|_2^\beta$, $x \in \mathbb{R}^d$, with $\beta > 0$, $\beta \notin 2\mathbb{N}$, has the generalized Fourier transform*
$$\widehat{\Phi}(\omega) = \frac{2^{\beta+\frac{d}{2}}\Gamma(\frac{d+\beta}{2})}{\Gamma(-\frac{\beta}{2})}\|\omega\|_2^{-\beta-d}, \qquad \omega \neq 0,$$
of order $m = \lceil \beta/2 \rceil$.

Proof Let us start with the function $\Phi_c(x) = (c^2 + \|x\|_2^2)^{\frac{\beta}{2}}$, $c > 0$. This function possesses a generalized Fourier transform of order $m = \lceil \beta/2 \rceil$ given by
$$\widehat{\Phi}_c(\omega) = \varphi_c(\omega) = \frac{2^{1+\beta/2}}{\Gamma(-\beta/2)}\|\omega\|_2^{-\beta-d}(c\|\omega\|_2)^{\frac{\beta+d}{2}}K_{\frac{\beta+d}{2}}(c\|\omega\|_2)$$

due to Theorem 1.5.3. Here, we use the subscript c instead of β, since β is fixed and we want to let c go to zero. Moreover, we can conclude from the proof of Theorem 1.5.3 that for $\gamma \in \mathcal{S}_m$ the product can be bounded by
$$|\varphi_c(\omega)\gamma(\omega)| \leq C_\gamma \frac{2^{\beta+d/2}\Gamma(\frac{\beta+d}{2})}{|\Gamma(-\beta/2)|}\|\omega\|_2^{2m-\beta-d}$$

for $\|\omega\|_2 \to 0$ and by
$$|\varphi_c(\omega)\gamma(\omega)| \leq C_\gamma \frac{2^{\beta+d/2}\Gamma(\frac{\beta+d}{2})}{|\Gamma(-\beta/2)|}\|\omega\|_2^{-\beta-d}$$

for $\|\omega\|_2 \to \infty$ independently of $c > 0$. Since $|\Phi_c(\omega)\widehat{\gamma}(\omega)|$ can also be bounded independently of c by an integrable function, we can use the convergence theorem of Lebesgue twice to derive

$$\begin{aligned}
\int_{\mathbb{R}^d} \|x\|_2^\beta \widehat{\gamma}(x)dx &= \lim_{c\to 0}\int_{\mathbb{R}^d}\Phi_c(x)\widehat{\gamma}(x)dx = \lim_{c\to 0}\int_{\mathbb{R}^d}\varphi_c(\omega)\gamma(\omega)d\omega \\
&= \frac{2^{1+\frac{\beta}{2}}}{\Gamma(-\frac{\beta}{2})}\int_{\mathbb{R}^d}\|\omega\|_2^{-\beta-d}\gamma(\omega)\lim_{c\to 0}(c\|\omega\|_2)^{\frac{\beta+d}{2}}K_{\frac{\beta+d}{2}}(c\|\omega\|_2)d\omega \\
&= \frac{2^{\beta+d/2}\Gamma(\frac{d+\beta}{2})}{\Gamma(-\beta/2)}\int_{\mathbb{R}^d}\|\omega\|_2^{-\beta-d}\gamma(\omega)d\omega
\end{aligned}$$

for $\gamma \in \mathcal{S}_m$. The last equality follows from

$$\lim_{r\to 0}r^\nu K_\nu(r) = \lim_{r\to 0}2^{\nu-1}\int_0^\infty e^{-t}e^{-\frac{r^2}{4t}}t^{\nu-1}dt = 2^{\nu-1}\Gamma(\nu);$$

see also the proof of Lemma 1.5.2. □

Theorem 1.5.5 *The function* $\Phi(x) = \|x\|_2^{2k}\log\|x\|_2$, $x \in \mathbb{R}^d$, $k \in \mathbb{N}$, *possesses the generalized Fourier transform*

$$\widehat{\Phi}(\omega) = (-1)^{k+1}2^{2k-1+\frac{d}{2}}\Gamma\left(k+\frac{d}{2}\right)k!\|\omega\|_2^{-d-2k}$$

of order $m = k + 1$.

Proof For fixed $r > 0$ and $\beta \in (2k, 2k+1)$ we expand the function $\beta \mapsto r^\beta$ using Taylor's theorem to obtain

$$r^\beta = r^{2k} + (\beta - 2k)r^{2k}\log r + \int_{2k}^\beta (\beta - t)r^t \log r\, dt. \quad (1.14)$$

From Theorem 1.5.4 we know the generalized Fourier transform of the function $x \mapsto \|x\|_2^\beta$ of order $m = \lceil \beta/2 \rceil = k+1$. From Proposition 1.4.3 we see that the generalized Fourier transform of order m of the function $x \mapsto \|x\|_2^{2k}$ equals zero. Thus we can conclude from (1.14) that for any test function $\gamma \in \mathcal{S}_m$

$$\begin{aligned}
\int_{\mathbb{R}^d}\|x\|_2^{2k}\log\|x\|_2\widehat{\gamma}(x)dx &= \frac{1}{\beta-2k}\int_{\mathbb{R}^d}\left(\|x\|_2^\beta - \|x\|_2^{2k}\right)\widehat{\gamma}(x)dx \\
&\quad -\frac{1}{\beta-2k}\int_{\mathbb{R}^d}\int_{2k}^\beta (\beta-t)\|x\|_2^t\log\|x\|_2\widehat{\gamma}(x)dt\,dx
\end{aligned}$$

$$= \frac{2^{\beta+\frac{d}{2}}\Gamma(\frac{d+\beta}{2})}{(\beta-2k)\Gamma(-\frac{\beta}{2})}\int_{\mathbb{R}^d}\|\omega\|_2^{-\beta-d}\gamma(\omega)d\omega$$
$$+ \mathcal{O}(\beta-2k)$$

for $\beta \to 2k$. Furthermore, we know from the property $\Gamma(z)\Gamma(1-z) = \pi/\sin(\pi z)$ that

$$\frac{1}{\Gamma(-\frac{\beta}{2})(\beta-2k)} = -\frac{\sin(\frac{\pi\beta}{2})\Gamma(1+\frac{\beta}{2})}{\pi(\beta-2k)}.$$

Because

$$\lim_{\beta \to 2k}\frac{\sin(\frac{\pi\beta}{2})}{\beta-2k} = \lim_{\beta \to 2k}\frac{\frac{\pi}{2}\cos(\frac{\pi\beta}{2})}{1} = \frac{\pi}{2}(-1)^k,$$

we see that

$$\lim_{\beta \to 2k}\frac{1}{\Gamma(-\frac{\beta}{2})(\beta-2k)} = (-1)^{k+1}k!/2.$$

Now we can apply the theorem of dominated convergence to get

$$\int_{\mathbb{R}^d}\|x\|_2^{2k}\log\|x\|_2\widehat{\gamma}(x)dx = 2^{2k+d/2}\Gamma(k+d/2)(-1)^{k+1}\frac{k!}{2}\int_{\mathbb{R}^d}\|\omega\|_2^{-d-2k}\gamma(\omega)d\omega$$

for all $\gamma \in \mathcal{S}_m$, which gives the stated generalized Fourier transform. □

Now it is easy to decide whether the functions just investigated are conditionally positive definite. As mentioned before, we state the result with the minimal m.

Corollary 1.5.6 *The following functions* $\Phi : \mathbb{R}^d \to \mathbb{R}$ *are conditionally positive definite of order* m:

- $\Phi(x) = (-1)^{\lceil\beta\rceil}(c^2 + \|x\|_2^2)^\beta$, $\beta > 0$, $\beta \notin 2\mathbb{N}$, $m = \lceil\beta\rceil$,
- $\Phi(x) = (c^2 + \|x\|_2^2)^\beta$, $\beta < 0$, $m = 0$,
- $\Phi(x) = (-1)^{\lceil\beta/2\rceil}\|x\|_2^\beta$, $\beta > 0$, $\beta \notin \mathbb{N}$, $m = \lceil\beta/2\rceil$,
- $\Phi(x) = (-1)^{k+1}\|x\|_2^{2k}\log\|x\|_2$, $k \in \mathbb{N}$, $m = k+1$.

1.6 Construction via dimension walk

So far we have seen that radial functions that work on $\mathbb{R}^d$ for all $d \geq 1$, are nicely characterized by the abstract results of Schoenberg and Micchelli, while translation-invariant functions for fixed dimensions are best handled via the Fourier transform, yielding explicit results for further use.

Here, we want to investigate radial functions for a fixed space dimension. Thus we have to take the Fourier transform, but we shall make use of radiality throughout, relying on ideas of Wu (1995) and Schaback and Wu (1996). Our main goal will be the construction of compactly supported positive definite radial basis functions for fixed space dimensions.

Theorem 1.6.1 *Suppose* $\Phi \in L_1(\mathbb{R}^d) \cap C(\mathbb{R}^d)$ *is radial, i.e.* $\Phi(x) = \phi(\|x\|_2)$, $x \in \mathbb{R}^d$. *Then its Fourier transform* $\widehat{\Phi}$ *is also radial, i.e.* $\widehat{\Phi}(\omega) = \mathcal{F}_d \phi(\|\omega\|_2)$ *with*

$$\mathcal{F}_d \phi(r) = r^{-\frac{d-2}{2}} \int_0^\infty \phi(t) t^{\frac{d}{2}} J_{\frac{d-2}{2}}(rt) dt,$$

and ϕ *satisfies* $\phi(t) t^{d-1} \in L_1[0, \infty)$.

Proof The case $d = 1$ follows immediately from

$$J_{-1/2}(t) = \left(\frac{2}{\pi t}\right)^{1/2} \cos t.$$

When $d \geq 2$, splitting the Fourier integral, and using the representation

$$\int_{S_{d-1}} e^{ix^T \xi} dS(\xi) = (2\pi)^{d/2} \|x\|_2^{-\frac{d-2}{2}} J_{\frac{d-2}{2}}(\|x\|_2)$$

of the classical Bessel function J_ν via an integral over the sphere $S_{d-1} \subset \mathbb{R}^d$ yields

$$\begin{aligned}
\widehat{\Phi}(x) &= (2\pi)^{-d/2} \int_{\mathbb{R}^d} \Phi(\omega) e^{-ix^T \omega} d\omega \\
&= (2\pi)^{-d/2} \int_0^\infty t^{d-1} \int_{S_{d-1}} \phi(t\|\omega\|_2) e^{-ix^T \omega} dS(\omega) dt \\
&= (2\pi)^{-d/2} \int_0^\infty \phi(t) t^{d-1} \int_{S_{d-1}} e^{-itx^T \omega} dS(\omega) dt \\
&= r^{-(d-2)/2} \int_0^\infty \phi(t) t^{d/2} J_{(d-2)/2}(rt) dt
\end{aligned}$$

The theorem's second assertion follows from the condition $\Phi \in L_1(\mathbb{R}^d) \cap C(\mathbb{R}^d)$, and the radiality of Φ. $\square$

Theorem 1.6.1 gives us the opportunity to interpret the d-variate Fourier transform of a radial function via $\mathcal{F}_d$ as an operator that maps univariate functions to univariate functions.

Now let us have a closer look at this operator with respect to the space dimension. If we use $\frac{d}{dz}\{z^\nu J_\nu(z)\} = z^\nu J_{\nu-1}(z)$ we get via integration by parts, for $d \geq 3$,

$$\begin{aligned}
\mathcal{F}_d\phi(r) &= r^{-d+2} \int_0^\infty \phi(t) t\, (rt)^{\frac{d-2}{2}} J_{\frac{d-2}{2}}(rt)\, dt \\
&= r^{-d+2} \left(-\int_t^\infty \phi(s)s\, ds\right)(rt)^{\frac{d-2}{2}} J_{\frac{d-2}{2}}(rt)\bigg|_{t=0}^{t=\infty} \\
&\quad + r^{-d+2}\int_0^\infty \left(\int_t^\infty \phi(s)s\, ds\right) r^{\frac{d}{2}} t^{\frac{d-2}{2}} J_{\frac{d-4}{2}}(rt)\, dt \\
&= \mathcal{F}_{d-2}\left(\int_\bullet^\infty \phi(s)s\, ds\right)(r)
\end{aligned}$$

whenever the boundary terms vanish. Thus if we define

$$\begin{aligned}
I\phi(r) &:= \int_r^\infty \phi(t) t\, dt \\
D\phi(r) &:= -\frac{1}{r}\frac{d}{dr}\phi(r)
\end{aligned}$$

we get the following result.

Theorem 1.6.2 *If $\phi \in C[0,\infty)$ satisfies $t \mapsto \phi(t)t^{d-1} \in L_1[0,\infty)$, for some $d \geq 3$, then we have $\mathcal{F}_d(\phi) = \mathcal{F}_{d-2}(I\phi)$. This means that ϕ is positive definite on $\mathbb{R}^d$ if and only if $I\phi$ is positive definite on $\mathbb{R}^{d-2}$. On the other hand, if for some $d \geq 1$, ϕ satisfies $t \mapsto \phi(t)t^{d-1} \in L_1[0,\infty)$ and $\phi(t) \to 0$ as $t \to \infty$, and if the even extension of ϕ to $\mathbb{R}$ is in $C^2(\mathbb{R})$, then $\mathcal{F}_d(\phi) = \mathcal{F}_{d+2}(D\phi)$. In this situation, the function ϕ is positive definite on $\mathbb{R}^d$ if and only if $D\phi$ is positive definite on $\mathbb{R}^{d+2}$.*

Since both operators I and D are easily computable and satisfy $I = D^{-1}$ and $D = I^{-1}$ wherever defined, this gives us a very powerful tool for constructing positive definite functions. For example, we could start with a very smooth compactly supported function on $\mathbb{R}^1$ and apply the operator D n times to get a positive definite and compactly supported function on $\mathbb{R}^{2n+1}$. Before we give an example, let us note that it is possible to generalize the operators $\mathcal{F}_d, I, D$ to step through the dimensions one by one and not two by two (Schaback and Wu (1996)).

Theorem 1.6.3 *Define $\phi_\ell(r) := (1-r)_+^\ell$ and $\phi_{d,k}$ by*

$$\phi_{d,k} = I^k \phi_{\lfloor d/2 \rfloor + k + 1}.$$

Then $\phi_{d,k}$ is compactly supported, a polynomial within its support, and pos-

itive definite on $\mathbb{R}^d$. In particular, the function $20\phi_{3,1}(r) = (1-r)_+^4(4r+1)$ is positive definite on $\mathbb{R}^3$.

Proof Since the operator I respects the polynomial structure and compact support, we only have to prove positive definiteness. Since

$$\mathcal{F}_d\phi_{d,k} = \mathcal{F}_d I^k \phi_{\lfloor d/2 \rfloor + k + 1} = \mathcal{F}_{d+2k}\phi_{\lfloor (d+2k)/2 \rfloor + 1}$$

it remains to show that $\mathcal{F}_d\phi_{\lfloor d/2 \rfloor + 1}$ is nonnegative for every space dimension d. We will follow ideas of Askey (1973) to do this. Let us start with an odd dimension $d = 2n + 1$. Then the Fourier transform is given by

$$r^{3n+2}\mathcal{F}_{2n+1}\phi_{n+1}(r) = \int_0^r (r-s)^{n+1} s^{n+\frac{1}{2}} J_{n-\frac{1}{2}}(s) ds.$$

Denoting the right-hand side of the last equation by $g(r)$, we see that g is the convolution $g(r) = \int_0^r g_1(r-s)g_2(s) ds$ of the functions $g_1(s) := (s)_+^{n+1}$ and $g_2(s) := s^{n+1/2} J_{n-1/2}(s)$. Thus its Laplace transform $\mathcal{L}g(r) = \int_0^\infty g(t) e^{-rt} dt$ is the product of the Laplace transforms of g_1 and g_2. These transforms can be computed for $r > 0$ as

$$\mathcal{L}g_1(r) = \frac{(n+1)!}{r^{n+2}}$$

and

$$\mathcal{L}g_2(r) \frac{n!\, 2^{n+1/2} r}{\sqrt{\pi}\,(1+r^2)^{n+1}}.$$

These combine to give

$$\mathcal{L}g(r) = \frac{2^{n+1/2} n! (n+1)!}{\sqrt{\pi}} \frac{1}{r^{n+1}(1+r^2)^{n+1}}.$$

On the other hand, it is well known that the function $1 - \cos r$ has the Laplace transform $\frac{1}{r(1+r^2)}$. Thus, if p denotes the n-fold convolution of this function with itself, we get

$$\mathcal{L}p(r) = \frac{1}{r^{n+1}(1+r^2)^{n+1}}.$$

By the uniqueness of the Laplace transform this leads to

$$g(r) = \frac{2^{n+1/2} n! (n+1)!}{\sqrt{\pi}} p(r),$$

which is clearly nonnegative and not identically zero. For even space dimension $d = 2n$ we need only to note that $\phi_{\lfloor \frac{2n}{2} \rfloor + 1} = \phi_{\lfloor \frac{2n+1}{2} \rfloor + 1}$. Hence $\phi_{\lfloor \frac{2n}{2} \rfloor + 1}$ induces a positive definite function on $\mathbb{R}^{2n+1}$ and therefore also on $\mathbb{R}^{2n}$. The function $\phi(r) = (1-r)_+^4(4r+1)$ is nothing but $20\phi_{3,1}$, and hence positive definite on $\mathbb{R}^3$. □

The parameter k in the last theorem controls the smoothness of the basis function. It can be shown (Wendland (1995)) that $\phi_{d,k}$ possesses $2k$ continuous derivatives as a radial function on $\mathbb{R}^d$ and is of minimal degree among all piecewise polynomial compactly supported functions that are positive definite on $\mathbb{R}^d$ and whose even extensions to $\mathbb{R}$ are in $C^{2k}(\mathbb{R})$. A different technique for generating compactly supported radial basis functions is due to Buhmann (1998), Buhmann (2000a), Buhmann (2000b).

1.7 Construction of general functions

So far we have only dealt with translation-invariant (conditionally) positive definite functions, and most of our work was even restricted to radial functions. As a consequence, we had to work with basis functions that are (conditionally) positive definite on all of $\mathbb{R}^d$. In this section we want to take a more general approach which allows us to construct positive definite functions on local domains Ω. Consequently, we have to drop Fourier and Laplace transforms, replacing them by expansions into orthogonal systems. As a by-product, this technique allows us to construct positive definite functions on manifolds, in particular on the sphere.

Theorem 1.7.1 *Suppose $\Omega \subseteq \mathbb{R}^d$ is measurable. Let $\varphi_1, \varphi_2, \ldots$ be an orthonormal basis for $L_2(\Omega)$ consisting of continuous and bounded functions. Suppose that the point evaluation functionals are linearly independent on* span$\{\varphi_j : j \in \mathbb{N}\}$. *Suppose ρ_n is a sequence of positive numbers satisfying*

$$\sum_{n=1}^{\infty} \rho_n \|\varphi_n\|_{L_\infty(\Omega)}^2 < \infty. \tag{1.15}$$

Then

$$\Phi(x,y) = \sum_{n=1}^{\infty} \rho_n \varphi_n(x) \overline{\varphi_n(y)}$$

is positive definite on Ω.

Proof Property (1.15) ensures that Φ is well-defined and continuous. Furthermore, we have, for $\alpha \in \mathbb{C}^N$ and distinct $x_1, \ldots, x_N \in \Omega$, that

$$\sum_{j,k=1}^{N} \alpha_j \overline{\alpha_k} \Phi(x_j, x_k) = \sum_{n=1}^{\infty} \rho_n \left| \sum_{j=1}^{N} \alpha_j \varphi_n(x_j) \right|^2 \geq 0.$$

Since the point evaluation functionals are linearly independent on span$\{\varphi_j : j \in \mathbb{N}\}$, the last expression can only vanish for $\alpha = 0$. □

Note that the condition on the point evaluation functionals is somewhat unnatural for the space $L_2(\Omega)$. It would be more natural to define Φ to be positive definite iff for every linear independent set $\Lambda = \{\lambda_1, \ldots \lambda_N\} \subseteq L_2(\Omega)^*$ and every $\alpha \in \mathbb{C}^N \setminus \{0\}$ the quadratic form

$$\sum_{j,k=1}^N \alpha_j \overline{\alpha_k} \lambda_j^x \lambda_k^y \Phi(x,y)$$

is positive. But we do not want to pursue this topic any further. Instead, we want to use Theorem 1.7.1 to give an example of a positive definite function on a restricted domain.

Our example deals with the space $L_2[0, 2\pi]^2$ which has the bounded and continuous orthogonal basis $\{\phi_{n,k}(x_1, x_2) = e^{i(nx_1 + kx_2)} : n, k \in \mathbb{Z}\}$ of functions with a 2π-periodic extension. Thus condition (1.15) is satisfied if the positive coefficients $\rho_{n,k}$ have the property

$$\sum_{n,k=-\infty}^{\infty} \rho_{n,k} < \infty.$$

In particular, the bivariate functions

$$\phi_{1,\ell}(x) = 1 + \sum_{(n,k) \in \mathbb{Z}^2 \setminus \{0\}} \frac{1}{(n^2 + k^2)^\ell} e^{i(nx_1 + kx_2)}$$

and

$$\phi_{2,\ell}(x) = 1 + \sum_{\substack{n=-\infty \\ n \neq 0}}^{\infty} \frac{1}{n^{2\ell}} e^{inx_1} + \sum_{\substack{k=-\infty \\ k \neq 0}}^{\infty} \frac{1}{k^{2\ell}} e^{ikx_2} + \sum_{\substack{n=-\infty \\ n \neq 0}}^{\infty} \sum_{\substack{k=-\infty \\ k \neq 0}}^{\infty} \frac{1}{(nk)^{2\ell}} e^{i(nx_1 + kx_2)}$$

generate positive definite 2π-periodic translation-invariant functions $\Phi(x, y) = \phi(x - y)$ on $[0, 2\pi]^2$, for sufficiently large ℓ. Due to their tensor product structure, the latter can be computed directly (see Narcowich and Ward (1996)). Some examples are:

$$\phi_{2,1}(x) = \prod_{j=1}^{2}\left(\frac{6 - \pi^2}{6} + \frac{1}{2}(x_j - \pi)^2\right);$$

$$\phi_{2,2}(x) = \prod_{j=1}^{2}\left(\frac{360 - 7\pi^4}{360} + \frac{\pi^2(x_j - \pi)^2}{12} - \frac{(x_j - \pi)^4}{24}\right).$$

For more examples, see Narcowich and Ward (1996). Of course, this tensor product approach generalizes to arbitrary space dimension, but the basic technique is much more general. See Schaback (1999) for the relation to positive integral operators.

References

Askey, R. (1973). Radial characteristic functions. MRC Technical Report Sum: Report No. 1262, University of Wisconsin.

Bochner, S. (1932). *Vorlesungen über Fouriersche Integrale*. Akademische Verlagsgesellschaft, Leipzig.

Bochner, S. (1933). Monotone funktionen, Stieltjes integrale und harmonische analyse. *Math. Ann.*, **108**, 378–410.

Buhmann, M.D. (1998). Radial functions on compact support. *Proceedings of the Edinburgh Mathematical Society*, **41**, 33–46.

Buhmann, M.D. (2000a). A new class of radial basis functions with compact support. To appear.

Buhmann, M.D. (2000b). This volume.

Chang, K.F. (1996a). Strictly positive definite functions. *J. Approx. Theory*, **87**, 148–158.

Chang, K.F. (1996b). Strictly positive definite functions II. Preprint.

Guo, K., Hu, S. and Sun, X. (1993). Conditionally positive definite functions and Laplace-Stieltjes integrals. *J. Approx. Theory*, **74**, 249–265.

Madych, W.R. and Nelson, S.A. (1983). Multivariate interpolation: a variational theory. Manuscript.

Micchelli, C.A. (1986). Interpolation of scattered data: distance matrices and conditionally positive definite functions. *Constr. Approx.* **2**, 11–22.

Narcowich, F.J. and Ward, J.D. (1996). Wavelets associated with periodic basis functions. *Appl. Comput. Harmonic Anal.*, **3**, 40–56.

Powell, M.J.D. (1987). Radial basis functions for multivariable interpolation: a review. In *Algorithms for Approximation*, ed. J.C. Mason and M.G. Cox., pp. 143–167. Clarendon Press, Oxford.

Schaback, R. (1999). A unified theory of radial basis functions (native Hilbert spaces for radial basis functions II). Preprint, Göttingen.

Schaback, R. and Wu, Z. (1996). Operators on radial functions. *J. Comp. Appl. Math.*, **73**, 257–270.

Schoenberg, I.J. (1938). Metric spaces and completely monotone functions. *Ann. of Math.*, **39**, 811–841.

Wendland, H. (1995). Piecewise polynomial, positive definite and compactly supported radial functions of minimal degree. *AiCM*, **4**, 389–396.

Wendland, H. (1999). On the smoothness of positive definite and radial functions. *Journal of Computational and Applied Mathematics*, **101**, 177–188.

Widder, D.V. (1946). *The Laplace Transform*. Princeton University Press, Princeton.

Wu, Z. (1995). Multivariate compactly supported positive definite radial functions. *AiCM*, **4**, 283–292.

2
Approximation and interpolation with radial functions

M.D. BUHMANN

Abstract

This chapter gives a short, up-to-date survey of some recent developments in the research on radial basis functions. Among its other new achievements, we consider results on convergence rates of interpolation with radial basis functions, and also recent contributions on approximation on spheres and on computation of interpolants with Krylov space methods.

2.1 Introduction

Research into radial basis functions is an immensely active and fruitful field at present and it is important and worthwhile to stand back and summarize the newest developments from time to time. In brief, this is the goal of this chapter, although we will by necessity be far from comprehensive. One of the most important aspects from the perspective of approximation theorists is the accuracy of approximation with radial basis functions when the centers are scattered. This is a subject quite suitable to begin this review with, as the whole development of radial basis functions was initiated by Duchon's contributions (1976,1978,1979) on exactly this question in a special context, especially for thin-plate splines approximation in $\mathbb{R}^2$.

Before we begin, we recall what is understood by approximation and interpolation by radial basis function. We always start with a univariate continuous function – the radial function – ϕ that is radialized by composition with the Euclidean norm on $\mathbb{R}^n$, or a suitable replacement thereof when we are working on an $(n-1)$ sphere in n-dimensional Euclidean space. Of course, due to radialization, we are no longer in a univariate setting and this is not a univariate theory. We are, in addition, given "centers" ξ from a finite set Ξ of distinct points in $\mathbb{R}^n$ (or in a compact subset Ω thereof, or from the sphere) which are simultaneously used for shifting the radial basis function and as interpolation (collocation) points. Our standard approximants therefore, have the form of linear combinations

$$s(x) = \sum_{\xi \in \Xi} \lambda_\xi \phi(\|x - \xi\|), \qquad x \in \mathbb{R}^n, \tag{2.1}$$

suitable adjustments being made when x is, e.g., from a compact Ω, and the coefficient vector $\lambda = (\lambda_\xi)_{\xi \in \Xi}$ is from $\mathbb{R}^\Xi$. It is important to know when such sums can be chosen to interpolate a given function at the centers. In many instances, especially when ϕ has compact support, the interpolation requirements

$$s\mid_\Xi = f\mid_\Xi \tag{2.2}$$

for given data $f\mid_\Xi$ lead to a positive definite interpolation matrix $A = \{\phi(\|\xi - \zeta\|)\}_{\xi,\zeta \in \Xi}$ and therefore the coefficients are uniquely determined by (2.2). In that case, we also call the radial basis function "positive definite". In other instances, when the radial function is only strictly *conditionally* positive definite of some order k, polynomials $p(x) \in \mathbb{P}_n^{k-1}$ of degree $k-1$ in n unknowns, are augmented to the right-hand side of (2.1) so as to render the interpolation problem again uniquely solvable. Of course, the linear system resulting from (2.2) is then under-determined. The extra degrees of freedom are taken up by requiring that the coefficient vector $\lambda \in \mathbb{R}^\Xi$ be orthogonal to the polynomial space $\mathbb{P}_n^{k-1}(\Xi)$, i.e., all polynomials of total degree less than k in n variables restricted to Ξ:

$$\lambda \perp \mathbb{P}_n^{k-1}(\Xi) \iff \sum_{\xi \in \Xi} \lambda_\xi q(\xi) = 0, \ \forall \ q \in \mathbb{P}_n^{k-1}. \tag{2.3}$$

In order to retain uniqueness, the set of centers Ξ must be $\mathbb{P}_n^{k-1}$-unisolvent in this case.

There are a number of important theoretical questions in this context. One is whether the coefficients in (2.1) are uniquely defined. Another is whether there are suitable radial basis functions which have compact support (see Schaback and Wendland (2000)). Finally, a central question is that of convergence and convergence rates of those interpolants to the function f that is being approximated by collocation to $f\mid_\Xi$ if f is in a suitable function space, usually a Sobolev space. Duchon (1976,1978,1979) has given answers to this question that address the important special case of thin-plate splines $\phi(r) = r^2 \log r$, especially in two dimensions, and its brethren. Recent research by Bejancu, Johnson, Powell, Schaback and Wu, has improved some of Duchon's 20-year-old results on L^p convergence orders in various directions, including inverse theorems and theorems about optimality of convergence orders. Several of these relevant results we address in the following section, and they hold in all dimensions.

Since this book also contains another chapter (by Schaback and Wendland) on radial basis functions, where much attention is given to those functions with compact support, we only devote a short section to compactly supported ϕs. They have many applications, for instance for numerical solutions of partial differential equations, when those radial functions are used because they can act as finite elements, and they are being tested at the moment by several researchers for this very purpose. (See, e.g., Fasshauer (1999), Franke and Schaback (1998), Pollandt (1997).)

In Section 2.4, several recent results about the efficient implementation of radial basis function interpolation by Krylov space methods are presented, notably iterative methods with guaranteed convergence, for the computation of interpolants when the number $|\Xi|$ of centers is very large. This is then necessary because no direct or simple iterative methods will work satisfactorily, the matrices resulting from (2.2) and (2.3) being large and ill-conditioned and sometimes exceptionally so with exponentially increasing condition numbers (Schaback (1994)).

The final section is devoted to convergence questions for radial basis functions on spheres. The multitude of possible applications (Hardy (1990)) of radial basis functions for interpolating data, especially on the earth's surface, be it potential or temperature data for instance, has led many researchers to consider the question of how to approximate accurately and efficiently when the data are from a sphere and when the whole idea of distance that goes into the definition of the above approximants is adjusted properly to distances on spheres, i.e. geodesic distances.

Of course, this brief review is not comprehensive and covers material from the author's perspective. However, in conjunction with the chapter by Schaback and Wendland, most issues on radial basis functions which are currently being investigated should be accessible to the reader.

2.2 Convergence rates

The paradigm of globally supported radial basis functions is the thin-plate spline $\phi(r) = r^2 \log r$ in two dimensions, and more generally, radial basis functions of thin-plate spline type in n dimensions, namely,

$$\phi(r) = \begin{cases} r^{2k-n} \log r & \text{if } 2k - n \text{ is an even integer,} \\ r^{2k-n} & \text{if } 2k - n \text{ is not an even integer,} \end{cases} \quad (2.4)$$

where we admit nonintegral k and always demand $k > \frac{n}{2}$. In this context, the approximants are usually taken from the "native spaces" X of distributions (see Jones (1982), for instance, on distributions in the Schwartz sense) in n unknowns whose total kth degree partial derivatives are square-integrable. We call this space $X := D^{-k}L^2(\mathbb{R}^n)$. The Sobolev embedding theorem tells us that this space consists of continuous functions so long as $k > \frac{n}{2}$. In fact it is continuously embedded therein.

As a first result we state a convergence result that holds in a domain Ω for general scattered sets of centers (Powell (1994)). In stating our convergence results we always take

$$h := \sup_{x \in \Omega} \inf_{\xi \in \Xi} \|x - \xi\|. \tag{2.5}$$

Theorem 2.2.1 *Let $k = n = 2$ and Ω be a bounded set which is not a subset of a straight line. Let s be the radial basis function interpolant to $f|_\Xi$ satisfying (2.2) for $\Xi \subset \Omega$, where k keeps the same meaning, i.e., linear polynomials are added to s and the appropriate side conditions (2.3) demanded. Then there is an h-independent C such that*

$$\|s - f\|_{\infty,\Omega} \leq Ch\sqrt{-\log h}, \qquad 0 < h < 1.$$

If domains as general as those in the statement of the theorem are admitted, then this is the best possible bound, i.e., the constant C on the right-hand side cannot be replaced by $o(1)$. Bejancu (1997) has generalized this result to arbitrary k and n, and his theorem includes the above result. There are no further restrictions on Ω except, in general, that it is $\mathbb{P}_n^{k-1}$-unisolvent (which we need for the following theorem).

Theorem 2.2.2 *Let Ω be bounded and $\mathbb{P}_n^{k-1}$-unisolvent. Let s be the radial basis function interpolant to $f|_\Xi$ for $\Xi \subset \Omega$ satisfying (2.2) where k keeps the same meaning, i.e., $\mathbb{P}_n^{k-1}$ polynomials are added to s with the appropriate side conditions (2.3) on the $\lambda \in \mathbb{R}^\Xi$. Then there is an h-independent C such that*

$$\|s - f\|_{\infty,\Omega} \leq C \begin{cases} h\sqrt{-\log h} & \text{if } 2k - n = 2, \\ \sqrt{h} & \text{if } 2k - n = 1, \text{ and} \\ h & \text{in all other cases,} \end{cases} \qquad 0 < h < 1.$$

Johnson (1998b) has proved uniform convergence orders of $O(h^{2k-n/2})$ that do require, however, Ω to be an open domain with Lipschitz continuous boundary $\partial \Omega$ and satisfying an interior cone condition (cf. Duchon (1976,

1978,1979)). Now (2.4) for general n and k are admitted, and H^{2k} denotes the usual Sobolev space.

Theorem 2.2.3 *Let Ω be open, bounded and satisfying the cone property, and let $f \in H^{2k}(\Omega)$ be supported in $\bar{\Omega}$. Then we have for any $\Xi \subset \Omega$ which is $\mathbb{P}_n^{k-1}$- unisolvent, the error estimate*

$$\|s - f\|_{p,\Omega} \leq C\|f\|_{H^{2k}} h^{2k+\min[n/p-n/2,0]}, \qquad 0 < h < 1, \qquad (2.6)$$

in the p-norm, where $1 \leq p \leq \infty$. Here s is the radial basis function interpolant to f on Ξ and C is some h-independent and f-independent constant.

Note that the optimal rate occurs in (2.6) when $1 \leq p \leq 2$.

It is of particular interest to have *upper bounds* on the obtainable convergence rate rather than the lower bounds thereon as above. Johnson (1998a) shows that the rates of Theorem 2.2.3 are almost optimal (as is well known, the optimal rates in case of $\Xi = h\mathbb{Z}^n$ and $p = \infty$ are h^{2k}, Buhmann (1990a,b), Johnson (1997)). We let the radial basis function still be from the above class (2.4) of thin-plate spline type functions.

Theorem 2.2.4 *Let $1 \leq p \leq \infty$ and let Ω be the unit ball. Suppose $\{\Xi_h \subset \Omega\}_{h>0}$ is a sequence of finite sets of centers with distance h when we set $\Xi = \Xi_h$ in (2.5) for each h. Then there exists an infinitely smooth f such that, for the best $L^p(\Omega)$-approximation s to f of the form (2.1) with the appropriate polynomials added, the error on the left-hand side of (2.6) is not $o(h^{2k+1/p})$ as h tends to zero.*

Comparing the last two results we observe that, in Theorem 2.2.3 we are not far from the optimal result. We do get the optimal rates of $O(h^{2k})$ for uniform convergence on grids $\Xi = h\mathbb{Z}^n$ as mentioned already, and also, as Bejancu (1997,2000) shows us, on finite grids $\Xi = \Omega \cap h\mathbb{Z}^n$, where we relax the interpolation conditions and only claim the existence of a suitable approximant of the form

$$s(x) = \sum_{\xi \in \Xi} \lambda_\xi \phi(\|x - \xi\|) + p(x).$$

Theorem 2.2.5 *Let Ω be compact, $k \in \mathbb{N}$, and let Ξ, $\tilde{\Omega}$ and the approximant s be as above. Then, for $f \in C^{2k}(\Omega)$, with $\operatorname{supp} f \subset \tilde{\Omega}$, there are λ as in (2.3) and $p \in \mathbb{P}^{k-1}$, such that*

$$\|s - f\|_{\infty,\Omega} \leq Ch^{2k}, \qquad 0 < h < 1.$$

Looking at Theorem 2.2.4 we see that this is the obtainable (saturation) order, and an inverse theorem of Schaback and Wendland (1998) tells us that all functions for which a better order is obtainable must be trivial in the sense of polyharmonic functions. That is, they are in the kernel of an iterated Laplace operator. In the following statement we use the standard notation Δ for the Laplace operator.

Theorem 2.2.6 *Let Ω be as in Theorem 2.2.3, $\Xi \subset \Omega$ a finite set of centers with distance h as in (2.5) and ϕ as in (2.4). If for a given $f \in C^{2k}(\Omega)$, and all compact $\tilde{\Omega} \subset \Omega$,*

$$\|s - f\|_{\infty, \tilde{\Omega}} = o(h^{2k}), \qquad 0 < h < 1,$$

then $\Delta^k f = 0$ on Ω.

Inverse theorems may be formulated in a variety of ways. Another way, from the same paper of Schaback and Wendland, is the following one with which we close this section. We recall that for a radial basis function with distributional Fourier transform that agrees with a positive function $\hat{\phi}(\|\cdot\|) : \mathbb{R}^n \setminus \{0\} \to \mathbb{R}$ (Estrada (1998)), the native space norm (which is, strictly speaking, usually a *semi*-norm) is

$$\|f\|_\phi^2 := \frac{1}{(2\pi)^n} \int_{\mathbb{R}^n} \frac{1}{\hat{\phi}(\|t\|)} |\hat{f}(t)|^2 \, dt. \tag{2.7}$$

The native space X is the space of all distributions f on $\mathbb{R}^n$ for which (2.7) is finite. In the case of the radial basis functions (2.4), X agrees with $D^{-k}L^2(\mathbb{R}^n)$, because $\hat{\phi}(\|t\|)^{-1}$ is a constant multiple of $\|t\|^{2k}$. Therefore if we take for simplicity $\hat{\phi}(\|t\|) = \|t\|^{-2k}$, then equation (2.7) becomes

$$\|f\|_\phi^2 := \frac{1}{(2\pi)^n} \int_{\mathbb{R}^n} \|t\|^{2k} |\hat{f}(t)|^2 \, dt.$$

We have already addressed the notion of conditionally positive functions in the introduction. The functions (2.4) are strictly conditionally positive definite of order k subject to a possible sign change. We use the notation $\tau(x) \sim t(x)$ if both $t(x)/\tau(x)$ and $\tau(x)/t(x)$ are uniformly bounded for the appropriate range of x.

Theorem 2.2.7 *Let Ω be a domain as in Theorem 2.2.3. Let ϕ be strictly conditionally positive definite of order k and*

$$\hat{\phi}(r) \sim r^{-2k}, \qquad r > 0.$$

If there is a $\mu > k$ such that, for some $f \in C(\Omega)$, and for the radial basis function interpolants s thereof on all finite sets of centers $\Xi \subset \Omega$ with distance (2.5), we have

$$\|s - f\|_{\infty,\Omega} \leq Ch^{\mu}, \qquad h \to 0,$$

with an f-dependent but h-independent C, then $f \in X$.

An example of an application of this result is the radial basis function (2.4) with the k there and in Theorem 2.2.7 being the same.

2.3 Compact support

There exist at present essentially two approaches to constructing univariate, compactly supported $\phi : \mathbb{R}_+ \to \mathbb{R}$, such that the interpolation problem is uniquely solvable with a positive definite collocation matrix $A = \{\phi(\|\xi - \zeta\|)\}_{\xi,\zeta\in\Xi}$. As usual for radial basis function interpolation, there will be no restriction on the geometry of the set Ξ of centers, but there are (there must be in fact) bounds on the maximal spatial dimension n which is admitted for each radial function ϕ so that positive definiteness is retained.

Following an initial idea of Wu, there is an approach, due both to Wu and Wendland, where the radial basis functions consist only of one polynomial piece on the unit interval $[0,1]$ and are otherwise zero. Although this means that the radial basis functions are piecewise polynomials as univariate functions, the resulting approximants are, of course, not. This whole idea begins with Askey (1973), see also Misiewicz and Richards (1994), who observed that $\phi(r) = \phi_0(r) = (1-r)_+^{\ell}$ gives rise to positive definite interpolation matrices A so long as $\ell \geq [n/2] + 1$. Already here we see, incidentally, a bound on the spatial dimension n if we fix the degree of the piecewise polynomial. The construction and examples are given in Schaback and Wendland (2000). They use the so-called *operator on radial functions*

$$If(x) := \int_x^{\infty} rf(r)\,dr, \qquad x > 0,$$

defined for suitably decaying f. The said interpolation matrix A remains positive definite if the basis function

$$\phi(r) = \phi_{n,k}(r) = I^k \phi_0(r), \qquad r \geq 0, \tag{2.8}$$

is used when $\ell \geq k + [n/2] + 1$. They take $\ell = k + [n/2] + 1$.

Theorem 2.3.1 *The radial function* (2.8) *gives rise to a positive definite* interpolation matrix A with centers Ξ in $\mathbb{R}^n$. Among all such functions for dimension n and smoothness C^{2k} it is of minimal polynomial degree.

Theorem 2.3.2 *Let ϕ be as in (2.8), let $f \in H^{k+(n+1)/2}(\mathbb{R}^n)$, and k be at least $\frac{1}{2}$ so long as $n = 1$ or $n = 2$. Then for a compact domain Ω with centers Ξ inside Ω, the interpolant (2.1) satisfies*

$$\|s - f\|_{\infty,\Omega} = O(h^{k+1/2}), \qquad h \to 0.$$

The definition of h is the same as in (2.5).

Another class of radial basis functions of compact support (Buhmann (1998, 1999)) is of a certain convolution form and contains, for example, the following cases. (We state the value of the function only on the unit interval, elsewhere it is zero. It can, of course, be suitably scaled.) Two examples that give twice and three-times continuously differentiable functions, respectively, in three and two dimensions are as follows. The parameter choices $\alpha = \delta = \frac{1}{2}$, $\rho = 1$, and $\lambda = 2$, in the theorem below, give $(n = 3)$

$$\phi(r) = 2r^4 \log r - \frac{7}{2}r^4 + \frac{16}{3}r^3 - 2r^2 + \frac{1}{6}, \qquad 0 \le r \le 1,$$

while the choices $\alpha = \frac{3}{4}$, $\delta = \frac{1}{2}$, $\rho = 1$, and $\lambda = 2$, provide, for $n = 2$,

$$\phi(r) = \frac{112}{45}r^{\frac{9}{2}} + \frac{16}{3}r^{\frac{7}{2}} - 7r^4 - \frac{14}{15}r^2 + \frac{1}{9}, \qquad 0 \le r \le 1.$$

Note that there is no log term in the above radial basis function of compact support.

Theorem 2.3.3 *Let $0 < \delta \le \frac{1}{2}$, $\rho \ge 1$ be reals, and suppose $\lambda \ne 0$ and α are also real quantities with*

$$\lambda \in \begin{cases} (-\frac{1}{2}, \infty), & -1 < \alpha \le \min[\frac{1}{2}, \lambda - \frac{1}{2}], & \text{if } n = 1 \text{ or} \\ [1, \infty), & -\frac{1}{2} < \alpha \le \frac{1}{2}\lambda, & \text{if } n = 1, \text{ and} \\ (-\frac{1}{2}, \infty), & -1 < \alpha \le \min[\frac{1}{2}(\lambda - \frac{1}{2}), \lambda - \frac{1}{2}], & \text{if } n = 2, \text{ and} \\ [0, \infty), & -1 < \alpha \le \frac{1}{2}(\lambda - 1), & \text{if } n = 3, \text{ and} \\ (\frac{1}{2}(n-5), \infty), & -1 < \alpha \le \frac{1}{2}(\lambda - \frac{1}{2}(n-1)), & \text{if } n > 3. \end{cases}$$

Then the radial basis function

$$\phi(r) = \int_0^\infty \left(1 - r^2/\beta\right)_+^\lambda \beta^\alpha (1 - \beta^\delta)_+^\rho \, d\beta, \qquad r \ge 0, \tag{2.9}$$

has a positive Fourier transform and therefore gives rise to positive definite interpolation matrices A with centers Ξ from $\mathbb{R}^n$. Moreover $\phi(\|\cdot\|) \in C^{1+\lceil 2\alpha \rceil}(\mathbb{R}^n)$.

Note that in the statement of the following theorem, the approximand f is continuous by the Sobolev embedding theorem since $1+\alpha$ is positive by the conditions in the previous theorem.

Theorem 2.3.4 *Let ϕ be as in the previous theorem and suppose additionally $\rho > 1$ and $2\alpha \leq \lambda - n/2 - 3 + \lfloor \rho \rfloor$. Let Ξ be a finite set of distinct centers in a compact domain Ω. Let s be the interpolant (2.1) to $f \in L^2(\mathbb{R}^n) \cap D^{-n/2-1-\alpha}L^2(\mathbb{R}^n)$, scaled of the form (for a positive scaling parameter η)*

$$s(x) = \sum_{\xi \in \Xi} \lambda_\xi \phi(\eta^{-1}\|x-\xi\|), \qquad x \in \mathbb{R}^n,$$

with the interpolation conditions $(s-f)|_\Xi = 0$ satisfied. Then the uniform convergence estimate

$$\|f - s\|_{\infty, \Omega} \leq Ch^{1+\alpha}\eta^{-n/2-1-\alpha}$$

holds for $h \to 0$ and positive bounded η, the positive constant C being independent of both h and η.

2.4 Iterative methods for implementation

The topic we wish to discuss in this section is the practical computation of radial basis function interpolation in the thin-plate spline case. For this there is a Krylov space method which uses local Lagrange functions, related to the so-called domain decomposition methods. Another highly relevant and useful class of methods are the fast multipole approaches which we do not describe here (but we recommend reading Beatson and Newsam (1992), Beatson and Light (1997) and Beatson et al. (1998)). Like the fast multipole methods, the Krylov space algorithm (Faul and Powell (1999), see also Saad and Schultz (1986)) is iterative and depends on an initial structuring of the centers Ξ prior to the start of the iteration. This initial structuring is less complicated than the fairly sophisticated hierarchical structure demanded for the multipole schemes, especially when $n > 2$.

We explain the method for $n = 2$ and thin-plate splines. Hence, the interpolant we wish to compute has the same form as in the second section for $k = n = 2$ and $\phi(r) = r^2 \log r$. But we denote by s^* the actual interpolant we seek, whereas s will now denote only the active approximation to s^* at each stage of the algorithm. The basic idea of the algorithm is to derive, from minimizing $\|s - s^*\|_\phi$, an s from the so-called Krylov space, by a line search method along mutually orthogonal directions. The directions depend on so-called local Lagrange functions. We recall the Lagrange formulation of the interpolant

$$s^*(x) = \sum_{\xi \in \Xi} f(\xi) L_\xi(x), \qquad x \in \mathbb{R}^n, \tag{2.10}$$

where each Lagrange function L_ξ satisfies the Lagrange conditions

$$L_\xi(\zeta) = \delta_{\zeta\xi}, \qquad \xi, \zeta \in \Xi, \tag{2.11}$$

and is of the form

$$L_\xi(x) = \sum_{\zeta \in \Xi} \lambda_{\zeta\xi} \phi(\|x - \zeta\|) + p_\xi(x), \qquad x \in \mathbb{R}^n. \tag{2.12}$$

Here $p_\xi \in \mathbb{P}_2^1$ and $\lambda_{\cdot,\xi} \perp \mathbb{P}_2^1(\Xi)$. Since the computation of such full Lagrange functions would be just as expensive as solving the full, usually linear interpolation system of equations, the idea here is to replace (2.11) by local Lagrange conditions which require for each ξ only that the identity holds for some $q = 30$, say, points, ζ that are near ξ. We therefore take Lagrange functions that are of the form (2.12), but all of whose coefficients vanish, except q of them from a certain set of "active points".

So we can associate with each ξ the Lagrange functions L_ξ and its active point set of centers, which we denote $\mathcal{L}_\xi \subset \Xi$, $|\mathcal{L}_\xi| = q$. In addition to having q elements, $\mathcal{L}_\xi$ must contain a unisolvent subset with respect to $\mathbb{P}_n^{k-1}$ as required previously. We also extract another set of centers Σ of approximately the same size q from Ξ for which no local Lagrange functions are computed because the set is sufficiently small to allow direct solution of the interpolation problem. That set must also contain a unisolvent subset.

In order that an iterative method can be applied, we order the centers $\Xi \setminus \Sigma = \{\xi_i\}_{i=1}^m$ for which local Lagrange functions are computed and include the final requirement that for all $i = 1, 2, \ldots, m$,

$$\{\xi_i\} \subset \mathcal{L}_{\xi_i} \subset \Xi \setminus \{\xi_1, \xi_2, \ldots, \xi_{i-1}\}.$$

All the elements of $\mathcal{L}_{\xi_i}$, except ξ_i itself, should be chosen so that they are from among all ξ_j, $j = i+1, i+2, i+3, \ldots, m$, closest to ξ_i. There may be equalities in this reordering that may be broken by random choice. The Lagrange conditions that must be satisfied by the local Lagrange functions are now

$$L_\xi(\zeta) = \delta_{\zeta\xi}, \qquad \zeta \in \mathcal{L}_\xi. \tag{2.13}$$

Using these local Lagrange functions, the right-hand side of (2.10) is not a representation of the exact approximant s^*, but only an approximation thereof. Here is where the iteration comes in. (Of course the Lagrange functions are computed in advance once and only once, and stored prior to the beginning of the iterations. This is an $O(|\Xi|)$ process.)

In the iterations, we refine the approximation at each step of the algorithm by correcting its residual through updates similar to a conjugate gradient method

$$s(x) \longrightarrow s(x) + \alpha \cdot d, \tag{2.14}$$

where d is a "search direction" defined as

$$d = \eta(s^* - s) + \beta \tilde{d}. \tag{2.15}$$

In (2.14) α is chosen so that $\|s - s^*\|_\phi$ becomes smaller, and in (2.15), $\tilde{d}$ is the search direction from the previous sweep, β is chosen so that $d \perp \tilde{d}$ with respect to the semi-inner product associated with $\|\cdot\|_\phi$, and η is a certain prescribed operator about which we shall comment below. The correction (2.15) is added in each sweep of the algorithm.

This completes the sweep of the algorithm. If we started with an approximation $s = s_0 = 0$ to s^*, the sweep replaces s_j by s_{j+1} that goes into the next sweep. The stopping criterion can be, for instance, that we terminate the algorithm if all the residuals on Ξ are sufficiently small. Let U be the linear space spanned by $\phi(\|\cdot - \xi\|)$, $\xi \in \Xi$, plus linear polynomials. Let U_j be the space spanned by $s^*, \eta(s^*), \eta(\eta(s^*)), \ldots, \eta^j(s^*)$. Those should not be used as bases for U_j as they are ill-conditioned. Then the goal of the above choices is to let, in the jth iteration, $s_{j+1} \in U_j$ be the element which minimizes $\|s - s^*\|_\phi$ among all $s \in U_j$. Now let $\eta : U \mapsto U$ be the map

$$s \mapsto \sigma + \sum_{k=1}^{m-q} \frac{(L_k, s) L_k}{\lambda_{\xi_k \xi_k}},$$

where $(g, g) = \|g\|_\phi^2$, and where σ is the full standard thin-plate spline solution of the interpolation problem

$$\sigma(\xi) = s(\xi), \quad \xi \in \Sigma. \tag{2.16}$$

Faul and Powell (1999) prove the following for the above method.

Theorem 2.4.1 *Let $\{s_j\}_{j=0}^\infty$ be a sequence of approximations to s^* generated by $s_0 = 0$ and the above algorithm, where the radial function is the thin-plate spline function and $n = 2$. Then $s_j \to s^*$, as $j \to \infty$.*

In fact, their proof shows that for *all* such operators η that recover elements of the nullspace of $\|\cdot\|_\phi$, and satisfy $(\eta(s), t) = (s, \eta(t))$ and $(s, \eta(s)) > 0$ if

$\|s\|_\phi > 0$, the above method enjoys guaranteed convergence with uniquely defined search directions. The semi-inner product $(\cdot,\cdot)$ is still the one associated with $\|\cdot\|_\phi$. It turns out in practice that this method often provides excellent accuracy with fewer than ten iterations.

Another approach to the iterative approximation and refinement of the radial basis function interpolants is that of *fast multipole methods*. We briefly outline these here. This is explained in detail in Buhmann (2000). However they are sufficiently important that we must, at least, outline the essentials.

These algorithms are based on analytic expansions of the underlying radial functions for large argument (see Greengard and Rokhlin (1987)).

The methods require that data be structured in a hierarchical way before the onset of the iteration, and also an initial computation of the so-called far-field expansions. The far-field expansions use the fact that radial basis functions of the form (2.4) are analytic except at the origin, even when made multivariate through composition with Euclidean norms, the derivative discontinuity still restricted to the origin, and can be approximated well away from the origin by a truncated Laurent expansion. The accuracy can be preset and be arbitrarily small, but of course the cost of the method (here the multiplier in the operation count), rises with higher accuracy. The multipliers in the operational count can be large and therefore the method is only worthwhile if $|\Xi|$ is large, say, 1000 or more. Using the hierarchical structure of the center set below, this method allows for simultaneous approximate evaluation of many radial basis function terms, whose centers are close to each other, by a single finite Laurent expansion. This is cheap, as long as the argument x is far from the "cloud" of centers. By contrast, all radial functions whose translates are close to x are computed exactly and explicitly, usually with a direct method.

The hierarchical structure of the centers is fundamental to the algorithm because it decides and orders what is far from and what is near any given x where we wish to evaluate our function. In a simple two-dimensional case it is built up as follows. We assume, for ease of exposition, that the ξ are fairly uniformly distributed in a square. We form a *tree* of centers which contains Ξ as a root, and as the next children, the intersections of the four quarter squares of the unit square with Ξ. These in turn are divided into four grandchildren, each in the same fashion, and so on. We stop this process at a predetermined level that in part determines the accuracy of the calculation in the end. Each member of this family is called a panel. Recall that our aim is to evaluate fast at a single point x. Given x, all centers ξ that are in the near-field give rise to explicit evaluations of $\phi(\|x-\xi\|)$. The so-called near-field consists simply of contributions from all points which are not "far", according to the following definition: we say x is far away from a panel T and therefore from all centers in that panel, if there is at least one more panel between x and T, and if this panel is on the same level of parenthood as T itself.

Having dealt with the near-field, what remains now is the far-field. We have to decide how to group the far points. All panels Q are in the *evaluation list* of a panel T if Q is either at the same or a coarser (that is, higher) level than T and every point in T is far away from Q, and if, finally, T contains a point that is *not* far away from the parent of Q. Thus the far field of an x, whose closest center is in a panel T, is the sum of all

$$s_Q(x) = \sum_{\xi \in Q} \lambda_\xi \phi(\|x - \xi\|) \tag{2.17}$$

such that Q is in the evaluation list of T. For each (2.17), a common Laurent series is computed. Since we do not know the value of x that is to be inserted into (2.17), we compute the *coefficients* of the Laurent series. These do not depend on x but on Ξ.

Recall that much of the work has to be done at the onset of the algorithm. In this set-up process, we compute the expansions of the radial function for large argument, i.e., their coefficients, and store them. When x is provided at the evaluation stage, we combine those expansions for all centers from each evaluation list, and finally approximate the whole far field by one Laurent series. To explain this in detail is long and technical because the various expansions which are used have to be stated and their accuracy estimated. We therefore refer, for example, to the paper by Beatson and Newsam (1992).

2.5 Interpolation on spheres

Beginning with Hardy (1990), there was a lot of interest in approximations on 2 and higher dimensional spheres due to the many applications, for instance in geodesy. In fact the work on radial basis functions on spheres, in itself, has developed into an important branch of approximation theory. Freeden and co-workers (1981,1986,1995,1998) have contributed much to this aspect of the theory. There are excellent and lengthy review papers available on the work of this group (see the cited references) and others. One of the main goals of that work is to determine useful classes of radial basis functions on spheres, e.g, such that interpolants exist and are unique. We no longer use the conventional Euclidean norm in connection with a univariate radial function when we approximate on the $(n-1)$ sphere S^{n-1}

within $\mathbb{R}^n$. Rather we apply geodesic distances $d(x,y) = \arccos(x^T y)$. The standard notions of positive definite functions and conditional positive definiteness no longer apply, and one has to study new concepts of (strictly conditionally) positive definite functions on the $(n-1)$ sphere. This study was, as so many other aspects of approximation theory, initiated long ago by Schoenberg (1942). He characterized positive definite functions on spheres as those whose expansions in the series of normalized Gegenbauer polynomials have nonnegative coefficients. Extending this work, Xu and Cheney (1992) characterized strict positive definiteness on spheres and gave necessary and sufficient conditions. It is sufficient that all of the aforementioned coefficients are positive. This was further generalized by Ron and Sun (1996). They showed that it suffices for strict positive definiteness for a fixed number m of centers, that the coefficients $a_k > 0$ for all $k \leq \frac{1}{2}m$. It is also sufficient that $\{k \mid a_k > 0\}$ contains arbitrarily long sequences of consecutive odd and consecutive even integers k. Quantitative results on existence and stability of interpolants can be found in Narcowich et al. (1998).

Papers on approximation orders by Jetter et al. (1999) and Levesley et al. (1999) (see also Dyn et al. (1997)), use native spaces (see below), (semi-)inner products and reproducing kernels (cf. Saitoh (1988) for the theory of reproducing kernels) to derive approximation orders in a way similar in fashion to that summarized in Section 2.2. They all use the *spherical harmonics* $\{Y_k^{(\ell)}\}_{k=1}^{d_\ell}$ that form an orthonormal basis for the d_ℓ dimensional space of polynomials on the sphere in $\mathbb{P}_n^\ell(S^{n-1}) \cap \mathbb{P}_n^{\ell-1}(S^{n-1})^\perp$. They are called spherical harmonics because they are the restrictions of polynomials of total degree ℓ to the sphere and are in the d_ℓ-dimensional kernel of the Laplace operator Δ.

A *native space* X is defined via the expansion of functions on the sphere in spherical harmonics. Namely,

$$X = \left\{ f \mid \sum_{\ell=0}^{\infty} \sum_{k=1}^{d_\ell} \frac{|\hat{f}_{\ell k}|^2}{a_{\ell k}} < \infty \right\}, \tag{2.18}$$

where

$$f(x) = \sum_{\ell=0}^{\infty} \sum_{k=1}^{d_\ell} \hat{f}_{\ell k} Y_k^{(\ell)}(x), \qquad x \in S^{n-1}, \tag{2.19}$$

and the $a_{\ell k}$ are prescribed positive real weights.

The native space has to be equipped both with an inner product and a (semi-)norm. In this case, the inner product can be described by the double sum

$$\langle f, g \rangle = \sum_{\ell=0}^{\infty} \sum_{k=1}^{d_\ell} \frac{1}{a_{\ell k}} \hat{f}_{\ell k} \hat{g}_{\ell k},$$

where the $a_{\ell k}$ are the coefficients that appear in (2.18). They are still assumed to be positive. The reproducing kernel that results from this Hilbert space X with the above inner product and that corresponds to the function of our previous radial basis functions in the native space is, when x and y are on the sphere, given by

$$\phi(x, y) = \sum_{\ell=0}^{\infty} \sum_{k=1}^{d_\ell} a_{\ell k} Y_k^{(\ell)}(x) Y_k^{(\ell)}(y), \qquad x, y \in S^{n-1}. \qquad (2.20)$$

Applying the famous addition theorem (Stein and Weiss (1971)) to $\phi(x, y) = \phi(x^T y)$, this can be written as

$$\phi(t) = \frac{1}{\omega_{n-1}} \sum_{\ell=0}^{\infty} d_\ell a_{\ell k} P_\ell(t), \qquad (2.21)$$

ω_{n-1} being the measure of the unit sphere, if the coefficients are constant with respect to k. Here, P_ℓ is the Gegenbauer polynomial (Abramowitz and Stegun (1972)) normalized to be one at one. We now use (2.20) or (2.21) for interpolation on the sphere, in the same place and with the same centers Ξ as before. Convergence estimates are available from the three articles mentioned above. Jetter et al. (1999) prove, using the mesh norm associated with the set of centers Ξ,

$$h = \sup_{x \in S^{n-1}} \inf_{\xi \in \Xi} \arccos(x^T \xi),$$

the following theorem.

Theorem 2.5.1 *Let X and Ξ be as above with the given mesh norm h. Let K be a positive integer such that $h \leq 1/(2K)$. Then, for any $f \in X$, there is a unique interpolant s in*

$$\mathrm{span}\left\{\phi(\xi, \cdot) \mid \xi \in \Xi\right\}$$

that interpolates f on Ξ and satisfies the error estimate

$$\|s - f\|_\infty^2 \leq \frac{5(|\Xi| + 1)}{\omega_{n-1}} \|f\|_\phi^2 \sum_{\ell = K+1}^{\infty} d_\ell \max_{1 \leq k \leq d_\ell} a_{\ell k}.$$

Corollary 2.5.2 *Let the assumptions of the previous theorem hold and suppose further that $|\Xi| + 1 \leq C_1 K^{n-1}$ and*

$$\frac{C_2}{1+K} \leq h \leq \frac{1}{2K}.$$

Then the said interpolant s provides

$$\|s - f\|_\infty = O\left(\left(\frac{h}{C_2}\right)^{(\alpha-n)/2}\right)$$

or

$$\|s - f\|_\infty = O\left(\frac{\exp(-\alpha C_2/2h)}{h^{(n-1)/2}}\right),$$

respectively, if $d_\ell \times \max_{1 \leq k \leq d_\ell} a_{\ell k}$ is bounded by a constant multiple of $(1+\ell)^{-\alpha}$ for an $\alpha > n$ or by a constant multiple of $\exp(-\alpha(1+\ell))$ for a positive α, respectively.

An error estimate due to Levesley et al. (1999) (see also Golitschek and Light (2000)) which includes strictly conditionally positive definite kernels for orders $\kappa = 1$ or $\kappa = 2$ is for $n = 2$ (see also Freeden and Hermann (1986), for a similar, albeit weaker result).

Theorem 2.5.3 *For X, Ξ, h and κ as above, let s be the minimal norm interpolant to $f \in X_\kappa$ on Ξ (so that to s of Theorem 2.5.1 in particular a polynomial $p \in \mathbb{P}_3^{\kappa-1}(S^2)$ is added). When ϕ is twice continuously differentiable on $[1-\epsilon, \epsilon]$ for some $\epsilon \in (0,1)$, then*

$$\|s - f\|_\infty \leq C h^2 \|f\|_\phi.$$

Acknowledgment

It is a pleasure to thank Oleg Davydov for several thoughtful comments.

References

Askey, R. (1973). Radial characteristic functions. MRC Report 1262, University of Wisconsin–Madison.

Abramowitz, M. and Stegun, I. (1972). *Handbook of Mathematical Functions*. National Bureau of Standards, Washington.

Beatson, R.K., Cherrie, J.B. and Moat, C.T. (1998). Fast fitting of radial basis functions: methods based on preconditioned GMRES iteration. Technical Report Univ., Canterbury, Christchurch.

Beatson, R.K., Goodsell, G. and Powell, M.J.D. (1995). On multigrid techniques for thin plate spline interpolation in two dimensions. *Lectures in Applied Mathematics*, Vol. 32, pp. 77–97.

Beatson, R.K. and Light, W.A. (1997). Fast evaluation of radial basis functions: methods for 2-dimensional polyharmonic splines. *IMA J. Numer. Anal.*, **17**, 343–372.

Beatson, R.K. and Newsam, G.N. (1992). Fast evaluation of radial basis functions: I. *Comput. Math. Appl.*, **24(12)**, 7–19.

Bejancu, A. (1997). The uniform convergence of multivariate natural splines. DAMTP Technical Report, Univ. Cambridge.

Bejancu, A. (2000). On the accuracy of surface spline approximation and interpolation to bump functions. DAMTP Technical Report, Univ. Cambridge.

Buhmann, M.D. (1990a). Multivariate interpolation in odd - dimensional Euclidean spaces using multiquadrics. *Constr. Approx.*, **6**, 21–34.

Buhmann, M.D. (1990b). Multivariate cardinal-interpolation with radial-basis functions. *Constr. Approx.*, **6**, 225–255.

Buhmann, M.D. (1998). Radial functions on compact support. *Proceedings of the Edinburgh Mathematical Society*, **41**, 33–46.

Buhmann, M.D. (1999). A new class of radial basis functions with compact support. Technical Report, University of Dortmund.

Buhmann, M.D. (2000). Radial basis functions. *Acta Numerica*, **9**, 1–37.

Duchon, J. (1976). Interpolation des fonctions de deux variables suivant le principe de la flexion des plaques minces. *Rev. Française Automat. Informat. Rech. Opér. Anal. Numer.*, **10**, 5–12.

Duchon, J. (1978). Sur l'erreur d'interpolation des fonctions de plusieurs variables pars les D^m–splines. *Rev. Française Automat. Informat. Rech. Opér. Anal. Numer.*, **12**, 325–334.

Duchon, J. (1979). Splines minimizing rotation–invariate semi–norms in Sobolev spaces. In *Constructive Theory of Functions of Several Variables*, ed. W. Schempp and K. Zeller, pp. 85–100. Springer, Berlin–Heidelberg.

Dyn, N., Narcowich, F.J. and Ward, J.D. (1997). A framework for interpolation and approximation on Riemannian manifolds. In *Approximation and Optimization, Tributes to M.J.D. Powell*, ed. M.D. Buhmann and A. Iserles, pp. 133–144. Cambridge University Press, Cambridge.

Estrada, R. (1998). Regularization of distributions. *Int. J. Math. Math. Sci.*, **21**, 625–636.

Fasshauer, G. (1999). Solving differential equations with radial basis functions: multilevel methods and smoothing. Technical Report, IIT Chicago.

Faul, A.C. and Powell, M.J.D. (1998). Proof of convergence of an iterative technique for thin plate spline interpolation in two dimensions. DAMTP Technical Report, Univ. Cambridge.

Faul, A.C. and Powell, M.J.D. (1999). Krylov subspace methods for radial basis function interpolation. DAMTP Technical Report, Univ. Cambridge.

Floater, M. and Iske, A. (1996). Multistep scattered data interpolation using compactly supported radial basis functions. *J. Comp. Appl. Math.*, **73**, 65–78.

Franke, C. and Schaback, R. (1998). Solving partial differential equations by collocation using radial basis functions. *Comput. Math. Appl.*, **93**, 72–83.

Freeden, W. (1981). On spherical spline interpolation and approximation. *Math. Meth. in Appl. Sci.*, **3**, 551–575.

Freeden, W. and Hermann, P. (1986). Uniform approximation by harmonic splines. *Math. Z.*, **193**, 265–275.

Freeden, W., Gervens, T. and Schreiner, M. (1998). *Constructive Approximation on the Sphere*. Oxford Science Publications, Clarendon Press.

Freeden, W., Schreiner, M. and Franke, R. (1995). A survey on spherical spline approximation. Technical Report 95–157, Univ. Kaiserslautern.

Golitschek, M.V. and Light, W.A. (2000). Interpolation by polynomials and radial basis functions on spheres. Preprint.

Greengard, L. and Rokhlin, V. (1987). A fast algorithm for particle simulations. *J. Comput. Physics*, **73**, 325–348.

Hardy, R.L. (1990). Theory and applications of the multiquadric-biharmonic method. *Comput. Math. Appl.*, **19**, 163–208.

Jetter, K., Stöckler J. and Ward, J.D. (1999). Error estimates for scattered data interpolation on spheres. *Math. Comp.*, **68**, 733–747.

Johnson, M.J. (1997). An upper bound on the approximation power of principal shift-invariant spaces. *Constr. Approx.*, **13**, 155–176.

Johnson, M.J. (1998a). A bound on the approximation order of surface splines. *Constr. Approx.*, **14**, 429–438.

Johnson, M.J. (1998b). On the error in surface spline interpolation of a compactly supported function. Manuscript, Univ. Kuwait.

Jones, D.S. (1982). *The Theory of Generalised Functions*. Cambridge University Press, Cambridge.

Levesley, J., Light, W.A., Ragozin, D. and Sun, X. (1999). A simple approach to the variational theory for interpolation on spheres. In *Approximation Theory Bommerholz 1998*, ed. M.D. Buhmann, M. Felten, D. Mache and M.W. Müller. Birkhäuser, Basel.

Micchelli, C.A. (1986). Interpolation of scattered data: distance matrices and conditionally positive definite functions. *Constr. Approx.*, **1**, 11–22.

Misiewicz, J.K. and Richards, D.St.P. (1994). Positivity of integrals of Bessel functions. *SIAM J. Math. Anal.*, **25**, 596–601.

Narcowich, F., Sivakumar, N. and Ward, J.D. (1998). Stability results for scattered data interpolation on Euclidean spheres. *Adv. Comp. Math.*, **8**, 137–163.

Pollandt, R. (1997). Solving nonlinear equations of mechanics with the boundary element method and radial basis functions. *Internat. J. for Numerical Methods in Engineering*, **40**, 61–73.

Pottmann, H. and Eck, M. (1999). Modified multiquadric methods for scattered data interpolation over the sphere. *Comput. Aided Geom. Design*, **7**, 313–322.

Powell, M.J.D. (1992). The theory of radial basis function approximation in 1990. In *Advances in Numerical Analysis II: Wavelets, Subdivision, and Radial Functions*, ed. W.A. Light, pp. 105–210. Oxford University Press, Oxford.

Powell, M.J.D. (1993). Truncated Laurent expansions for the fast evaluation of thin plate splines. *Numer. Algorithms*, **5**, 99–120.

Powell, M.J.D. (1994). The uniform convergence of thin-plate spline interpolation in two dimensions. *Numer. Math.*, **67**, 107–128.

Ron, A. and Sun, X. (1996). Strictly positive definite functions on spheres in Euclidean spaces. *Math. Comp.*, **65**, 1513–1530.

Saad, Y. and Schultz, M.H. (1986). GMRES: a generalized minimum residual algorithm for solving nonsymmetric linear systems. *SIAM J. Sci. Stat. Comp.*, **7**, 856–869.

Saitoh, S. (1988). *Theory of Reproducing Kernels and its Applications*. Longman, Harlow.

Schaback, R. (1994). Lower bounds for norms of inverses of interpolation matrices for radial basis functions. *J. Approx. Theory*, **79**, 287–306.

Schaback, R. and Wendland, H. (1998). Inverse and saturation theorems for radial basis function interpolation. Technical Report, Univ. Göttingen.

Schaback, R. and Wendland, H. (2000). Characterization and construction of radial basis functions. This volume.

Schaback, R. and Wu, Z. (1996). Operators on radial basis functions. *J. Comp. Appl. Math.*, **73**, 257–270.

Schoenberg, I.J. (1942). Positive definite functions on spheres. *Duke Math. J.*, **9**, 96–108.

Schreiner, M. (1997). On a new condition for strictly positive definite functions on spheres. *Proc. AMS*, **125**, 531–539.

Sibson, R. and Stone, G. (1991). Computation of thin plate splines. *SIAM J. Scient. Stat. Comput.*, **12**, 1304–1313.

Stein, E.M. and Weiss, G. (1971). *Introduction to Fourier Analysis on Euclidean Spaces*. Princeton University Press, Princeton.

Steward, J. (1976). Positive definite functions and generalizations, a historical survey. *Rocky Mountains Math. J.*, **6**, 409–434.

Wendland, H. (1995). Piecewise polynomial, positive definite and compactly supported radial functions of minimal degree. *Advances in Computational Mathematics*, **4**, 389–396.

Wendland, H. (1997). Sobolev-type error estimates for interpolation by radial basis functions. In *Surface Fitting and Multiresolution Methods*, ed. A. LeMéhauté and L.L. Schumaker, pp. 337–344. Vanderbilt University Press, Nashville.

Wendland, H. (1998). Error estimates for interpolation by radial basis functions of minimal degree. *J. Approx. Theory*, **93**, 258–272.

Wu, Z. (1992). Hermite–Birkhoff interpolation of scattered data by radial basis functions. *Approx. Theory Appl.*, **8**, 1–10.

Wu, Z. (1995). Multivariate compactly supported positive definite radial functions. *Advances in Computational Mathematics*, **4**, 283–292.

Wu, Z. and Schaback, R. (1993). Local error estimates for radial basis function interpolation of scattered data. *IMA J. Numer. Anal.*, **13**, 13–27.

Xu, Y. and Cheney, E.W. (1992). Strictly positive definite functions on spheres. *Proc. Amer. Math. Soc.*, **116**, 977-981.

3
Representing and analyzing scattered data on spheres

H.N. MHASKAR, F.J. NARCOWICH and J.D. WARD

Abstract

Geophysical or meteorological data collected over the surface of the earth via satellites or ground stations will invariably come from scattered sites. There are two extremes in the problems one faces in handling such data. The first is representing *sparse* data by fitting a surface to it. This arises in geodesy in conjunction with measurements of the gravitation field from satellites, or meteorological measurements – temperature, for example – made at ground stations. The second is analyzing *dense* data to extract features of interest. For example, one may wish to process satellite images for mapping purposes. Between these two extremes there are many other problems. We will review various aspects of fitting surfaces to scattered data, addressing problems involving interpolation and order of approximation, and quadratures. Analyzing data is a more recent problem that is currently being addressed via various spherical wavelet schemes, which we will review, along with multilevel schemes. We close by discussing quadrature methods, which arise in many of the wavelet schemes as well as some interpolation methods.

3.1 Introduction

3.1.1 Overview

In this survey, we discuss recent progress in the representation and analysis of scattered data on spheres. As is the case with $\mathbb{R}^s$, many practical problems have stimulated interest in this direction. More and more data is taken from satellites each year. This, in turn, requires for example, improved image processing techniques for fault detection and for generation of maps. Meteorological readings (temperature, pressure, etc.) as well as geophysical data (gravitational field) are a few examples where enhanced representation and analysis of data capabilities prove useful.

Our domain of consideration is the q-sphere S^q rather than $\mathbb{R}^s$. While the q-sphere, $\mathbf{S}^q$, shares many similarities with $\mathbb{R}^s$, there are also significant differences. Since $\mathbf{S}^q$ is compact, there is no need for "infinite" amounts of data for the analysis, while if one restricts oneself to compact subdomains of $\mathbb{R}^s$, then the boundary of such sets introduces many difficulties. On the other hand, all large data sets on the sphere are necessarily scattered. In other words, there is no analogue to "gridded" data (for data sets larger than 20). This fact makes the standard $\mathbb{R}^s$ construction of wavelets on the sphere inapplicable.

A brief outline of this article is as follows. In Section 3.2 we discuss the representation of data on $\mathbf{S}^q$. Here we are primarily concerned with interpolation and quasi-interpolation/approximation. Recent progress along these lines involves families based on *spherical basis functions*, which are discussed below. Section 3.3 deals with the analysis of data on $\mathbf{S}^q$. Various constructs of wavelets, including continuous wavelet transforms as well as discrete wavelets, are mentioned here. In addition, we discuss localization and uncertainty principles on $\mathbf{S}^q$. Also included in this section is a brief introduction to multilevel interpolation. Finally, in Section 3.4, various numerical schemes for implementing the ideas in the previous sections are detailed. In both the representation and analysis of scattered data on spheres, efficient computations of Fourier coefficients and convolutions are needed. A quick tour of quadrature rules to estimate integrals over the q-sphere as well as related topics, such as sampling theorems and discrete Fourier transforms, are also given.

Many authors have contributed valuable insights into the topics discussed here. In this regard, we wish to apologize in advance for any omissions, and also for any other mistakes that we may have made in the preparation of this article.

3.1.2 Background – spherical harmonics

There are a number of books that treat spherical harmonics and their approximation properties. For classical results on spherical harmonics, there is the book of Hobson (1965). Later, very useful references are Müller's book (Müller (1966)) and, of course, the book by Stein and Weiss (1971). To a large extent, we will follow Müller's notation. Let $q \geq 1$ be an integer which will be fixed throughout the rest of this article, and let $\mathbf{S}^q$ be the unit sphere in the Euclidean space $\mathbb{R}^{q+1}$, with $d\mu_q$ being its usual area element, which is invariant under arbitrary coordinate changes. The volume of $\mathbf{S}^q$ is

$$\omega_q := \int_{\mathbf{S}^q} d\mu_q = \frac{2\pi^{(q+1)/2}}{\Gamma((q+1)/2)}.$$

Corresponding to $d\mu_q$, we have the inner product,

$$\langle f, g \rangle_{\mathbf{S}^q} := \int_{\mathbf{S}^q} f(x)\overline{g(x)} d\mu_q(x),$$

and $L^p(\mathbf{S}^q)$ norms,

$$\|f\|_{\mathbf{S}^q, p} := \left(\int_{\mathbf{S}^q} |f(x)|^p d\mu_q(x) \right)^{1/p}.$$

The class of all measurable functions $f : \mathbf{S}^q \to \mathbb{C}$ for which $\|f\|_{\mathbf{S}^q,p} < \infty$ will be denoted by $L^p(\mathbf{S}^q)$, with the usual understanding that functions that are equal almost everywhere are considered equal as elements of $L^p(\mathbf{S}^q)$. All continuous complex valued functions on $\mathbf{S}^q$ will be denoted by $C(\mathbf{S}^q)$.

For an integer $\ell \geq 0$, the restriction to $\mathbf{S}^q$ of a homogeneous harmonic polynomial of degree ℓ is called *a spherical harmonic of degree ℓ*. The class of all spherical harmonics of degree ℓ will be denoted by H_ℓ^q, and the class of all spherical harmonics of degree $\ell \leq n$ will be denoted by Π_n^q. Of course, $\Pi_n^q = \bigoplus_{\ell=0}^n H_\ell^q$, and it comprises the restriction to $\mathbf{S}^q$ of all algebraic polynomials in $q+1$ variables of total degree not exceeding n. The dimension of H_ℓ^q is given by Müller (1966), p. 11

$$d_\ell^q := \dim H_\ell^q = \begin{cases} \frac{(2\ell+q-1)}{(\ell+q-1)} \binom{\ell+q-1}{\ell}, & \text{if } \ell \geq 1, \\ 1, & \text{if } \ell = 0 \end{cases}$$

and that of Π_n^q is $\sum_{\ell=0}^n d_\ell^q$.

The spherical harmonics have an intrinsic characterization as the eigenfunctions of $\Delta_{\mathbf{S}^q}$, the Laplace–Beltrami operator on $\mathbf{S}^q$. $\Delta_{\mathbf{S}^q}$ is an elliptic, (unbounded) selfadjoint operator on $L^2(\mathbf{S}^q)$, invariant under arbitrary coordinate changes; its spectrum comprises discrete eigenvalues $\lambda_\ell := -\ell(\ell+q-1)$, $\ell = 0, 1, \ldots$, each having finite multiplicity d_ℓ^q. The space H_ℓ^q can be characterized intrinsically as the eigenspace of $\Delta_{\mathbf{S}^q}$ corresponding to λ_ℓ. Hence, for distinct ℓ's, the H_ℓ^q's are mutually orthogonal relative to the inner product $\langle \cdot, \cdot \rangle_{\mathbf{S}^q}$; in addition, the closure of their direct sum is all of $L^2(\mathbf{S}^q)$. That is, closure$\{\bigoplus_\ell H_\ell^q\} = L^2(\mathbf{S}^q)$. If we choose an orthonormal basis $\{Y_{\ell,k} : k = 1, \ldots, d_\ell^q\}$ for each H_ℓ^q, then we automatically have that the set $\{Y_{\ell,k} : \ell = 0, 1, \ldots; k = 1, \ldots, d_\ell^q\}$ is an orthonormal basis for $L^2(\mathbf{S}^q)$. (See Narcowich (1995), §IV for a more complete discussion and further references.)

The spherical harmonics satisfy the well-known addition formula (Müller (1966), Stein and Weiss (1971))

$$\sum_{k=1}^{d_\ell^q} Y_{\ell,k}(\mathbf{x})\overline{Y_{\ell,k}(\mathbf{y})} = \frac{d_\ell^q}{\omega_q}\mathcal{P}_\ell(q+1;\mathbf{x}\cdot\mathbf{y}), \qquad \ell = 0, 1, \ldots,$$

where $\mathcal{P}_\ell(q+1;t)$ is the degree ℓ Legendre polynomial in t, associated with $\mathbf{S}^q$. The Legendre polynomials are normalized so that $\mathcal{P}_\ell(q+1;1) = 1$, and satisfy the orthogonality relations

$$\int_{-1}^{1} \mathcal{P}_\ell(q+1;x)\mathcal{P}_k(q+1;x)(1-x^2)^{\frac{q}{2}-1}dx = \frac{\omega_q}{\omega_{q-1}d_\ell^q}\delta_{\ell,k},$$

and, up to normalization, are the same as the ultraspherical polynomials $P_\ell^{((q-1)/2)}$ (Müller (1966), p. 33) and the Jacobi polynomials $P_\ell^{(q/2-1,q/2-1)}$ (Szegö (1975)).

The addition formula allows us to express a variety of projection operators directly in terms of simple kernels. For example, the projection of $f \in L^2(\mathbf{S}^q)$ onto H_ℓ^q is just

$$P_\ell f(\mathbf{x}) := \text{Proj}_{H_\ell^q} f(\mathbf{x}) = \frac{d_\ell^q}{\omega_q} \int_{\mathbf{S}^q} \mathcal{P}_\ell(q+1;\mathbf{x}\cdot\mathbf{y})f(\mathbf{y})d\mu_q(\mathbf{y}),$$

with the kernel being $\frac{d_\ell^q}{\omega_q}\mathcal{P}_\ell(q+1;\mathbf{x}\cdot\mathbf{y})$. For the sequence $\{P_\ell\}_{\ell=0}^n$ of these mutually orthogonal projections, we define the order k Cesàro means via the standard formula,

$$\Sigma_n^{(k)} := \binom{n+k}{n}^{-1} \sum_{\ell=0}^{n} \binom{n-\ell+k}{n-\ell} P_\ell.$$

Replacing the projections above by their kernels yields kernels for $\Sigma_n^{(k)}$, which we denote by $\sigma_n^{(k)}(\mathbf{x}\cdot\mathbf{y})$. For

$$k \geq k_q := \left\lfloor \frac{q-1}{2} \right\rfloor + 1, \tag{3.1}$$

k fixed, these kernels have L^1 norms uniformly bounded in n (Szegö (1975)), say by C.

The Cesàro operators are needed to construct the *delayed mean* operators introduced by Stein (1957). These operators have the form

$$T_n^{(k)} := \sum_{j=1}^{k+1} a_j^{k,n} \Sigma_{jn-1}^{(k)}, n \geq 1, \tag{3.2}$$

and reduce to the identity on Π_n^q, provided n is larger than some integer N_q, that is,

$$T_n^{(k)}|_{\Pi_n^q} = \mathrm{Id}_{\Pi_n^q}, n \geq N_q. \tag{3.3}$$

Moreover, the coefficients are bounded independently of n. A simple example is the de la Vallée Poussin operator,

$$T_n^{(1)} := 2\Sigma_{2n-1}^{(1)} - \Sigma_{n-1}^{(1)} = \sum_{\ell=0}^{n} P_\ell + \sum_{\ell=n+1}^{2n-1} \frac{2n-\ell}{n} P_\ell, \; n \geq 1. \tag{3.4}$$

The first sum on the right in (3.4) is the orthogonal projection onto Π_n^q and the second maps Π_n^q to 0. If q is 1 or 2, then because the Cesàro operators are bounded, the sequence of operators $T_n^{(1)} : L^\infty(\mathbf{S}^q) \to C(\mathbf{S}^q)$ satisfies $\|T_n^{(1)}\| \leq 3C$ for all $n \geq 1$. We remark that if $q \geq 3$ and k satisfies the inequality (3.1), then $T_n^{(k)}$ has similar properties (cf. Mhaskar et al. (2000), Theorem 2.1 and Stein (1957), Theorem 1).

3.2 Representing data

In this section, we wish to discuss some past and present developments concerning scattered-data surface fitting and approximation on the sphere. Many articles have been written on the subject, including several good, recent survey articles (Cheney (1995), Fasshauer and Schumaker (1998), Freeden et al. (1997), Narcowich (1998)). We will focus on approximation/surface-fitting by positive definite (and related) functions.

Approximation on the sphere has evolved in the same way as approximation on $\mathbb{R}^s$. The early results focused on approximation with polynomials (or spherical harmonics) but approximation with other families of functions were later studied in order to address the shortcomings of polynomial approximation, such as instability or inability of polynomials to interpolate at scattered sites, Gibbs' phenomena, the computational expense of evaluating polynomials of high degree, and global support.

There have been at least three approaches for circumventing some of the problems mentioned above. One of them concerns the spherical splines originally investigated about 20 years ago by Freeden (1981), Wahba (1981) and others. These functions have the form $\phi(\mathbf{x} \cdot \mathbf{y})$ where $\mathbf{x}$ and $\mathbf{y}$ are unit vectors and ϕ is a spline-like function defined on $[-1, 1]$. Much has been written along these lines and for the interested reader we refer to the recent survey paper (Freeden et al. (1997)). In another direction, locally supported piecewise polynomials built on triangulated spherical patches and described in terms of spherical Bezier nets have recently been investigated by Alfeld et al. (1996). An excellent review in this direction has been given in Fasshauer and Schumaker (1998). The third approach originated in a celebrated paper of Schoenberg (1942), in which the key ideas for interpolation and approximation on the sphere were introduced.

3.2.1 Interpolation on the sphere

Schoenberg (1942) introduced functions that he termed *positive definite* on $\mathbf{S}^q$. Let $\theta(\mathbf{x}, \mathbf{y})$ be the geodesic distance between two points $\mathbf{x}$ and $\mathbf{y}$ on $\mathbf{S}^q$. A continuous function $G(\theta(\mathbf{x}, \mathbf{y}))$ is said to be a *positive definite function* on $\mathbf{S}^q$ if, for every set of points $\mathcal{C} = \{\mathbf{x}_1, \ldots, \mathbf{x}_N\}$, the $N \times N$ matrix with entries $[G(\theta(\mathbf{x}_j, \mathbf{x}_k))]$ is self adjoint and positive semi-definite. Schoenberg (1942) completely classified all such functions in terms of expansions in ultraspherical (Gegenbauer) polynomials.

Theorem 3.2.1 (Schoenberg (1942)) *A continuous function $G(\theta(\mathbf{x}, \mathbf{y}))$ is a positive definite function on $\mathbf{S}^q$ if and only if*

$$G(\theta(\mathbf{x}, \mathbf{y})) = G(\arccos(\mathbf{x} \cdot \mathbf{y})) = \sum_{\ell=0}^{\infty} a_\ell \mathcal{P}_\ell(q+1; \mathbf{x} \cdot \mathbf{y}),$$

where $\mathcal{P}_\ell(q+1; \mathbf{x} \cdot \mathbf{y})$ is the degree ℓ Legendre polynomial associated with $\mathbf{S}^q$, and the $a_\ell \geq 0$ decay sufficiently fast for $G(\theta(\mathbf{x}, \mathbf{y}))$ to be continuous in $\mathbf{x} \cdot \mathbf{y}$.

Since the invertibility of the matrix $A_{j,k} := [G(\theta(\mathbf{x}_j, \mathbf{x}_k))]$ is necessary to interpolate data $f(\mathbf{x}_j), j = 1, \ldots, N$, from the space span$\{G(\theta(\mathbf{x}, \mathbf{x}_j))\}$, the representation theorem of Schoenberg suggests that positive definite functions on spheres would be an ideal class of functions for interpolation, provided the interpolation matrices were invertible for distinct points.

Xu and Cheney (1992) gave a number of interesting conditions on G guaranteeing that the matrix $[G(\theta(\mathbf{x}_j, \mathbf{x}_k))]$ will be positive definite, and therefore invertible, for classes of finite subsets $\mathcal{C} \subset \mathbf{S}^q$ in which the points are distinct. We say that G is *strictly positive definite of order N* if the matrix $[G(\theta(\mathbf{x}_j, \mathbf{x}_k))]$ is positive definite whenever the set $\mathcal{C}$ has cardinality less than or equal to N. When this happens for all N, we will say that G is *strictly positive definite*. One particular condition in Xu and Cheney (1992) sufficient for G to be strictly positive definite, is that the coefficients in the expansion of G satisfy $a_\ell > 0$ for every ℓ. Ron and Sun (1996) shed further light on the invertibility of the interpolation matrices.

Theorem 3.2.2 (Ron and Sun (1996)) *A function G is strictly positive definite of order N if the coefficients $a_1, \ldots, a_{\lfloor N/2 \rfloor}$ are positive. Moreover, G is strictly positive definite for all orders if the set $K := \{k \in Z_+ : a_k > 0\}$ contains arbitrarily long sequences of consecutive even integers, as well as arbitrarily long sequences of consecutive odd integers.*

It is still an open question as to what conditions are both necessary *and* sufficient for G to be strictly positive definite, either for finite order or all orders.

Narcowich (1995) investigated solving very general Hermite interpolation problems on compact Riemannian manifolds, with data "sampled" by applying arbitrary distributions to smooth functions. One consequence was that if G is C^∞ and satisfies the condition that $a_\ell > 0$ for all ℓ, then it is strictly positive definite in an even stronger sense than what is discussed above. The set of point evaluations corresponding to $\mathcal{C}$ can be replaced by a finite set of arbitrary, linearly independent distributions, and the (generalized) interpolation matrices that result will be strictly positive definite. Analogues hold when G is not C^∞, but is in a Sobolev space. (See the discussion below.)

The functions G for which $a_\ell > 0$ for all ℓ are quite useful; we call them **spherical basis functions** (SBFs).

The work described above was directed toward the solvability of various interpolation problems; it set the stage for the next round of investigations, which were aimed at answering the question of how well the interpolants approximate functions from a given smoothness class. Such questions were investigated in the $\mathbb{R}^s$ case (for scattered data) in a series of papers starting with Madych and Nelson (1983). A convenient setting for analyzing such questions is a reproducing kernel Hilbert space whose relevance to interpolation problems was first pointed out in Golomb and Weinberger (1959).

An early work concerning error estimates on the sphere was that of Freeden. For a given scattered, finite set of points $\mathcal{C}$ on the sphere, one defines the mesh norm $\delta_\mathcal{C}$ of $\mathcal{C}$ by

$$\delta_\mathcal{C} := \max_{\mathbf{x} \in \mathbf{S}^q} \text{dist}(\mathbf{x}, \mathcal{C}) = \max_{\mathbf{x} \in \mathbf{S}^q} \min_{\mathbf{y} \in \mathcal{C}} \text{dist}(\mathbf{x}, \mathbf{y}), \qquad (3.5)$$

where the distance is the geodesic distance on the sphere. Freeden showed that interpolants formed from spherical splines converge at rates $|\delta_\mathcal{C}|^\tau$, where $\tau \in [0, 1]$. The value of τ depends on a number of factors; see Freeden (1986).

In subsequent years, further application of the Golomb–Weinberger framework (Golomb and Weinberger (1959)) in the case of the sphere produced results for which the rates of approximation reflected the smoothness of both the approximating kernel as well as the function to be approximated. We give a brief description of these results next. A more complete explanation of relevant material can be found in Dyn et al. (1997,1999).

We will work in the same setting as Narcowich (1995); namely, we take the underlying space to be a compact, C^∞ Riemannian manifold $\mathcal{M}$. Spheres and tori are of course examples.

We will need to introduce some notation. We will denote the Riemannian metric on $\mathcal{M}$ by g_{ij} and its inverse by g^{ij}. The corresponding volume element, inner product, Laplace–Beltrami operator and its eigenvalues and eigenfunctions, are

$$\left.\begin{aligned}
d\mu(\mathbf{x}) &:= g(\mathbf{x})^{1/2}d\mathbf{x}, \text{ where } g(\mathbf{x}) := \det g_{ij}(\mathbf{x}) > 0 \\
\langle f,g\rangle_\mathcal{M} &:= \int_\mathcal{M} f(\mathbf{x})\overline{g(\mathbf{x})}\, d\mu(\mathbf{x}) \\
\Delta w &:= g^{-\frac{1}{2}} \sum_{i,j} \frac{\partial}{\partial x^i}\left(g^{\frac{1}{2}}g^{ij}\frac{\partial w}{\partial x^j}\right) \\
\lambda_k &:= \text{k}^{th} \text{ eigenvalue of } \Delta \\
F_k &:= \text{eigenfunction corresponding to } \lambda_k.
\end{aligned}\right\} \quad (3.6)$$

Finally, we let $\mathcal{D}'$ denote the set of all distributions on $\mathcal{M}$, and define the Sobolev space

$$H_s(\mathcal{M}) = \left\{u \in \mathcal{D}' : \sum_{j=1}^\infty \lambda_j^s|\hat{u}(j)|^2 < \infty\right\}, \quad \hat{u}(j) := \langle u, F_j\rangle_\mathcal{M}.$$

The (Sobolev) norm for this space is given by

$$\|u\|_s := \left(\sum_{j=1}^\infty \lambda_j^s|\hat{u}(j)|^2\right)^{1/2}.$$

One can integrate the product of a kernel $\kappa(\mathbf{x},\mathbf{y})$ in $H_{2s}(\mathcal{M}\times\mathcal{M})$ with a distribution $u(\mathbf{y}) \in H_{-s}(\mathcal{M})$, with the resulting distribution being a function in $H_s(\mathcal{M})$. In particular, we define

$$\kappa \star u(\mathbf{x}) = \int_\mathcal{M} \kappa(\mathbf{x},\mathbf{y})u(\mathbf{y})d\mu(\mathbf{y}).$$

Note that "$\star$" is *not* commutative. In the special case in which u is a linear combination of point evaluations, $u = \sum_{j=1}^N a_j\delta_{\mathbf{x}_j}$, then $\kappa \star u(\mathbf{x}) = \sum_{j=1}^N a_j\kappa(\mathbf{x},\mathbf{x}_j)$.

A function $\kappa \in H_{2s}(\mathcal{M}\times\mathcal{M})$ is a *strictly positive definite kernel* if for every nonzero distribution $u \in H_{-s}(\mathcal{M})$ we have that $\langle u, \kappa \star u\rangle_\mathcal{M} > 0$. Consequently, we can use a strictly positive definite kernel κ to define a norm on $H_{-s}(\mathcal{M})$; namely,

$$\|u\| := \sqrt{\langle u, \kappa \star u\rangle_\mathcal{M}}. \quad (3.7)$$

All kernels of the form

$$\kappa(\mathbf{x}, \mathbf{y}) = \sum_{k=1}^{\infty} a_k F_k(\mathbf{x}) \overline{F_k(\mathbf{y})}, \quad a_k > 0. \tag{3.8}$$

are strictly positive definite, provided that the a_ks decay sufficiently fast for them to be in one of the Sobolev spaces defined above.

In practical applications one often wants interpolants that reproduce exactly some fixed, finite dimensional space of functions – for example, polynomials on $\mathbb{R}^s$ or spherical harmonics on $\mathbf{S}^q$. In Dyn et al. (1999), *conditionally positive definite* kernels were introduced to deal with such problems for general manifolds. Let $\mathcal{I}$ be a finite set of indices in $\mathbb{Z}_+$, and consider the finite dimensional space defined by $\mathcal{S}_\mathcal{I} := \mathrm{span}\{F_i : i \in \mathcal{I}\}$. In addition, define the space of distributions $\mathcal{S}_\mathcal{I}^\perp := \{u \in H_{-s}(\mathcal{M}) : \langle u, F \rangle_\mathcal{M} = 0, \ F \in \mathcal{S}_\mathcal{I}\}$. We will say that a selfadjoint, continuous kernel κ in $H_{2s}(\mathcal{M} \times \mathcal{M})$ is *conditionally* positive definite (CPD) with respect to a finite set of indices $\mathcal{I} \subset \mathbb{Z}_+$ if for every $u \in \mathcal{S}_\mathcal{I}^\perp$ we have $\langle u, \kappa \star u \rangle_\mathcal{M} \geq 0$. Moreover, if for all such $u \neq 0$ we have $\langle u, \kappa \star u \rangle_\mathcal{M} > 0$, then we will say that κ is **conditionally strictly positive definite** (CSPD) on $\mathcal{M}$ with respect to $\mathcal{S}_\mathcal{I}$. All kernels of the form (3.8) are CSPD for $\mathcal{S}_\mathcal{I}$, provided only that $a_k > 0$ for all $k \in \mathbb{Z}_+ \backslash \mathcal{I}$.

The next result gives error estimates for interpolation using distributions and a strictly positive definite kernel $\kappa \in H_{2s}(\mathcal{M} \times \mathcal{M})$. Let $\{u_j : j = 1, \ldots, N\}$ be a linearly independent set of distributions in $H_{-s}(\mathcal{M})$, and define the space $\mathcal{U} := \mathrm{span}\{u_j : j = 1, \ldots, N\}$. The problem is that if $f \in H_s(\mathcal{M})$ and if we are given data $d_k = \langle u_k, f \rangle_\mathcal{M}$, then we want to find $u \in \mathcal{U}$ such that $\tilde{f} := \kappa \star u$ satisfies $d_k = \langle u_k, \tilde{f} \rangle_\mathcal{M}$ for $k = 1, \ldots, N$. Equivalently, we want to find $c_1, \ldots, c_N$ such that $\tilde{f} = \sum_{j=1}^{N} c_j \kappa \star u_j$ satisfies $d_k = \sum_{j=1}^{N} c_j \langle u_k, \kappa \star u_j \rangle_\mathcal{M}$ for $k = 1, \ldots, N$. Because κ is strictly positive definite, the interpolation matrix $A_{j,k} := \langle u_j, \kappa \star u_k \rangle_\mathcal{M}$ is positive definite and invertible, so we can always solve for the c_js and find a unique distribution $u \in \mathcal{U}$ such that $\tilde{f} := \kappa \star u$ solves the interpolation problem.

How close are f and $\tilde{f}$? Let $f = \kappa \star v$, with $v \in H_{-s}(\mathcal{M})$. The following result is a version of the hypercircle inequality (Golomb and Weinberger (1959)).

Theorem 3.2.3 (Dyn et al. (1999), Proposition 3.2) *Let $f = \kappa \star v$ and let $\tilde{f} = \kappa \star u$ be the interpolant described above. If w is an arbitrary distribution in H_{-s}, then*

$$|\langle w, f - \tilde{f} \rangle_\mathcal{M}| \leq \mathrm{dist}(v, \mathcal{U}) \mathrm{dist}(w, \mathcal{U}),$$

where the distances are computed relative to the norm (3.7).

Of course, $|\langle w, f - \tilde{f}\rangle_{\mathcal{M}}|$ defined above represents the interpolation error relative to the distribution $w \in H_{-s}$. In order for this formulation to be useful, one must estimate the distances given on the right hand side of the inequality. Using the notation defined in (3.6), we can state the next result, which gives a framework for making such estimates.

Theorem 3.2.4 (Dyn et al. (1999), Proposition 3.6) *Let n be a positive integer, $s > 0$, and let κ have the form (3.8). If there are coefficients $c_1, \ldots, c_N$ such that for $k = 1, \ldots, n$, $|\langle w - \sum_{j=1}^{N} c_j u_j, F_k\rangle_{\mathcal{M}}| = 0$, then*

$$\mathrm{dist}(w, \mathcal{U}) \leq \|w - \sum_{j=1}^{N} c_j u_j\| \leq \left(\sum_{k=n+1}^{\infty} a_k b_k\right)^{1/2},$$

where $|\langle w - \sum_{j=1}^{N} c_j u_j, F_k\rangle_{\mathcal{M}}|^2 \leq b_k$ for $k \geq n+1$.

What is apparent from the theorem above is that if the b_ks are bounded or have controlled growth, then the interpolation error estimates will reflect the rate of decay of the coefficients a_k. In Dyn et al. (1997), estimates were obtained for point evaluations and derivatives of point evaluations at scattered points on the torus. In the case of the sphere, estimates were obtained for point evaluation functionals located at gridded sites on the rectangle $[0, 2\pi] \times [0, \pi]$; the b_ks were estimated by means of the quadrature formulae developed in Driscoll and Healy (1994).

We now restrict our attention to scattered data interpolation problems. In these problems the u_js are point evaluations at scattered sites on the underlying manifold, which we now take to be $\mathbf{S}^q$.

In Golitschek and Light (2000) and Jetter et al. (1999), a more complete picture of scattered data interpolation on the sphere emerged. The thrust and scope of the two papers differ somewhat, so we will discuss each briefly.

We begin with Golitschek and Light (2000). This paper obtains error estimates for problems involving interpolation with reproduction of spherical harmonics, and supersedes earlier results in Levesley et al. (1999). For now, we ignore the reproduction of spherical harmonics and concentrate on simple interpolation. Let $\mathcal{C} = \{\mathbf{x}_1, \mathbf{x}_2, \ldots, \mathbf{x}_N\} \subset \mathbf{S}^q$ be a finite set of centers with mesh norm $\delta_{\mathcal{C}}$, and consider interpolating data using an SBF kernel of the form

$$\phi(\mathbf{x} \cdot \mathbf{y}) = \sum_{\ell=0}^{\infty} \frac{d_\ell^q a_\ell}{\omega_q} \mathcal{P}_\ell(q+1; \mathbf{x} \cdot \mathbf{y}),$$

where the $\mathcal{P}_k(q+1;t)$s are the Legendre polynomials described in Section

3.1.2 and $a_\ell > 0$ for all $\ell \geq 0$. The function $\phi(\mathbf{x} \cdot \mathbf{y})$ is the reproducing kernel for the Hilbert space

$$\mathcal{H}_\phi := \left\{ f \in L^2(\mathbf{S}^q) : \sum_{\ell=0}^\infty \sum_{m=1}^{d_\ell^q} \frac{|\langle f, Y_{\ell,m}\rangle_{\mathbf{S}^q}|^2}{a_\ell} < \infty \right\}.$$

Let f be in $\mathcal{H}_\phi$. Either from Theorem 3.2.3 or by standard reproducing-kernel estimates for the interpolation error at a given point $\mathbf{x}$, one has

$$|f(\mathbf{x}) - \tilde{f}(\mathbf{x})| \leq Q_{\mathbf{x}} \|f\|_{\mathcal{H}_\phi}$$

where

$$Q_{\mathbf{x}}^2 = \phi(1) - 2\sum_{j=1}^N c_j \phi(\mathbf{x} \cdot \mathbf{x}_j) + \sum_{j,k=1}^N c_j c_k \phi(x_j \cdot \mathbf{x}_k).$$

The c_js are all real and arbitrary. The function $\tilde{f} \in \text{span}\{\phi(\mathbf{x} \cdot \mathbf{x}_j)\}_{j=1}^N$ is the interpolant for f.

Suppose that ϕ is in $C^{(\nu)}([0,1])$, and that for each $\mathbf{x} \in \mathbf{S}^q$ there is a geodesic disk $D_{\mathbf{x}}^h$ with radius h and center $\mathbf{x}$ for which the centers in $\mathcal{C} \cap D_{\mathbf{x}}^h$ form a unisolvent set for spherical harmonics of degree less than ν. Using only centers in $\mathcal{C} \cap D_{\mathbf{x}}^h$, we can choose the c_js so that they reproduce all spherical harmonics through degree $\ell = \nu - 1$. Applying Taylor's formula with remainder to ϕ, we have $\phi(t) = (T_{\nu-1}\phi)(t) + \frac{\phi^{(\nu)}(z)}{\nu!}(t-1)^\nu$, where $T_{\nu-1}\phi$ is the Taylor polynomial of degree $\nu-1$ about $t=1$, and z satisfies $\cos(2h) \leq t \leq z \leq 1$. Inserting this in the expression for $Q_{\mathbf{x}}^2$ and noting that in $Q_{\mathbf{x}}^2$ the polynomial part of ϕ vanishes by our choice of c_js, we obtain the bound

$$Q_{\mathbf{x}}^2 \leq C\left(1 + \sum_j |c_j|\right)^2 h^{2\nu}, \text{ where } C := \sup_{z \in [0,1]} \frac{|\phi^{(\nu)}(z)|}{\nu!}.$$

The bound is not necessarily uniform in h, because $1 + \sum_j |c_j|$ may depend on h or some other local quantities. In Golitschek and Light (2000), a difficult, technical argument is used to show that the factor $1 + \sum_j |c_j|$ can be bounded *uniformly* in h. Here is the result.

Theorem 3.2.5 (Golitschek and Light (2000), Corollary 3.3) *Let $\mathcal{C} \subset \mathbf{S}^q$ be finite, with $h = \delta_\mathcal{C}$ being the mesh norm given in (3.5), and let $\nu \in \mathbb{Z}_+$. If $\phi \in C^{(\nu)}([0,1])$, $f \in \mathcal{H}_\phi$, then there exist $\epsilon > 0$ and $C > 0$ such that whenever $h < \epsilon$,*

$$\|f - \tilde{f}\|_\infty < Ch^\nu \|f\|_{\mathcal{H}_\phi},$$

where $\tilde{f} \in \operatorname{span}\{\phi(\mathbf{x} \cdot \mathbf{x}_j)\}_{j=1}^N$ interpolates f on $\mathcal{C}$. Moreover, with appropriate changes in ϕ and $\mathcal{H}_\phi$, the bounds above hold even if the interpolant is also required to reproduce spherical harmonics up to order n, where $n \leq \nu - 1$.

The methods used in the proof are based on polynomial reproduction, so the additional statements concerning such reproduction are easily established. One thing that is not clear, however, is how many points one needs to get the error rate. The function of ϵ in the statement is to force the mesh norm $h = \delta_\mathcal{C}$ to be small enough for the set $\mathcal{C}$ to be unisolvent with respect to the spherical harmonics of order $\nu - 1$. This is not a difficulty for small ν, but it becomes problematic as ν gets large. In that case, one may need to take ϵ to be quite small, which implies that the number of points required will be large.

A completely different approach to error estimates was used in Jetter et al. (1999). Let $\Lambda > 0$ be an integer. The starting point for Jetter et al. (1999) was the condition from Theorem 3.2.4 that $|\langle w - \sum_{j=1}^N c_j \delta_{\mathbf{x}_j}, Y_{\ell,m}\rangle_{\mathbf{S}^q}| = 0$ for $\ell = 0, \ldots, \Lambda$ and $m = 1, \ldots, d_\ell^q$. In functional analytic terms, this condition will be satisfied whenever the set $\{\delta_{\mathbf{x}_j}\}_{j=1}^N$ spans the space dual to Π_Λ^q, which is the span of the spherical harmonics $\{Y_{\ell,m}\}$ with $\ell \leq \Lambda$ and $m = 0, \ldots, d_\ell^q$. One would also like that N be no larger than a constant multiple of the dimension of Π_Λ^q. In Jetter et al. (1999), the notion of norming sets and some recently discovered Markov inequalities (Bos et al. (1995)) for spherical harmonics were used to show the following error estimates

Theorem 3.2.6 (Jetter et al. (1999)) Let $\kappa(\mathbf{x}, \mathbf{y}) = \sum_{\ell,m} a_{\ell,m} Y_{\ell,m}(\mathbf{x}) \overline{Y_{\ell,m}(\mathbf{y})}$ be in $C(\mathbf{S}^q \times \mathbf{S}^q)$ and have all $a_{\ell,m} > 0$, and let $\mathcal{H}$ be the Hilbert space

$$\mathcal{H} := \left\{ f \in L^2(\mathbf{S}^q) \,:\, \sum_{\ell,m} \frac{|\langle f, Y_{\ell,m}\rangle_{\mathbf{S}^q}|^2}{a_{\ell,m}} < \infty \right\},$$

which has κ as a reproducing kernel. If Λ is chosen to be a positive integer such that $\frac{1}{2\Lambda+2} < \delta_\mathcal{C} \leq \frac{1}{2\Lambda}$, then for each $f \in \mathcal{H}$, the interpolation error is bounded by

$$\|f - \tilde{f}\|_\infty^2 \leq \frac{5(N+1)}{\omega_q} \|f\|_\mathcal{H}^2 \sum_{\ell > \Lambda} \hat{a}_\ell d_\ell^q$$

where $\tilde{f} \in \operatorname{span}\{\kappa(\mathbf{x}, \mathbf{x}_j)\}_{j=1}^N$ is the interpolant for f and $\hat{a}_\ell := \max_{1 \leq m \leq d_\ell^q} a_{\ell,m}$.

3.2.2 Approximation on the sphere

Interpolation by SBFs, which are the positive definite spherical functions whose Legendre expansions have only positive coefficients, is always possible because SBFs are strictly positive definite. It is also an implementable approximation scheme and is quite useful in representing data in many cases. However, as the data sites become more dense, interpolation becomes quite unstable. There have been investigations into how the condition number of the interpolation matrix depends on the separation distance of the data as well as the smoothness of the SBF. However, in practice, one typically uses least squares approximation to handle dense data sets, which leads naturally to theoretical questions concerning L^2 approximation on the q-sphere. The corresponding question on $\mathbb{R}^s$ was investigated in de Boor et al. (1994), where optimal rates of approximation were derived for various classes of RBFs in case the shifts of the RBF corresponded to a scaled lattice. Since the shift points on the sphere are necessarily scattered, optimal rates of approximation for general SBFs remain an elusive goal. Nevertheless, in Mhaskar et al. (1999a) nearly optimal L_p rates of approximation have been obtained for a general class of spherical functions. This class includes SBFs, and also similar functions that have non-vanishing Legendre coefficients.

The general approach follows along the lines of Mhaskar and Micchelli (1995) where near-best rates of approximation on the circle were obtained when the shift sites corresponded to the nth roots of unity. The case for the q-sphere, using scattered sites is more complex. We next give a brief overview of these results. Full details can be found in Mhaskar et al. (1999a).

The delayed mean operators $T_n^{(k)}$ defined by (3.2) are basic in this approach. For example, in the case of the circle the classical de la Vallée Poussin operators $T_n^{(1)}$ reproduce trigonometric polynomials of degree n or less. The norm $\|f - T_n^{(1)}(f)\|$ is thus comparable to the distance of f to trigonometric polynomials of degree n in any L^p norm. They also reproduce Π_n^2 in the case of $\mathbf{S}^2$. For $q \geq 3$ and $k \geq k_q = \lfloor \frac{q-1}{2} \rfloor + 1$ the delayed mean operators $T_n^{(k)}$ in (3.2) have similar properties.

These operators figure prominently in the linear approximation scheme given in Mhaskar et al. (1999a). This scheme can be summarized as follows. For a given function f defined on $\mathbf{S}^q$, one first approximates f by $T_n^{(k)}(f) := P$ where $k \geq k_q$. Let $\phi : [-1,1] \to \mathbb{R}$ be continuous and have the Legendre expansion $\phi(\cdot) = \sum_{l=0}^{\infty} \hat{\phi}(l) \mathcal{P}(q+1; \cdot)$, where $\hat{\phi}(l) \neq 0$. We want to get L^p-rates of approximation for the space $\text{span}\{\phi(\mathbf{x} \cdot \mathbf{y}) : \mathbf{y} \in \mathcal{C}\}$, where $\mathcal{C}$ is a finite subset of distinct points in $\mathbf{S}^q$. Using the Legendre expansion of ϕ, one finds a polynomial $Q \in \Pi_n^q$ for which $P(\mathbf{x}) = \int_{\mathbf{S}^q} \phi(\mathbf{x} \cdot \mathbf{y}) Q(\mathbf{y}) d\mu_q(\mathbf{x})$.

Finally discretizing this integral by means of an appropriate quadrature formula yields the desired approximant, $R_\mathcal{C}^\phi(f) := \sum_{\mathbf{y} \in \mathcal{C}} a_\mathbf{y} \phi(\mathbf{x} \cdot \mathbf{y})$. These quadrature formulae associated with scattered points are central in proving the rates below, and will be discussed in Section 3.4.3.

To simplify notation, we will state the result for L^∞, although the results described in Mhaskar et al. (1999a) are valid – with some details regarding formulae – for all L^p, $1 \leq p \leq \infty$. In the theorem, the mesh norm $\delta_\mathcal{C}$ is defined by (3.5), and the integers k_q and N_q are defined by (3.1) and (3.3), respectively. The constant α_q is the one needed for the Marcinkiewicz–Zygmund inequality (3.9) in Section 3.4.3 to hold. The distances are taken relative to L^∞. Finally, π_n is the set of univariate polynomials of degree n or less. Here is the result.

Theorem 3.2.7 (Mhaskar et al. (1999a), Theorem 4.2) *Let n, m be integers, with $n \geq N_q$; set $N = (k_q + 1)n - 1$, and take $\mathcal{C}$ to be a set of N distinct points on $\mathbf{S}^q$ such that $(k_q + 2)n + m \leq \alpha_q \delta_\mathcal{C}^{-1}$. If $f \in L^\infty(\mathbf{S}^q)$, then*

$$\|f - R_\mathcal{C}^\phi f\|_{\mathbf{S}^q,\infty} \leq c\left(\mathrm{dist}(f, \Pi_n^q) + \frac{\mathrm{dist}(\phi, \pi_m)}{r_N}\|f\|_{\mathbf{S}^q,\infty}\right),$$

where

$$r_N := \min_{0 \leq \ell \leq N} \frac{|\hat{\phi}(\ell)|}{d_\ell^q}.$$

We now wish to apply this theorem to the important case in which $\phi(x) = e^x$. This example was done in detail in Mhaskar et al. (1999a), §4. The error estimate consists of the sum of two terms. The first is a constant multiple of $\mathrm{dist}(f, \Pi_n^q)$; and the second is a constant multiple of a ratio of $\mathrm{dist}(\phi, \pi_m)$ to the smallest Legendre coefficient of ϕ. Using a Taylor series argument, one can easily show that $\mathrm{dist}(\phi, \pi_m) = \mathcal{O}(1/m!)$. Also, from the specific form of the Legendre expansion for e^x, one has that $r_N = 2^{-N}(\Gamma(N + \frac{q+1}{2}))^{-1}$. If we choose $m = \lfloor (1+t)N \rfloor$, $t > 0$, then

$$\frac{\mathrm{dist}(\phi, \pi_m)}{r_N} = \mathcal{O}(N^{-tN/2}).$$

Next, pick $\mathcal{C}$ so that $\delta_\mathcal{C}$ is small enough for the inequality in the theorem to be satisfied; this can be done in such a way that the cardinality of $\mathcal{C}$ is $\mathcal{O}(\delta_\mathcal{C}^{-q})$, which when coupled with the inequality $\delta_\mathcal{C} < \frac{\alpha_q}{n}$, implies that N, the cardinality of $\mathcal{C}$, is $\mathcal{O}(n^q)$. For functions f in $C^r(\mathbf{S}^q)$, the rate of approximation from Π_n^q is $\mathcal{O}(n^{-r})$. This is comparable to the best n-width (Mhaskar et al. (1999a), §3.1) from spaces of the same dimension as Π_n^q, which is also $\mathcal{O}(n^q)$. Thus the overall rate for this example is the same as the polynomial rate, and so is also close to optimal.

Petrushev (1998) has recently obtained interesting, similar results for ridge functions. Let X_n be an n-dimensional linear space of univariate functions in $L^2(I)$, $I = [-1, 1]$, and let $\Omega \subset \mathbf{S}^q$ have (finite) cardinality m. The space of ridge functions associated with X_n and Ω is defined as

$$Y_n := \{\rho(x \cdot \omega) : \rho \in X_n, \omega \in \Omega\}.$$

His results concern how well linear combinations of ridge functions approximate in the metric $L^2(B^{q+1})$, where B^{q+1} is the unit ball in $\mathbb{R}^{q+1}$. If X_n provides approximation order $\mathcal{O}(n^{-r})$ for univariate functions with r derivatives in $L^2(I)$, and if Ω is properly chosen, with cardinality $m = \mathcal{O}(n^q)$, then, in $L^2(B^{q+1})$, Y_n will provide approximation of order $\mathcal{O}(n^{-r-q/2})$ for every function $f \in L^2(B^{q+1})$ with smoothness of order $r + q/2$.

Sloan (1995,1997) takes a different approach to approximation on $\mathbf{S}^2$, which he calls *hyperinterpolation*. In hyperinterpolation, one uses an operator H_L that truncates the spherical harmonic expansion of f at order L, and replaces the coefficients there with ones computed using quadrature rules. These rules are postulated to satisfy a regularity condition; namely, an infinite family of positive-weight m-point quadrature rules $\{Q_m\}$, where

$$Q_m(f) := \sum_{j=1}^m a_j^m f(\xi_j^m), \qquad f \in C(\mathbf{S}^2),$$

is said to satisfy the *quadrature regularity assumption* if there exists a constant $c_1 > 0$, with c_1 independent of m, such that for every spherical cap $A_{\mathbf{x}} := \{\mathbf{y} \in \mathbf{S}^2 : \arccos(\mathbf{y} \cdot \mathbf{x}) \leq 1/\sqrt{m}\}$ (i.e., a geodesic disk with radius $1/\sqrt{m}$ and center $\mathbf{x}$) one has

$$\sum_{\xi_j^m \in A_{\mathbf{x}}} a_j^m \leq c_1 |A_{\mathbf{x}}|,$$

where $|A_{\mathbf{x}}| \approx \pi/m$ is the surface area of $A_{\mathbf{x}}$.

Theorem 3.2.8 (Sloan and Womersley (2000), Theorem 5.4) *If H_L is the hyperinterpolation operator that is computed using rules that satisfy the quadrature regularity assumption, then $\|H_L\|_{C \to C} \asymp L^{1/2}$.*

3.3 Analyzing data
3.3.1 Wavelets

Wavelets on the sphere have been a topic of recent interest. As is the case for Euclidean space, there are two types of transforms, continuous and discrete.

Continuous case. Continuous wavelet transforms (CWTs) on spheres have been developed by Torrésani (1995), Dahlke and Maass (1996), Freeden and Windheuser (1996), Holschneider (1996) and Antoine and Vandergheynst (1998).

Torrésani (1995), and Dahlke and Maass (1996) approach the construction of a CWT from a tangent bundle viewpoint, with the wavelet transforms being functions of position and two frequency variables.

Another approach is to use a wavelet transform that involves a scaling variable. For analyzing wavelets, Freeden and Windheuser (1996) use a family of zonal functions $\{\Psi_{\rho;\eta}(\mathbf{x}) = \Psi_\rho(\mathbf{x} \cdot \eta)\}$ that depend on a scaling parameter $\rho > 0$ and where $\eta \in \mathbf{S}^2$. Holschneider starts with a family of the form $\{g_a(\mathbf{x})\}$, where $a > 0$ is also a scaling parameter, and uses the family $\{U(\xi)g_a(\mathbf{x}) = g_a(\xi^{-1}\mathbf{x})\}$, where $\xi \in \mathrm{SO}(3)$, as the analyzing wavelet. Although it is far from obvious, if one specializes Holschneider's choice of g_a to $\Psi_{\rho;\eta}(\mathbf{x})$, with η fixed, then the resulting transforms are essentially the same. It should be mentioned that either type of wavelet family has to satisfy extra conditions, such as "finite energy" and vanishing moments. The CWT developed by Antoine and Vandergheynst uses generalized coherent states, with scaling coming via various group representations (not only for $\mathrm{SO}(n)$, but, surprisingly, for the Lorentz group). This approach carries over to the more general case of Riemannian symmetric spaces. It should be mentioned that Antoine and Vandergheynst also show that their CWT reduces to the standard flat-space one in the Euclidean limit. Holschneider also shows that this is true for his CWT. By virtue of the remarks just made, it is thus true for the Freeden–Windheuser CWT.

Discrete case. On spheres (or non-flat manifolds) there are no natural grids on which to do a discrete wavelet analysis. Thus, constructing a discrete wavelet transform (DWT) requires approaches that at the very least modify the usual ones for Euclidean wavelets.

The first of these is a method that starts with some initial grid and wavelets, and then refines them both (and scaling function, too) via an adaptive method. These wavelets are naturally bi-orthogonal. Such methods were independently introduced by two groups of researchers: Dahlke, Dahmen, Schmidt, and Weinrich (Dahlke et al. (1995)); and Schroeder and Sweldens

(Schröder and Sweldens (1995), Sweldens (1996), Schröder and Sweldens (1997), Sweldens (1998)).

Another approach is that of Tasche and Potts (1995). In it, one uses the standard latitude–longitude grid, and then one constructs wavelets by taking tensor products of Euclidean wavelets designed for intervals (e.g., Cohen et al. (1993)) with periodic wavelets. Of course, doing this results in singularities and distortions near the poles, and these must be dealt with.

Freeden et al. (Freeden et al. (1998), Chapter 11) develop a discrete wavelet transform based on their version of the CWT mentioned above. The wavelets and scaling functions are constructed via spherical harmonics, on the discrete "frequency" (or angular momentum) variable ℓ. The construction gives rise to a multiresolution analysis well suited to analyzing data from the earth's gravitational field. See Bayer et al. (1998) for results from such an analysis.

Narcowich and Ward (1996b) take an approach that constructs a multiresolution analysis from an SBF, G, and $N+M$ distinct, scattered sites on the sphere, $\{\mathbf{x}_1, \ldots, \mathbf{x}_{N+M}\}$. In this approach, the highest level sampling space is $\mathcal{V}_{N+M} = \text{span}\{G(\mathbf{x} \cdot \mathbf{x}_1), \ldots, G(\mathbf{x} \cdot \mathbf{x}_{N+M})\}$. To get the wavelets, pick a set of N distinct points from among the $N+M$ original points and label the N points by $\{\mathbf{x}_1, \ldots, \mathbf{x}_N\}$. This yields a second sampling space, $\mathcal{V}_N$, which itself is a subspace of $\mathcal{V}_{N+M}$, and is the "next level down" from $\mathcal{V}_{N+M}$. Note that N is not predetermined; one is free to choose it. The wavelet space $\mathcal{W}_N$ at the same level as $\mathcal{V}_N$ is the orthogonal complement of $\mathcal{V}_N$ in $\mathcal{V}_{N+M}$, relative to $\langle f, g \rangle_{\mathbf{S}^2}$, the usual inner product on $L^2(\mathbf{S}^2)$. In Narcowich and Ward (1996b), explicit bases – both orthogonal and biorthogonal – for $\mathcal{W}_N$ are given. These are the wavelets. Using these, one obtains a highly flexible framework for an analysis/synthesis algorithm on the sphere.

3.3.2 Localization and uncertainty principles

The chief measure of localization for a function f defined on the line is the variance (or its square-root: standard deviation). The notion of variance for functions defined on $\mathbf{S}^q$ is more subtle than variance on the line or in $\mathbb{R}^s$. Adapting coherent state results obtained by Carruthers and Nieto (1965), Breitenberger (1983) discusses both a notion of variance and an uncertainty principle for the circle $\mathbf{S}^1$. For the circle, these have been adapted to deal with localization and uncertainty for the periodic wavelets discussed in Narcowich and Ward (1996a).

The definition and interpretation of variance used in Breitenberger (1983) and in Narcowich and Ward (1996a) can easily be extended to $\mathbf{S}^q$. Let f be in $L^2(\mathbf{S}^q)$, and let $\|f\|_{L^2} = 1$; set

$$\mathbf{r}_f := \int_{\mathbf{S}^q} \eta |f(\eta)|^2 d\omega(\eta).$$

We remark that we regard $\mathbf{S}^q$ as embedded in $\mathbb{R}^{q+1}$, and η as both the normal to the sphere and a point on the sphere. We have *not* chosen coordinates on $\mathbf{S}^q$. One may interpret $|f(\eta)|^2 d\omega(\eta)$ as a mass distribution in $\mathbb{R}^{q+1}$ that is confined to $\mathbf{S}^q$, with $\mathbf{r}_f$ being the center of mass of the distribution. The corresponding variance is easily seen to be

$$\mathrm{var}_f(\eta) = \int_{\mathbf{S}^q} \|\eta - \mathbf{r}_f\|^2 |f(\eta)|^2 d\omega(\eta) = 1 - \|\mathbf{r}_f\|^2,$$

which is defined on $\mathbf{S}^q$ and is invariant under both choice of coordinates and rotations. Note that in the above equation, $\|\cdot\|$ stands for the Euclidean norm.

We will discuss an uncertainty relation only for $q = 2$. Infinitesimal rotations about an axis pointing along a unit vector $\mathbf{n}$ are generated by the differential operator $\mathbf{n} \cdot \Omega$, where

$$\Omega := -i\vec{x} \times \nabla, \qquad \vec{x} \in \mathbb{R}^3,$$

is a vector operator obtained by formally calculating the cross product. In quantum mechanics, the angular momentum operator $\mathbf{L}$ is proportional to Ω (see, e.g. Merzbacher (1961)). We remark that for $\mathbf{S}^2$ there exist quantum mechanical uncertainty principles, involving trade-offs between components of the angular momentum operator relative to different axes (Merzbacher (1961), p. 167). The uncertainty principle that we discuss here measures the trade-off between localization on $\mathbf{S}^2$ and a "total" angular frequency, which is related to the Laplace–Beltrami operator $\Delta_{\mathbf{S}^2}$.

We first define the expected value of Ω and the corresponding variance. If $f(\mathbf{x})$ is a twice-continuously differentiable complex-valued function on $\mathbf{S}^2$ for which $\int_{\mathbf{S}^q} |f(\eta)|^2 d\omega(\eta) = 1$, then the expected vectorial value and variance are

$$\mathbf{o}_f := \int_{\mathbf{S}^2} \overline{f(\eta)} (\Omega f)(\eta) d\omega(\eta) \quad \text{and} \quad \mathrm{var}_f(\Omega) := \int_{\mathbf{S}^2} \|(\Omega - \mathbf{o}_f) f(\eta)\|^2 d\omega(\eta).$$

With this notation in hand, we have the following.

Theorem 3.3.1 (Uncertainty Principle, Narcowich and Ward (1996b), Theorem 5.1)

$$\|\mathbf{r}_f\|^2 \leq \mathrm{var}_f(\eta) \mathrm{var}_f(\Omega).$$

This inequality can be put in a form more closely resembling the usual uncertainty relation. If we define $\Delta\Omega := \sqrt{\text{var}_f(\Omega)}$ and $\Delta\eta := \sqrt{\text{var}_f(\eta)}/\|\mathbf{r}_f\|$, then it takes the familiar form

$$\Delta\Omega\Delta\eta \geq 1.$$

With some work (cf. Narcowich and Ward (1996b), §5), one can show that

$$\Delta\Omega = \sqrt{\langle -\Delta_{\mathbf{S}^2}f, f\rangle - \|\mathbf{o}_f\|^2},$$

which expresses $\Delta\Omega$ in terms of the expected value of the Laplace–Beltrami operator (total angular "frequency") and $\|\mathbf{o}_f\|$. It is in this sense that $\Delta\eta\Delta\Omega \geq 1$ expresses the trade-off between localization and frequency.

Rösler and Voit (1997) have shown that the lower bound of 1 is sharp, and they also developed a similar uncertainty principle for general ultraspherical expansions. See Rösler and Voit (1997) for additional references related to uncertainty principles on the sphere. We close by mentioning that Freeden et al. (Freeden et al. (1998), Corollary 5.5.3) have modified the uncertainty principle above to fit the needs of their expansions.

3.3.3 Multilevel

We want to briefly describe a method for performing pointwise interpolation of a large, dense set of scattered data on the sphere, Euclidean space, or some other manifold that has basis functions (BFs) (Narcowich (1995)) similar to RBFs. We assume that we have available the data set $\{(\mathbf{x}_1, f(\mathbf{x}_1)), \ldots, (\mathbf{x}_N, f(\mathbf{x}_N))\}$, where N is large and f is smooth, but unknown. The data sites $\mathcal{C} = \{\mathbf{x}_1, \ldots, \mathbf{x}_N\}$ serve to provide the sets of "centers" to be used at the various levels. In general, $\mathcal{C}$ comprises distinct, scattered points. For the important case of least-squares surface fitting, one could simply choose $\mathcal{C}$ to be a subset of the data sites, or some other suitable point set.

A direct approach using a single BF Φ runs into difficulties: full interpolation matrices, a goodness-of-fit vs. stability tradeoff, and function evaluation problems. To circumvent them, Floater and Iske (1996), using RBFs in $\mathbb{R}^2$, came up with the following multilevel method. The method proceeds from coarse to fine, precisely the reverse of wavelet methods.

The central idea is to extract from the original set $\mathcal{C}$ a finite sequence of m nested sets, $\mathcal{C}_1 \subset \cdots \subset \mathcal{C}_m := \mathcal{C}$, each $\mathcal{C}_j$ corresponding to a level j. The $\mathcal{C}_j$s are chosen so that at any level j one has points that are neither too far apart, nor too close together. (In Floater and Iske (1996), a thinning algorithm was used for this purpose.) For the jth level, one picks a basis of the form $\{\Phi_j(\mathbf{x}\cdot\mathbf{y})\}_{\mathbf{y}\in\mathcal{C}_j}$, where Φ_j is a BF appropriate for the level. The interpolant is then found as follows.

1. Find $f_1^\star \in \mathrm{span}\{\Phi_1(\mathbf{x}\cdot\mathbf{y})\}_{\mathbf{y}\in\mathcal{C}_1}$ that interpolates f on $\mathcal{C}_1$.
2. Find $g_2^\star \in \mathrm{span}\{\Phi_2(\mathbf{x}\cdot\mathbf{y})\}_{\mathbf{y}\in\mathcal{C}_2}$ that interpolates the residual $g_2 := f - f_1^\star$ on $\mathcal{C}_2$. The level 2 interpolant is then $f_2^\star = f_1^\star + g_2^\star$.
3. For $j = 3,\ldots,m$, find $g_j^\star \in \mathrm{span}\{\Phi_j(\mathbf{x}\cdot\mathbf{y})\}_{\mathbf{y}\in\mathcal{C}_j}$ that interpolates the residual $g_j := g_{j-1} - g_{j-1}^\star$ on $\mathcal{C}_j$. The level j interpolant is $f_j^\star = f_1^\star + \sum_{\ell=2}^{j} g_\ell^\star$.

We assume that f belongs to a space $\mathcal{W}_0$ of smooth functions, a Sobolev space for example. The reproducing kernel Hilbert space $\mathcal{W}_j$ associated with each Φ_j changes with the level; smoother spaces capture trends, and "spikier" ones, details. These spaces are nested, with $\mathcal{W}_0$ being a subspace of them all.

$$\mathcal{W}_0 \subset \underbrace{\mathcal{W}_1}_{\text{smooth}} \subset \mathcal{W}_2 \subset \cdots \subset \mathcal{W}_{m-1} \subset \underbrace{\mathcal{W}_m}_{\text{spikey}}$$

The Floater–Iske technique is to pick the lowest (smoothest) level basis function Φ, and scale it: $\Phi_j(r) := \Phi(\alpha^{j-1}r)$, where $\alpha > 0$. Narcowich et al. (1999) replaced scaling with convolution, and were then able to apply results from Dyn et al. (1999) to obtain rates of convergence on the torus and sphere. It would be of interest to obtain rates of convergence under less restrictive conditions than those employed in Narcowich et al. (1999).

3.4 Continuous to discrete – computational issues

3.4.1 Problems and approaches

Numerical computation of Fourier coefficients, convolutions, and other integrals on the 2-sphere are needed in many applications – geodesy, geophysics, and meteorology, to name a few. Developing good quadrature rules to approximate integrals over the sphere is central to all of these numerical problems.

Quadrature rules are "good" if they satisfy most (and, one hopes, all) of the following requirements. The integration of spherical harmonics up to some fixed order should be exact, and the number of quadrature nodes needed to do this should be proportional to the dimension of the space of spherical harmonics that is integrated exactly. The quadrature weights should be positive and computable, and finally the magnitude of the weights should reflect the spacing of the nodes.

There are three approaches to quadrature on the sphere. One concerns *spherical designs* (Delsarte et al. (1977), Goethals and Seidel (1982)), where the aim is to determine good quadrature rules that use one weight. This approach amounts to finding a minimal set of points on $\mathbf{S}^q$ for which there is a one-weight quadrature formula that is exact for spherical harmonics up to some fixed degree. Such quadrature rules are very efficient, since the only weight is ω_q divided by the number of nodes. As Maier (2000) recently pointed out, there exist spherical designs for all $\mathbf{S}^q$; however, neither a general algorithm for finding them nor even the number of points required is known as yet.

Driscoll and Healy (1994), and Potts et al. (1996,1998) developed a second approach. One uses nodes that are selected points on a latitude–longitude grid, and then finds corresponding weights that will integrate spherical harmonics up to some fixed order. The number of nodes is comparable to the dimension of the space of spherical harmonics. The aim here is to construct sampling theorems and Fourier transforms for the 2-sphere that are fast and stable.

In many applications, the nodes are sites where a function is sampled. They will be scattered, and will not be equiangular or coincide with some predetermined set. The third approach takes these scattered sites, and uses some of them as nodes. The problem is then to find weights that give good quadrature rules. A method for doing this has recently been developed. We will discuss it in Section 3.4.3.

3.4.2 Quadrature using the latitude–longitude grid on $\mathbf{S}^2$

Driscoll and Healy (1994) investigated FFTs, quadrature rules and sampling theorems on the 2-sphere. For nodes on the latitude–longitude grid (LLG), they derived a sampling theorem that enabled fast, discrete numerical calculation of Fourier coefficients and convolutions involving band-limited functions (i.e., spherical harmonics of some fixed degree).

Recall that the Fourier decomposition of functions on the 2-sphere into spherical harmonics is simply the expansion of a function $f \in L^2(\mathbf{S}^2)$ in the orthonormal basis provided by spherical harmonics. Thus for $0 \leq \theta \leq \pi$, $0 \leq \phi \leq 2\pi$

$$f(\theta, \phi) = \sum_{l=0}^{\infty} \sum_{|m| \leq l} \hat{f}(l,m) Y_{l,m}(\theta, \phi), \text{ where } \hat{f}(l,m) = \int_{\mathbf{S}^2} f \bar{Y}_{l,m} d\omega.$$

The following sampling theorem was established in Driscoll and Healy (1994).

Theorem 3.4.1 (Driscoll and Healy (1994)) *Let $f(\theta, \phi)$ be a "band-limited" function on $\mathbf{S}^2$ such that $\hat{f}(l, m) = 0$ for $l \geq \Lambda$. Then*

$$\hat{f}(l,m) = \frac{\sqrt{2}\pi}{2\Lambda} \sum_{j=0}^{2\Lambda-1} \sum_{k=0}^{2\Lambda-1} a_j^{(\Lambda)} f(\theta_j, \phi_k) \bar{Y}_{l,m}(\theta_j, \phi_k)$$

for $0 \leq l < \Lambda$ and $|m| \leq l$. Here $\theta_j = \frac{\pi j}{2\Lambda}$, $\phi_k = \frac{\pi k}{\Lambda}$ and the coefficients $a_k^{(\Lambda)}$ are suitably defined.

It is clear that this sampling theorem immediately yields a quadrature rule which is exact for spherical harmonics of order Λ since for a given function g on $\mathbf{S}^2$, its integral is simply $\hat{g}(0,0)$ which is equal to the right-hand side of the equality given above. Moreover, Driscoll and Healy gave a formula for computing the coefficients $a_k^{(\Lambda)}$. For general Λ, the general sequence $\{a_k^{(\Lambda)}\}$ was shown to be the unique solution vector for the system of equations

$$a_0^{(\Lambda)} \mathcal{P}_\Lambda\big(\cos(\theta_0)\big) + a_1^{(\Lambda)} \mathcal{P}_\Lambda\big(\cos(\theta_1)\big) + \cdots + a_{2\Lambda-1}^{(\Lambda)} \mathcal{P}_\Lambda\big(\cos(\theta_{2\Lambda-1})\big) = \sqrt{2}\delta_l^0,$$

where $l = 0, \ldots, 2\Lambda - 1$. For the case of n samples, n a power of two, more can be said.

Proposition 3.4.2 (Driscoll and Healy (1994)) *For n samples, n a power of two, the sampling weights are*

$$A_j = \frac{2\sqrt{2}}{n} \sin\left(\frac{\pi j}{n}\right) \sum_{l=0}^{n/2-1} \frac{1}{2l+1} \sin\left([2l+1]\frac{\pi j}{n}\right), \qquad j = 0, \ldots, n-1.$$

Although it is not explicitly mentioned in Driscoll and Healy (1994), these weights are positive.

Again when the number of samples is a power of two, Potts et al. (1996) gave another set of quadrature rules. The derivation of weights given by Potts was more elementary than that given by Driscoll and Healy. They also derived bounds on these weights.

The original Driscoll–Healy algorithm becomes unstable when one tries an FFT on the 2-sphere for spherical harmonics with order 360 or higher. To remedy this, Moore (1994) used the method of *stable by-pass operations* introduced earlier in Moore et al. (1993). This, however, sacrifices speed. In Potts et al. (1998), the authors take a different approach, using Clenshaw weights. The result is a fast stable algorithm, though the weights are no longer positive.

Petrushev (1998) gave another quadrature formula which he used to obtain rates of approximation on the sphere (see Section 2). The nodes used were not on the LLG, but still coalesced at the poles of the sphere. His idea was to use two types of quadrature rules to exactly integrate trigonometric polynomials of fixed degree, one for $[0, 2\pi]$ and another for $[0, \pi]$. For the interval $[0, 2\pi]$, a standard one sufficed. The nontrivial, novel part of this program was deriving the one for $[0, \pi]$. Here is the result.

Proposition 3.4.3 (Petrushev (1998)) *For any $k = 1, 2, \ldots$, there exists a quadrature formula,*

$$Q_k(g) = \sum_{j=0}^{2k} \lambda_j g(\beta_j) \sim \int_0^\pi g(\theta) d\theta,$$

with the following properties:

(a) *$Q_k(g)$ is exact for all trigonometric polynomials of degree k.*

(b) *Let $\beta_{-1} := 0, \beta_{2k+1} := \pi$. For $0 < \beta_0 < \beta_1 < \cdots < \beta_{2k} < \pi$,*

$$\beta_j - \beta_{j-1} \leq \pi k^{-1}, \qquad j = 0, 1, \ldots, 2k + 1.$$

(c) *There exists an absolute constant c such that*

$$0 < \lambda_j \leq c(\beta_{j+1} - \beta_{j-1}), \qquad j = 0, 1, \ldots, 2k.$$

(d) *The exact values of the nodes β_j and the coefficients λ_j of the quadrature are computable.*

3.4.3 Quadrature formulae for scattered sites

The emphasis in the spherical design approach and in the Driscoll–Healy approach is on weights. The points used for nodes are unrelated to any sites for actual sampled data. Moreover, for the various approaches discussed in the previous section, nodes coalesce at the poles. In order to be useful when data comes from scattered sites, these methods will have to be augmented by some kind of surface fit, simply because data values may not be available at predetermined nodes. Recently, quadrature formulae that make use of nodes at scattered sites have been developed (Jetter et al. (1998), Mhaskar et al. (2000)).

In Jetter et al. (1998), the formulae, which are exact for spherical harmonics of some prescribed order L or less, were constructed by applying the notion of norming sets to spaces of spherical harmonics. They used a set of scattered nodes with cardinality comparable to that of the spherical harmonics they exactly integrate. These formulae are almost "good", though the weights involved have indeterminate sign. Moreover, bound estimates on these weights are poor.

In Mhaskar et al. (2000), "good" quadrature formulae for scattered sites were given. The main technical device which allowed one to establish the positivity of the quadrature coefficients was a *Marcinkiewicz–Zygmund* inequality which we now discuss.

For a finite subset $\mathcal{C} \subset \mathbf{S}^q$, we may decompose the q-sphere into a finite collection $\mathcal{R}$ of regions that are spherical simplices each of which contains at least one point from $\mathcal{C}$ in its interior. The union of these regions is $\mathbf{S}^q$, and any two regions intersect only at the boundary. The norm $\|\mathcal{R}\|$ is the maximum among the diameters of the regions. Once we pick $\mathcal{R}$, it simplifies matters if we discard points in $\mathcal{C}$ that are on boundaries or are "duplicates" in a region. The result is that there is now a one-to-one correspondence between regions in $\mathcal{R}$ and points in $\mathcal{C}$, and so we can label a region by a point $\mathbf{y}$; that is, $\mathcal{R} = \{R_\mathbf{y}\}_{\mathbf{y} \in \mathcal{C}}$. We will call such a decomposition $\mathcal{C}$-compatible. It is important to note that we may choose $\mathcal{R}$ so that the quantities $\|\mathcal{R}\|$, $\delta_\mathcal{C}$, $|\mathcal{C}|^{-\frac{1}{q}}$ are all comparable. Details may be found in Mhaskar et al. (2000). For such a decomposition, there is a constant α_q, which depends only on q, such that whenever the mesh norm $\delta_\mathcal{C}$ satisfies $\delta_\mathcal{C} < \frac{\eta \alpha_q}{n}$, with $0 < \eta < 1$ we have that, for all $P \in \Pi_n^q$,

$$(1-\eta)\|P\|_{\mathbf{S}^q,p} \leq \|P\|_{\mathcal{C},p} \leq (1+\eta)\|P\|_{\mathbf{S}^q,p}, \qquad (3.9)$$

where

$$\|f\|_{\mathcal{C},p} := \begin{cases} \left(\sum_{\mathbf{x} \in \mathcal{C}} |f(\mathbf{x})|^p \mu_q(R_\mathbf{x})\right)^{1/p} & \text{if } 1 \leq p < \infty, \\ \sup_{\mathbf{x} \in \mathcal{C}} \{|f(\mathbf{x})|\} & \text{if } p = \infty. \end{cases}$$

This Marcinkiewicz–Zygmund inequality is the key to establishing the result on quadrature rules for scattered nodes.

Theorem 3.4.4 (Mhaskar et al. (2000), Theorem 4.1) *Let $1 \leq p \leq \infty$. If in (3.9), $\eta < \frac{1}{2}$, then there exist nonnegative numbers $\{c_\mathbf{y}\}_{\mathbf{y} \in \mathcal{C}}$, such that for every $P \in \Pi_n^q$, we have*

$$\int_{\mathbf{S}^q} P(\mathbf{x}) d\mu_q(\mathbf{x}) = \sum_{\mathbf{y} \in \mathcal{C}} a_\mathbf{y} P(\mathbf{y}),$$

and

$$\left\| \left(\frac{a_\mathbf{y}}{\mu_q(R_\mathbf{y})} \right) \right\|_{\mathcal{C},p'} \leq \omega_q^{1/p'} (1-\eta)^{-1}, \quad 1 \leq p' := \frac{p}{p-1} \leq \infty.$$

If we choose $\frac{1}{2} \leq \eta < 1$, then the quadrature rule continues to hold, although the coefficients $\{a_\mathbf{y}\}$ are no longer guaranteed to be positive.

The theorem is in fact a feasibility result. Computing the weights requires solving a quadratic programming problem. The theorem states that the program has a solution, and so the standard algorithms will converge to the solution.

There is an application of the results above to hyperinterpolation, which we discussed at the end of Section 3.2.2. Quadrature plays an important role in this method of approximation. In particular, the approximation rates mentioned in Theorem 3.2.8 postulate quadrature rules that satisfy the regularity assumption discussed prior to Theorem 3.2.8. The quadrature rules constructed in Theorem 3.4.4 satisfy this regularity assumption, with $c_1 \approx (1-\eta)^{-1}$ where $\eta < 1/2$.

The techniques used to prove Marcinkiewicz–Zygmund inequality (3.9), were employed in Mhaskar et al. (1999b), to obtain rates of approximation and quasi-interpolants constructed from shifts of a compactly supported function, with data supplied by a continuous function sampled at scattered sites in the unit cube in $\mathbb{R}^s$. This extends results of Chui and Diamond (1990).

Acknowledgments

Research of the authors was sponsored by the Air Force Office of Scientific Research, Air Force Materiel Command, USAF, under grant numbers F49620-97-1-0211 and F49620-98-1-0204. The US Government is authorized to reproduce and distribute reprints for Governmental purposes notwithstanding any copyright notation thereon. The views and conclusions contained herein are those of the authors and should not be interpreted as necessarily representing the official policies or endorsements, either expressed or implied, of the Air Force Office of Scientific Research or the US Government.

The authors also wish to thank Professor Rudolph A. Lorentz for many helpful suggestions.

References

Alfeld, P., Neamtu, M. and Schumaker, L.L. (1996). Bernstein–Bezier polynomials on spheres and sphere-like surfaces. *Comput. Aided Geom. Design*, **13**, 333–349.

Antoine, J.-P. and Vandergheynst, P. (1998). Wavelets on the n-sphere and related manifolds. *J. Math. Phys.*, **39**, 3987–4008.

Bayer, M., Beth, S. and Freeden, W. (1998). Geophysical field modelling by multiresolution analysis. *Acta Geod. Geoph. Hung.*, **33**, 289–319.

Boor, C. de, DeVore, R. and Ron, A. (1994). Approximation from shift-invariant subspaces of $L_2(\mathbb{R}^d)$. *Trans. Amer. Math. Soc.*, **341**, 787–806.

Bos, L., Levenberg, N., Milman, P. and Taylor, B.A. (1995). Tangential Markov inequalities characterize algebraic submanifolds of $\mathbb{R}^N$. Ind. Univ. Math. J., **44**, 115–138.

Breitenberger, E. (1983). Uncertainty measures and uncertainty relations for angle observables. Found. Phys., **15**, 353–364.

Carruthers, P. and Nieto, M.M (1965). Coherent states and the number-phase uncertainty relation. Phys. Rev. Lett., **14**, 387–389.

Cheney, W. (1995). Approximation using positive definite functions. In *Approximation Theory VIII, Vol. 1: Approximation and Interpolation*, ed. C.K. Chui and L.L. Schumaker, pp. 145–168. World Scientific Publishing Co., Inc., Singapore.

Chui, C.K. and Diamond, H. (1990). A characterization of multivariate quasi-interpolation formulas and its applications. Numer. Math., **57**, 105–121.

Cohen, A., Daubechies, I. and Vial, P. (1993). Wavelets on the interval and fast wavelet transforms. Appl. Comp. Harm. Anal., **1**, 54–81.

Dahlke, S., Dahmen, W., Schmidt, E. and Weinrich, I. (1995). Multiresolution analysis on $\mathbf{S}^2$ and $\mathbf{S}^3$. Numer. Funct. Anal. Optimiz., **16**, 19–41.

Dahlke, S. and Maass, P. (1996). Continuous wavelet transforms with applications to analyzing functions on spheres. J. Fourier Anal. Appl., **2**, 379–396.

Delsarte, Ph., Goethals, J.M. and Seidel, J.J. (1977). Spherical codes and designs. *Geometriae Dedicata*, **6**, 363–388.

Driscoll, J.R. and Healy, D.M. (1994). Computing Fourier transforms and convolutions for the 2-sphere. Adv. in Appl. Math. , **15**, 202–250.

Dyn, N., Narcowich, F.J. and Ward, J.D. (1997). A framework for interpolation and approximation on Riemannian manifolds. In *Approximation Theory and Optimization*, ed. M.D. Buhmann and A. Iserles, pp. 133–144. Cambridge University Press, Cambridge.

Dyn, N., Narcowich, F.J. and Ward, J.D. (1999). Variational principles and Sobolev-type estimates for generalized interpolation on a Riemannian manifold. Constr. Approx., **15**, 175–208.

Fasshauer, G.E. and Schumaker, L.L. (1998). Scattered data fitting on the sphere. In *Mathematical Methods for Curves and Surfaces II*, ed. M. Dæhlen, T. Lyche and L.L. Schumaker, pp. 117–166. Vanderbilt University Press, Nashville & London.

Floater, M.S. and Iske, A. (1996). Multistep scattered data interpolation using compactly supported radial basis functions. J. Comput. Appl. Math., **73(1-2)**, 65–78.

Freeden, W. (1981). On spherical spline interpolation and approximation. Math. Meth. Appl. Sci., **3**, 551–575.

Freeden, W. (1984). Spherical spline approximation: basic theory and computational aspects. J. Comput. Appl. Math., **11**, 367–375.

Freeden, W. (1986). Uniform approximation by spherical spline interpolation: basic theory and computational aspects. Math. Z., **193**, 265–275.

Freeden, W., Gervens, T. and Schreiner, M. (1998). *Constructive Approximation on the Sphere: with Applications to Geomathematics*. Clarendon Press, Oxford.

Freeden, W., Schreiner, M. and Franke, R. (1997). A survey on spherical spline approximation. Surveys Math. Indust., **7**, 29–85.

Freeden, W. and Windheuser, U. (1996). Spherical wavelet transform and its discretization. Adv. Comput. Math., **5**, 51–94.

Goethals, J.M. and Seidel, J.J. (1982). Cubature formulae, polytopes, and spherical designs. In *The Geometric Vein, Coexeter Festschrift*, ed. C. Davis, B. Grunbaum and F.A. Sherk, pp. 203–218. Springer, New York.

Golitschek, M. von and Light, W.A. (2000). Interpolation by polynomials and radial basis functions. Preprint.

Golomb, M. and Weinberger, H.F. (1959). Optimal approximation and error bounds. In *On Numerical Approximation*, ed. R.E. Langer, pp. 117–190. The University of Wisconsin Press, Madison.

Hobson, E.W. (1965). *The Theory of Spherical and Ellipsoidal Harmonics*, 2nd Ed. Cambridge University Press. Reprinted by Chelsea Publishing, New York.

Holschneider, M. (1996). Continuous wavelet transforms on the sphere. *J. Math. Phys.*, **37**, 4156–4165.

Jetter, K., Stöckler, J. and Ward, J.D. (1998). Norming sets and spherical cubature formulas. In *Computational Mathematics*, ed. Z. Chen, Y. Li, C.A. Micchelli and Y. Xu, pp. 237–245. Marcel Decker, New York.

Jetter, K., Stöckler, J. and Ward, J.D. (1999). Error estimates for scattered data interpolation on spheres. *Math. Comp.*, **68**, 743–747.

Levesley, J., Light, W., Ragozin, D. and Sun, X. (1999). A simple approach to the variational theory for interpolation on spheres. In *New Developments in Approximation Theory*, ed. M.W. Müller, M.D. Buhmann, D.H. Mache, M. Felten, pp. 117–143. ISNM, vol. 132, Birkhaüser, Basel.

Madych, W.R. and Nelson, S.A. (1983). Multivariate interpolation: a variational approach. Preprint.

Maier, U. (2000). Numerical calculation of spherical designs. In *Proceedings of the International Conference on Multivariate Approximation*, ed. W. Haussmann, K. Jetter and M. Reimer. John Wiley, to appear.

Merzbacher, E. (1961). *Quantum Mechanics*. Wiley, New York.

Mhaskar, H.N. and Micchelli, C.A. (1995). Degree of approximation by neural and translation networks with a single hidden layer. *Adv. in Appl. Math.*, **16**, 151–183.

Mhaskar, H.N., Narcowich, F.J. and Ward, J.D. (1999a). Approximation properties of zonal function networks using scattered data on the sphere. *Adv. in Comp. Math.*, **11**, 121–137.

Mhaskar, H.N., Narcowich, F.J. and Ward, J.D. (1999b). Quasi-interpolation in shift invariant spaces. Center for Approximation Theory Report # 396, Department of Mathematics, Texas A&M University.

Mhaskar, H.N., Narcowich, F.J. and Ward, J.D. (2000). Spherical Marcinkiewicz–Zygmund inequalities and positive quadrature. *Math. Comp.*, to appear.

Moore, S.S.B. (1994). Efficient stabilization methods for fast polynomial transforms. Thesis, Dartmouth College.

Moore, S.S.B., Healy, D. and Rockmore, D.N. (1993). Symmetry stabilization for fast discrete monomial transforms and polynomial evaluation. *Linear Algebra Appl.*, **192**, 249–299.

Müller, C. (1966). *Spherical Harmonics*. Lecture Notes in Mathematics, **17**, Springer Verlag, Berlin.

Narcowich, F.J. (1995). Generalized Hermite interpolation and positive definite kernels on a Riemannian manifold. *J. Math. Anal. Applic.*, **190**, 165–193.

Narcowich, F.J. (1998). Recent developments in approximation via positive definite functions. In *Approximation IX, Vol. II: Computational Aspects*, ed.

C.K. Chui and L. Schumaker, pp. 221-242. Vanderbilt University Press, Nashville.

Narcowich, F.J., Schaback, R. and Ward, J.D. (1999). Multilevel interpolation and approximation. *Appl. Comp. Harm. Anal.*, **7**, 243–261.

Narcowich, F.J. and Ward, J.D. (1996a). Wavelets associated with periodic basis functions. *Appl. Comp. Harm. Anal.*, **3**, 40–56.

Narcowich, F.J. and Ward, J.D. (1996b). Nonstationary wavelets on the m-sphere for scattered data. *Appl. Comp. Harm. Anal.*, **3**, 324–336.

Pawelke, S. (1972). Uber die approximationsordnung bei kugelfunktionen und algebraischen polynomen. *Tôhoku Math. J.*, **24**, 473–486.

Petrushev, P. (1998). Approximation by ridge functions and neural networks. *SIAM J. Math. Anal.*, **30**, 155–189.

Potts, D., Steidl, G. and Tasche, M. (1996). Kernels of spherical harmonics and spherical frames. In *Advanced Topics in Multivariate Approximation*, ed. F. Fontanella, K. Jetter, P.-J. Laurent, pp. 287–301. World Scientific Publishing.

Potts, D., Steidl, G. and Tasche, M. (1998). Fast and stable algorithms for discrete spherical Fourier transforms. *Linear Algebra Appl.*, **275-276**, 433–450.

Rösler, M. and Voit, M. (1997). An uncertainty principle for ultraspherical expansions. *J. Math. Anal. Applic.*, **209**, 624–634.

Ron, A. and Sun, X. (1996). Strictly positive definite functions on spheres on spheres in Euclidean spaces. *Math. Comp.*, **65**, 1513–1530.

Schoenberg, I.J. (1942). Positive definite functions on spheres. *Duke Math. J.*, **9**, 96–108.

Schröder, P. and Sweldens, W. (1995). Efficiently representing functions on the sphere. *Computer Graphics Proceedings (SIGGRAPH 95)* (ACM Siggraph, Los Angeles), 161–172.

Schröder, P. and Sweldens, W. (1997). Spherical wavelets: Texture processing. Preprint.

Sloan, I.H. (1995). Polynomial interpolation and hyperinterpolation over general regions. *J. Approx. Theory*, **83**, 238–254.

Sloan, I.H. (1997). Interpolation and hyperinterpolation on the sphere. In *Multivariante approximation: recent trends and results*, ed. W. Haussmann, K. Jetter and M. Reimer, pp. 255–268. Akademie Verlag, Berlin.

Sloan, I.H. and Womersley, R.S. (2000). Constructive polynomial approximation on the sphere. *J. Approx. Theory*, **103**, 91–118.

Stein, E.M. (1957). Interpolation in polynomial classes and Markoff's inequality. *Duke Math. J.*, **24**, 467–476.

Stein, E.M. and Weiss, G. (1971). *Fourier Analysis on Euclidean Spaces*. Princeton University Press, Princeton, New Jersey.

Sweldens, W. (1996). A custom-design construction of biorthogonal wavelets. *Appl. Comp. Harm. Anal.*, **3**, 186–200.

Sweldens, W. (1998). The lifting scheme: a construction of second generation wavelets. *SIAM J. Math. Anal.*, **29**, 511–546.

Szegö, G. (1975). *Orthogonal Polynomials*. Amer. Math. Soc. Colloq. Publ., **23**, Amer. Math. Soc., Providence.

Tasche, M. and Potts, D. (1995). Interpolatory wavelets on the sphere. In *Approximation Theory VIII, Vol. 2: Wavelets and Multilevel Approximation*, ed. C.K. Chui and L.L. Schumaker, pp. 335–342. World Scientific, Singapore.

Torrésani, B. (1995). Position-frequency analysis for functions defined on spheres. *Signal Process.*, **43**, 341–346.

Wahba, G. (1981). Spline interpolation and smoothing on the sphere. *SIAM J. Sci. Statist. Comput.*, **2**, 5–16.

Wahba, G. (1984). Surface fitting with scattered noisy data on Euclidean d-space and on the sphere. *Rocky Mountain J. Math.*, **14**, 281–299.

Xu, Y. and Cheney, E.W. (1992). Strictly positive definite functions on spheres. *Proc. Amer. Math. Soc.*, **116**, 977–981.

4
A survey on L_2-approximation orders from shift-invariant spaces

K. JETTER and G. PLONKA

Abstract

This chapter aims at providing a self-contained introduction to notions and results connected with the L_2-approximation order of finitely generated shift-invariant (FSI) spaces $S_\Phi \subset L_2(R^d)$. Here, the approximation order is with respect to a scaling parameter and to the usual scaling of the L_2-projector onto S_Φ, where $\Phi = \{\phi_1, \ldots, \phi_n\} \subset L_2(R^d)$ is a given set of functions, the so-called generators of S_Φ. Special attention is given to the principal shift-invariant (PSI) case, where the shift-invariant space is generated from the multi-integer translates of just one generator. This case is interesting in itself because of its possible applications in wavelet methods. The general FSI case is considered subject to a stability condition being satisfied, and the recent results on so-called superfunctions are developed. For the case of a refinable system of generators the sum rules for the matrix mask and the zero condition for the mask symbol, as well as invariance properties of the associated subdivision and transfer operator are discussed. References to the literature and further notes are extensively given at the end of each section. In addition, the list of references has been enlarged in order to provide a rather comprehensive overview of the existing literature in the field.

4.1 Introduction

In this chapter we give an overview on recent results concerning the L_2-approximation order of so-called shift-invariant subspaces of $L_2(\mathbb{R}^d)$. We are going to consider only specific shift-invariant spaces, namely principal shift-invariant (PSI) spaces S_ϕ, generated by the multi-integer translates of just one single function $\phi \in L_2(\mathbb{R}^d)$, and more generally finitely generated shift-invariant (FSI) spaces S_Φ with a finite set $\Phi = \{\phi_1, \ldots, \phi_n\} \subset L_2(\mathbb{R}^d)$

of generators. This leads to the following notion of approximation order. Let $P_\Phi : L_2(\mathbb{R}^d) \to S_\Phi$ be the L_2-projector onto the shift invariant space, and let $P_{\Phi,h}$ be its scaled version, i.e., $P_{\Phi,h}(f) := \{P_\Phi(f_h)\}(\frac{\cdot}{h})$ with $f_h(x) := f(h\cdot x)$, for the scale parameter $0 < h \in \mathbb{R}$. Then S_Φ is said to have L_2-approximation order $0 < m \in \mathbb{R}$ for the subspace $W \subset L_2(\mathbb{R}^d)$ if

$$\|f - P_{\Phi,h}(f)\|_2 = \mathcal{O}(h^m) \quad \text{as } h \to 0 \, ,$$

for every $f \in W$. When W is the Sobolev space $W_2^m(\mathbb{R}^d)$, this definition can be replaced by the following condition on the unscaled operator

$$\|f - P_\Phi(f)\|_2 \leq \text{const.} \, |f|_{m,2} \quad \text{for any } f \in W_2^m(\mathbb{R}^d) \, ,$$

where $|f|_{m,2}^2 := (2\pi)^{-d} \int_{\mathbb{R}^d} |\xi|^{2m} \, |f^\wedge(\xi)|^2 \, d\xi$ denotes the usual Sobolev seminorm of order m. Here $f^\wedge$ denotes the Fourier transform of f and $|\xi|^2 = \xi_1^2 + \cdots + \xi_d^2$ is the Euclidian norm. It is this notion of approximation order which we will be referring to.

One essential concern of this chapter is to give a self-contained summary of the subject, where the PSI case and the FSI case are developed independently. This concept aims at simplifying the account for those readers only interested in the PSI case. At present, this circle of readers is certainly the larger part of the community interested in the approximation-theoretic aspects of wavelets and other related multiresolution methods. Furthermore, we are able to point out the close connections to, and the basic ideas of, the generalization to the FSI case. The main results will be worked out in a form which is perhaps not always the most general, but hopefully highly readable. They can be understood without referring back to the original literature. The interested reader, however, will find remarks and extensions at the end of each section, including explicit references to the original literature. The bibliography of this survey is intended to be even more comprehensive. However, we are aware of the fact that we have been somewhat selective. While we will only consider L_2-projectors as approximation methods, the reference list provides information about various papers treating other linear approximation processes which are quasi-optimal, i.e., having the same approximation order as the L_2-projectors.

This chapter is organized as follows. It consists of two main sections. The first deals with general shift-invariant spaces, while the second provides more details in the case where the system of generators is refinable. Both sections are very much influenced by some basic material from the list of references. Section 4.2 could not have been written without recourse to the fundamental work of de Boor, DeVore and Ron. Section 4.3 frequently uses Jia's important contributions. We believe that here and there we have added our own viewpoint. We would like to stress that the results of the FSI case in Section 4.3 are new.

Notation will be given in the text at appropriate places. We note that we use the standard multi-index notation. A multi-index is a d-tuple $\mu = (\mu_1, \ldots, \mu_d)$ whose components are nonnegative integers. Further, $|\mu| := \mu_1 + \cdots + \mu_d$, and $\mu! := \mu_1! \cdots \mu_d!$. For two multi-indices $\mu = (\mu_1, \ldots, \mu_d)$ and $\nu = (\nu_1, \ldots, \nu_d)$, we write $\nu \leq \mu$ if $\nu_j \leq \mu_j$ for $j = 1, \ldots, d$. In addition, $\binom{\mu}{\nu} := \frac{\mu!}{\nu!(\mu-v)!}$ for $\nu \leq \mu$, and D^μ is shorthand for the differential operator $\frac{\partial^{|\mu|}}{\partial x_1^{\mu_1} \cdots \partial x_d^{\mu_d}}$.

4.2 L_2-projectors onto FSI spaces

4.2.1 Shift-invariant spaces

A *shift-invariant space* S is a subspace of $L_2(\mathbb{R}^d)$ which is invariant under multi-integer shifts,

$$s \in S \implies s(\cdot - \alpha) \in S \text{ for all } \alpha \in \mathbb{Z}^d .$$

We shall deal with specific shift-invariant spaces spanned by the multi-integer translates of given "basis functions", or "generators". A *principal shift-invariant space* (or *PSI space*) S_ϕ is determined through a *single* generator $\phi \in L_2(\mathbb{R}^d)$ as the closure (with respect to the topology of $L_2(\mathbb{R}^d)$) of

$$S_\phi^0 := \text{span}\langle \phi(\cdot - \alpha); \alpha \in \mathbb{Z}^d \rangle.$$

Similarly, given a set of *finitely many* generators $\Phi = \{\phi_1, \ldots, \phi_n\} \subset L_2(\mathbb{R}^d)$ its associated *finitely generated shift-invariant space* (or *FSI space*) S_Φ is the closure of

$$S_\Phi^0 := \sum_{i=1}^n S_{\phi_i}^0 .$$

Some preliminary notations follow. Given $f, g \in L_2(\mathbb{R}^d)$, their scalar product can be expressed as

$$(f|g) := \int f(x) \, \overline{g(x)} \, dx = (2\pi)^{-d} \, (f^\wedge | g^\wedge) = (2\pi)^{-d} \int_C [f^\wedge | g^\wedge](\xi) \, d\xi$$

where

$$[f^\wedge | g^\wedge] := \sum_{\alpha \in \mathbb{Z}^d} f^\wedge(\cdot + 2\pi\alpha) \overline{g^\wedge(\cdot + 2\pi\alpha)}$$

is the 2π-periodization of $f^\wedge \overline{g^\wedge}$, now often called the *bracket product* of $f^\wedge$

and $g^\wedge$. Here, we have used Parseval's identity and the Fourier transform with the normalization

$$f^\wedge(\xi) = \int f(x)\, e^{-ix\cdot\xi}\, dx\,,$$

where $x \cdot \xi$ denotes the scalar product of the two vectors in $\mathbb{R}^d$. Unindexed integrals are taken with respect to the full space $\mathbb{R}^d$, and C stands for the d-dimensional fundamental cube,

$$C := [-\pi, +\pi]^d\,.$$

It is not hard to verify that $[f^\wedge | g^\wedge] \in L_1(C)$. Hence its Fourier coefficients can be expressed as

$$\begin{aligned}(2\pi)^{-d} \int_C [f^\wedge|g^\wedge](\xi)\, e^{+i\alpha\cdot\xi}\, d\xi &= (2\pi)^{-d}\, (f^\wedge | e^{-i\alpha\cdot\xi} g^\wedge) \\ &= (f|g(\cdot - \alpha)) \\ &= \int f(x)\, \overline{g(-(\alpha-x))}\, dx \\ &=: (f * g^*)(\alpha)\,, \qquad \alpha \in \mathbb{Z}^d\,,\end{aligned}$$

with $g^*(x) := \overline{g(-x)}$ denoting the involution of g, and $f * g^*$ the convolution of f and g^* (which is a continuous function). This shows that the bracket product has the Fourier series

$$[f^\wedge | g^\wedge] \sim \sum_{\alpha \in \mathbb{Z}^d} (f * g^*)(\alpha)\, e^{-i\alpha\cdot\xi}\,, \qquad (4.1)$$

and at the same time verifies the useful fact that f is orthogonal to the PSI space S_g if and only if $[f^\wedge | g^\wedge] = 0$ as an identity in $L_1(C)$. Specializing to the case where $f = g$, $F := f * f^*$ is called the auto-correlation of f, and

$$[f^\wedge | f^\wedge] = \sum_{\alpha \in \mathbb{Z}^d} |f^\wedge(\cdot + 2\pi\alpha)|^2 \sim \sum_{\alpha \in \mathbb{Z}^d} F(\alpha)\, e^{-i\alpha\cdot\xi}\,.$$

The following useful characterization of FSI spaces in the Fourier domain holds true.

Lemma 4.2.1 *For a finite set of generators $\Phi = \{\phi_1, \ldots, \phi_n\} \subset L_2(\mathbb{R}^d)$ and $f \in L_2(\mathbb{R}^d)$ the following are equivalent:*

(i) $f \in S_\Phi$;

(ii) *there are 2π-periodic functions $\tau_1, \ldots, \tau_n$ such that $f^\wedge = \sum_{i=1}^n \tau_i\, \phi_i^\wedge$.*

With this characterization in hand the following is easy to see.

Lemma 4.2.2 *A function $f \in L_2(\mathbb{R}^d)$ is orthogonal to S_Φ, with $\Phi = \{\phi_1, \ldots, \phi_n\} \subset L_2(\mathbb{R}^d)$, if and only if $[f^\wedge | \phi_i^\wedge] = 0$, $i = 1, \ldots, n$, a.e. on C.*

4.2.2 L_2-projectors onto S_Φ

The L_2-projector onto S_Φ is the linear (continuous) operator $P_\Phi : L_2(\mathbb{R}^d) \to S_\Phi$ characterized by

$$f - P_\Phi(f) \perp S_\Phi \quad \text{for any } f \in L_2(\mathbb{R}^d).$$

We give a representation for this projector according to Lemma 4.2.1, in the various cases.

4.2.2.1 The PSI case

We first assume that for the single generator ϕ the translates are *orthonormal*, i.e., $(\phi | \phi(\cdot - \alpha)) = (\phi * \phi^*)(\alpha) = \delta_{0,\alpha}$ or equivalently,

$$[\phi^\wedge | \phi^\wedge] = 1 \quad \text{in } L_1(C).$$

In this case it is clear that

$$P_\phi(f) = \sum_{\alpha \in \mathbb{Z}^d} \left(f | \phi(\cdot - \alpha) \right) \phi(\cdot - \alpha),$$

whence $P_\phi(f)^\wedge = \tau_f \, \phi^\wedge$ with $\tau_f = \sum_{\alpha \in \mathbb{Z}^d} (f | \phi(\cdot - \alpha)) \, e^{-i\alpha \cdot \xi} \in L_2(C)$ the Fourier series of $[f^\wedge | \phi^\wedge] \in L_1(C)$, and therefore

$$P_\phi(f)^\wedge = \tau_f \, \phi^\wedge \quad \text{with} \quad \tau_f = [f^\wedge | \phi^\wedge]. \tag{4.2}$$

Next we deal with the case where the translates of ϕ form a *Riesz basis* for S_ϕ, i.e., for some constants $0 < A \leq B < \infty$ we have

$$A \sum_{\alpha \in \mathbb{Z}^d} |c_\alpha|^2 \leq \left\| \sum_{\alpha \in \mathbb{Z}^d} c_\alpha \phi(\cdot - \alpha) \right\|^2 \leq B \sum_{\alpha \in \mathbb{Z}^d} |c_\alpha|^2$$

for any $\ell_2(\mathbb{Z}^d)$-sequence $c = (c_\alpha)$, where $\|f\| := (f|f)^{1/2}$ denotes the usual $L_2(\mathbb{R}^d)$-norm. Letting $\tau(\xi) = \sum_{\alpha \in \mathbb{Z}^d} c_\alpha \, e^{-i\alpha \cdot \xi}$, it follows that

$$\left\| \sum_{\alpha \in \mathbb{Z}^d} c_\alpha \, \phi(\cdot - \alpha) \right\|^2 = (2\pi)^{-d} \, (\tau \, \phi^\wedge | \tau \, \phi^\wedge) = (2\pi)^{-d} \int_C |\tau(\xi)|^2 \, [\phi^\wedge | \phi^\wedge](\xi) \, d\xi$$

and $\sum_{\alpha \in \mathbb{Z}^d} |c_\alpha|^2 = (2\pi)^{-d} \int_C |\tau(\xi)|^2 \, d\xi$. Hence the Riesz basis property is equivalent to

$$A \leq [\phi^\wedge | \phi^\wedge](\xi) \leq B \quad \text{a.e. in } C.$$

By performing the *orthogonalization process*

$$(\phi^\perp)^\wedge := \phi^\wedge / \sqrt{[\phi^\wedge|\phi^\wedge]} \qquad (4.3)$$

we see that $\phi^\perp \in S_\phi$, hence $S_{\phi^\perp} \subset S_\phi$. From the preceding orthonormal case, $f^\wedge = P_{\phi^\perp}(f)^\wedge$ for $f \in S_\phi$, whence $S_{\phi^\perp} = S_\phi$, and

$$P_\phi(f)^\wedge = \tau_f \, \phi^\wedge \quad \text{with} \quad \tau_f = \frac{[f^\wedge|\phi^\wedge]}{[\phi^\wedge|\phi^\wedge]}. \qquad (4.4)$$

In the *general PSI case* we use the same formulas with the modification

$$(\phi^\perp)^\wedge(\xi) := 0 =: \tau_f(\xi) \quad \text{if} \quad [\phi^\wedge|\phi^\wedge](\xi) = 0. \qquad (4.5)$$

Again, $\phi^\perp \in S_\phi$ and $S_{\phi^\perp} \subset S_\phi$. From

$$[f^\wedge - P_\phi(f)^\wedge | \phi^\wedge] = 0 \quad \text{a.e. in } C , \quad \text{for any } f \in L_2(\mathbb{R}^d) ,$$

we conclude that P_ϕ is indeed the orthogonal projector onto S_ϕ.

In order to obtain an explicit error formula, we mention that

$$\|f - P_\phi(f)\|^2 = \|f\|^2 - \|P_\phi(f)\|^2$$

by orthogonality. Therefore, using Parseval's identity, we find

$$\|f - P_\phi(f)\|^2 = (2\pi)^{-d} \int \left\{ |f^\wedge(\xi)|^2 - \left| \frac{[f^\wedge|\phi^\wedge](\xi)}{[\phi^\wedge|\phi^\wedge](\xi)} \right|^2 |\phi^\wedge(\xi)|^2 \right\} d\xi ,$$

from which we have

Theorem 4.2.3 *Let P_ϕ denote the orthogonal projector onto the PSI space S_ϕ. Then, for $f \in L_2(\mathbb{R}^d)$ such that $\operatorname{supp} f^\wedge \subset C$, we have the error formula*

$$\|f - P_\phi(f)\|^2 = (2\pi)^{-d} \int_C |f^\wedge(\xi)|^2 \left\{ 1 - \frac{|\phi^\wedge(\xi)|^2}{[\phi^\wedge|\phi^\wedge](\xi)} \right\} d\xi.$$

Remark 4.2.4 As previously noted, in this theorem we must set

$$\Lambda_\phi{}^\wedge(\xi) := \frac{|\phi^\wedge(\xi)|^2}{[\phi^\wedge|\phi^\wedge](\xi)} := 0 \quad \text{if} \quad [\phi^\wedge|\phi^\wedge](\xi) = 0.$$

This implies that $0 \leq \Lambda_\phi{}^\wedge(\xi) \leq 1$, and

$$\sum_{\alpha \in \mathbb{Z}^d} \Lambda_\phi{}^\wedge(\xi + 2\pi\alpha) = 1 - \chi_{Z_\phi}(\xi),$$

with Z_ϕ the set of all $\xi \in \mathbb{R}^d$ for which $[\phi^\wedge|\phi^\wedge](\xi) = 0$, i.e.,

$$Z_\phi = \left\{ \xi \in \mathbb{R}^d;\ \phi^\wedge(\xi + 2\pi\alpha) = 0 \text{ for all } \alpha \in \mathbb{Z}^d \right\}.$$

In the case Z_ϕ has measure 0, we see that Λ_ϕ has the fundamental interpolation property

$$\Lambda_\phi(\alpha) = \delta_{0,\alpha} \quad \text{for} \quad \alpha \in \mathbb{Z}^d.$$

4.2.2.2 The FSI case

Here we deal with the stable case only when the translates of the system $\Phi = \{\phi_1, \ldots, \phi_n\}$ form a *Riesz basis* of S_Φ, i.e.,

$$A \sum_{i=1}^n \sum_{\alpha \in \mathbb{Z}^d} |c_\alpha^{(i)}|^2 \leq \left\| \sum_{i=1}^n \sum_{\alpha \in \mathbb{Z}^d} c_\alpha^{(i)} \phi_i(\cdot - \alpha) \right\|^2 \leq B \sum_{i=1}^n \sum_{\alpha \in \mathbb{Z}^d} |c_\alpha^{(i)}|^2$$

for some constants $0 < A \leq B < \infty$. It is convenient to introduce the Gramian $\mathbf{G}_\Phi$ for the system Φ as given by

$$\begin{aligned}
\mathbf{G}_\Phi^t &= \begin{pmatrix}
[\phi_1^\wedge, \phi_1^\wedge] & [\phi_1^\wedge, \phi_2^\wedge] & \cdots & [\phi_1^\wedge, \phi_n^\wedge] \\
[\phi_2^\wedge, \phi_1^\wedge] & [\phi_2^\wedge, \phi_2^\wedge] & \cdots & [\phi_2^\wedge, \phi_n^\wedge] \\
\vdots & \vdots & \ddots & \vdots \\
[\phi_n^\wedge, \phi_1^\wedge] & [\phi_n^\wedge, \phi_2^\wedge] & \cdots & [\phi_n^\wedge, \phi_n^\wedge]
\end{pmatrix} \\
&= \sum_{\alpha \in \mathbb{Z}^d} \Phi^\wedge(\cdot + 2\pi\alpha)\left(\Phi^\wedge(\cdot + 2\pi\alpha)\right)^* ;
\end{aligned}$$

here we are using the vector notation

$$\Phi^\wedge := (\phi_1^\wedge \; \phi_2^\wedge \; \cdots \; \phi_n^\wedge)^t.$$

The superscripts t and $*$ denote the transpose and the conjugate-transpose, respectively. This Gramian is a 2π-periodic (Hermitian) matrix function. It can be shown that the Riesz basis condition is equivalent to requiring that the spectrum $\{\sigma_1(\xi), \ldots, \sigma_n(\xi)\}$ of $\mathbf{G}_\Phi^t(\xi)$ satisfies

$$A \leq \min_{i=1,\ldots,n} \sigma_i(\xi) \leq \max_{i=1,\ldots,n} \sigma_i(\xi) \leq B \quad \text{a.e. on } C.$$

Without loss of generality we therefore assume that

$$B \geq \sigma_1(\xi) \geq \sigma_2(\xi) \geq \cdots \geq \sigma_n(\xi) \geq A , \tag{4.6}$$

and that the Gramian has the spectral decomposition

$$\mathbf{G}_\Phi(\xi) = \sum_{i=1}^n \sigma_i(\xi) \, \mathbf{u}_i(\xi) \, \mathbf{u}_i^*(\xi)$$

with $\mathbf{U} := (\mathbf{u}_1 \; \mathbf{u}_2 \; \cdots \; \mathbf{u}_n)$ a unitary, 2π-periodic matrix function.

Theorem 4.2.5 *The orthogonal projector* $P_\Phi : L_2(\mathbb{R}^d) \to S_\Phi$ *takes the form*
$$P_\Phi(f)^\wedge = (\Phi^\wedge)^t [f^\wedge | \tilde{\Phi}^\wedge]$$
with
$$\tilde{\Phi}^\wedge = \begin{pmatrix} \tilde{\phi}_1^\wedge \\ \vdots \\ \tilde{\phi}_n^\wedge \end{pmatrix} := (\mathbf{G}_\Phi^t)^{-1} \Phi^\wedge \quad and \quad [f^\wedge | \tilde{\Phi}^\wedge] := \begin{pmatrix} [f^\wedge | \tilde{\phi}_1^\wedge] \\ \vdots \\ [f^\wedge | \tilde{\phi}_n^\wedge] \end{pmatrix}.$$

Proof Assuming for the moment that $P_\Phi(f)^\wedge \in L_2(\mathbb{R}^d)$ it is easy to verify that $f - P_\Phi(f)$ is orthogonal to S_Φ, for any $f \in L_2(\mathbb{R}^d)$, viz.

$$\begin{aligned}[P_\Phi(f)^\wedge | \Phi^\wedge] &= \sum_{\alpha \in \mathbb{Z}^d} \overline{\Phi^\wedge(\cdot + 2\pi\alpha)} \, (\Phi^\wedge(\cdot + 2\pi\alpha))^t \, [f^\wedge | \tilde{\Phi}^\wedge] \\ &= \mathbf{G}_\Phi^t \, [f^\wedge | \tilde{\Phi}^\wedge] = [f^\wedge | \mathbf{G}_\Phi^t \, \tilde{\Phi}^\wedge] \\ &= [f^\wedge | \Phi^\wedge].\end{aligned}$$

Hence $[f^\wedge - P_\Phi(f)^\wedge | \Phi^\wedge]$ is the zero vector a.e. on C, and the orthogonality follows from Lemma 4.2.2.

In order to prove the above assumption, we use the equivalent representation

$$P_\Phi(f)^\wedge = (\Phi^{\perp \wedge})^t [f^\wedge | \Phi^{\perp \wedge}] \quad \text{with} \quad \Phi^{\perp \wedge} = \begin{pmatrix} \phi_1^{\perp \wedge} \\ \vdots \\ \phi_n^{\perp \wedge} \end{pmatrix} := (\mathbf{G}_\Phi^t)^{-1/2} \Phi^\wedge. \tag{4.7}$$

This is an extension of the orthogonalization process (4.3). Using the spectral decomposition of the Gramian it is not too hard to see that $\phi_i^{\perp \wedge} \in L_2(\mathbb{R}^d)$, hence $\phi_i^\perp \in S_\Phi$ by Lemma 4.2.1, $i = 1, \ldots, n$. But the Gramian of $\Phi^{\perp \wedge}$ is the identity matrix, whence the sum $\sum_{i=1}^n S_{\phi_i^\perp}$ is an orthogonal sum of PSI subspaces of S_Φ. Since $P_\Phi = \sum_{i=1}^n P_{\phi_i^\perp}$, the proof is complete. $\square$

We remark that this proof yields the following consequence. By the orthogonalization process (4.7) we have the orthogonal decomposition

$$S_\Phi = \sum_{i=1}^n S_{\phi_i^\perp}$$

into PSI spaces. As an easy consequence of Theorem 4.2.5 we get the following analogue of Theorem 4.2.3.

Theorem 4.2.6 Let P_Φ denote the orthogonal projector onto the stable FSI space S_Φ. Then for $f \in L_2(\mathbb{R}^d)$ such that $\operatorname{supp} f^\wedge \subset C$, we have the error formula

$$\|f - P_\Phi(f)\|^2 = (2\pi)^{-d} \int_C |f^\wedge(\xi)|^2 \left\{ 1 - (\Phi^\wedge)^t(\xi)\, \mathbf{G}_\Phi^{-1}(\xi)\, \overline{\Phi^\wedge(\xi)} \right\} d\xi.$$

Remark 4.2.7 We may add the useful information that the function

$$A^\wedge(\xi) := (\Phi^\wedge)^t(\xi)\, \mathbf{G}_\Phi^{-1}(\xi)\, \overline{\Phi^\wedge(\xi)}$$

satisfies

$$0 \leq \frac{1}{B}(\Phi^\wedge)^t(\xi)\, \overline{\Phi^\wedge(\xi)} \leq A^\wedge(\xi) = \sum_{i=1}^n |(\phi_i^\perp)^\wedge(\xi)|^2 \leq 1,$$

where B is the Riesz constant in (4.6). Furthermore

$$\sum_{\alpha \in \mathbb{Z}^d} A^\wedge(\xi + 2\pi\alpha) = \sum_{i=1}^n \left[(\phi_i^\perp)^\wedge \overline{(\phi_i^\perp)^\wedge} \right] = n,$$

and we arrive at the following *fundamental interpolation property*:

$$A(\alpha) = n\delta_{0,\alpha}, \quad \alpha \in \mathbb{Z}^d.$$

4.2.3 Approximation order

We first consider the PSI space generated by the famous sinc function

$$\phi(x) := \prod_{i=1}^d \frac{\sin \pi x_i}{\pi x_i} \quad \text{and} \quad \phi^\wedge(\xi) = \chi_C(\xi) = \begin{cases} 1, & \text{if } \xi \in C, \\ 0, & \text{otherwise.} \end{cases} \quad (4.8)$$

Here, by (4.2), $P_\phi(f)^\wedge = f^\wedge \chi_C$, whence

$$\|f - P_\phi(f)\|^2 = (2\pi)^{-d} \int_{\mathbb{R}^d \setminus C} |f^\wedge(\xi)|^2 \, d\xi \leq (2\pi)^{-d} \int |\xi|^{2m} \, |f^\wedge(\xi)|^2 \, d\xi$$

for any $m \in \mathbb{R}_+$, with $|\xi|$ the Euclidian norm of $\xi \in \mathbb{R}^d$. This implies

Lemma 4.2.8 *The PSI space generated by the sinc function (4.8) has approximation order m, for $f \in W_2^m(\mathbb{R}^d)$, where m is any positive real.*

We next prove the important fact that the approximation order of an FSI space is already given by the approximation order of a PSI subspace S_ψ as follows.

Theorem 4.2.9 *Let ψ denote the orthogonal projection of the* sinc *function onto the (stable) FSI space S_Φ, i.e., $\psi = P_\Phi(\text{sinc})$. Then the approximation order of S_ψ and S_Φ are the same.*

Proof From Theorem 4.2.5 and (4.2) we find that $P_\Phi \circ P_{\text{sinc}} = P_\psi \circ P_{\text{sinc}}$. Writing

$$f - P_\psi(f) = f - P_\Phi(f) + P_\Phi(f) - P_\Phi \circ P_{\text{sinc}}(f) + P_\psi \circ P_{\text{sinc}}(f) - P_\psi(f)$$

we see that (since $P_\psi(f) \in S_\Phi$)

$$\|f - P_\Phi(f)\| \leq \|f - P_\psi(f)\| \leq \|f - P_\Phi(f)\| + 2\|f - P_{\text{sinc}}(f)\|.$$

The theorem now follows from Lemma 4.2.8. $\square$

A function $\psi \in S_\Phi$ having the property that S_ψ has the same approximation order as the larger space S_Φ, is called a *superfunction* in the FSI space. Finding a superfunction with specific properties is the general aim of so-called *superfunction theory*. One such property is ψ being *compactly supported* whenever Φ is. This question will be addressed later.

Remark 4.2.10 By Theorem 4.2.5, the superfunction ψ of Theorem 4.2.9 is given by

$$\psi^\wedge = (\Phi^\wedge)^t [\chi_C | \tilde{\Phi}^\wedge],$$

hence satisfies

$$\psi^\wedge = (\Phi^\wedge)^t \,\overline{\tilde{\Phi}^\wedge} = A(\xi) \quad \text{for} \quad \xi \in C .$$

Also, since $\text{sinc} - \psi$ is orthogonal to S_ψ, we have $[\chi_C - \psi^\wedge | \psi^\wedge] = 0$, whence

$$[\psi^\wedge | \psi^\wedge](\xi) = [\chi_C | \psi^\wedge](\xi) = \overline{\psi^\wedge(\xi)} \quad \text{for} \quad \xi \in C .$$

Theorem 4.2.11 *For the (stable) FSI space S_Φ the following are equivalent, for given $0 < m \in \mathbb{R}$:*

(i) *S_Φ has approximation order m, for $f \in W_2^m(\mathbb{R}^d)$;*
(ii) *the function*

$$\xi \;\mapsto\; |\xi|^{-2m}\Big\{1 - (\Phi^\wedge)^t(\xi)\, \mathbf{G}_\Phi^{-1}(\xi)\, \overline{\Phi^\wedge(\xi)}\Big\}$$

is in $L_\infty(C)$;
(iii) *for $\psi := P_\Phi(\text{sinc})$ we have that $|\xi|^{-2m}\Big\{1 - \psi^\wedge(\xi)\Big\} \in L_\infty(C)$.*

Proof With $f = f_1 + f_2$, where $f_1 = f^\wedge \chi_C$ is the orthogonal projection of f onto S_{sinc}, we have

$$\|f - P_\Phi(f)\| \leq \|f_1 - P_\Phi(f_1)\| + \|f_2 - P_\Phi(f_2)\| \leq \|f_1 - P_\Phi(f_1)\| + \|f_2\|.$$

Moreover, for $f \in W_2^m(\mathbb{R}^d)$,

$$\|f_2\|^2 = (2\pi)^{-d} \int_{\mathbb{R}^d \setminus C} |f^\wedge(\xi)|^2 \, d\xi \leq |f|_{m,2}^2.$$

The theorem now follows from Theorems 4.2.3 and 4.2.6, in view of Remark 4.2.10. □

Remark 4.2.12 The second statement of this theorem is equivalent to the order relation

$$1 - A(\xi) := 1 - (\Phi^\wedge)^t(\xi) \, \mathbf{G}_\Phi^{-1}(\xi) \, \overline{\Phi^\wedge}(\xi) \quad = \quad \mathcal{O}(|\xi|^{2m}) \quad \text{as} \quad \xi \to 0 \,,$$

see Remark 4.2.7. Also, statement (iii) of the theorem is equivalent to the order relation

$$1 - \psi^\wedge(\xi) \quad = \quad \mathcal{O}(|\xi|^{2m}) \quad \text{as} \quad \xi \to 0 \,.$$

4.2.3.1 The (stable) compactly supported case

We now assume that S_Φ is generated by compactly supported functions. In this case, the bracket products $[\phi_i^\wedge | \phi_j^\wedge]$, $i,j = 1, \ldots, n$, all have finitely many Fourier coefficients (since the functions $\phi_i * \phi_j^*$ are compactly supported as well); i.e., we have identity in (4.1), and the entries of the Gramian are all trigonometric polynomials.

In particular, if $\mathbf{G}_\Phi(0)$ is regular, and this is certainly true under the Riesz basis condition assumed here, then the function A is holomorphic in a neighborhood of the origin. Hence the order relations in Remark 4.2.12, at least for $2m \in \mathbb{N}$, can be checked by looking at the power series expansion of $1 - A(\xi)$ at the origin.

4.2.3.2 The compactly supported PSI case

This case can be dealt with without recourse to the assumption of stability. Here, the zero set of the trigonometric polynomial $[\phi^\wedge | \phi^\wedge]$ in C has (d-dimensional Lebesgue) measure 0, and Λ_ϕ as defined in Remark 4.2.4 has the fundamental interpolation property. Indeed, let us set

$$\Omega(\xi) := 1 - \frac{|\phi^\wedge(\xi)|^2}{[\phi^\wedge | \phi^\wedge](\xi)} \quad = \sum_{0 \neq \alpha \in \mathbb{Z}^d} \frac{|\phi^\wedge(\xi + 2\pi\alpha)|^2}{[\phi^\wedge | \phi^\wedge](\xi)} \,. \tag{4.9}$$

From Theorem 4.2.3 we can deduce

Theorem 4.2.13 Let $\phi \in L_2(\mathbb{R}^d)$ be compactly supported, and $0 < m \in \mathbb{R}$. Then S_ϕ has approximation order m for $f \in W_2^m(\mathbb{R}^d)$ if and only if

$$\Omega(\xi) = \mathcal{O}(|\xi|^{2m}) \quad \text{as} \quad \xi \to 0.$$

This order relation can be expressed in another equivalent way, viz. from (4.9) we see that it is equivalent to requiring that

$$|\phi^\wedge(\xi + 2\pi\alpha)|^2 = \mathcal{O}(|\xi|^{2m}[\phi^\wedge|\phi^\wedge](\xi)) \quad \text{as} \quad \xi \to 0, \quad \text{for all } 0 \neq \alpha \in \mathbb{Z}^d.$$

A special case of this is

Theorem 4.2.14 Let $\phi \in L_2(\mathbb{R}^d)$ be compactly supported, and assume that $[\phi^\wedge|\phi^\wedge](0) \neq 0$. Then for $m \in \mathbb{N}$ the following are equivalent:

(i) S_ϕ has approximation order m, for $f \in W_2^m(\mathbb{R}^d)$;
(ii) ϕ satisfies the conditions

$$D^\beta \phi^\wedge(2\pi\alpha) = 0 \quad \text{for } 0 \neq \alpha \in \mathbb{Z}^d \text{ and } |\beta| < m .$$

4.2.3.3 Compactly supported superfunction

In the case 4.2.3.1 it is interesting to search for a compactly supported superfunction. With ψ as in Theorem 4.2.9, according to Remark 4.2.10, we see that on the fundamental cube C, ψ coincides with A as given in Remark 4.2.7. Since the components of $\Phi^\wedge$ are entire functions, and since the coefficient functions of the Gramian $\mathbf{G}_\Phi$ are trigonometric polynomials (with determinant non-vanishing due to the Riesz basis property) we see that A (hence ψ) is C^∞ in a neighborhood of the origin. Also, the representation of $\psi^\wedge \in S_\Phi$ according to Lemma 4.2.1 is determined in Remark 4.2.10 as $\psi^\wedge = (\tilde{\Phi}^\wedge)^* \Phi^\wedge \chi_C = (\Phi^\wedge)^* (\mathbf{G}_\Phi^t)^{-1} \Phi^\wedge \chi_C$, i.e.,

$$\psi^\wedge = \sum_{i=1}^n \tau_i \, \phi_i^\wedge \quad \text{with} \quad \tau_i|_C = \overline{\phi_i^\wedge}|_C \, , \, i = 1, \ldots, n.$$

From this it also follows that the 2π-periodic functions τ_i are C^∞ in a neighborhood of the origin (in fact everywhere, according to our strong assumptions).

The idea is now to mimic the behavior of $\psi^\wedge$ at the origin by a function

$$\tilde{\psi}^\wedge = \sum_{i=1}^n \tilde{\tau}_i \, \phi_i^\wedge \tag{4.10}$$

with *trigonometric polynomials* $\tilde{\tau}_i$, in order to have $\tilde{\psi}$ as a compactly supported function. This can be done, *at least for integral m*, by forcing the trigonometric polynomials to satisfy the interpolatory conditions

$$D^\beta \tau_i(0) = D^\beta \tilde{\tau}_i(0) \quad \text{for} \quad 0 \leq |\beta| < m \quad \text{and} \quad i = 1, \ldots, n.$$

Using Remarks 4.2.12 and 4.2.10 for ψ, we see that

$$D^\beta \tilde{\psi}^\wedge(0) = D^\beta \psi^\wedge(0) = \delta_{0\beta} \quad \text{and}$$
$$D^\beta \tilde{\psi}^\wedge(2\pi\alpha) = D^\beta \psi^\wedge(2\pi\alpha) = 0, \quad \text{for} \quad 0 \neq \alpha \in \mathbb{Z}^d \text{ and } |\beta| < m.$$

Hence $S_{\tilde{\psi}}$ has approximation order m for $f \in W_2^m(\mathbb{R}^d)$. We obtain

Theorem 4.2.15 *Given a stable FSI space S_Φ with compactly supported generators $\phi_1, \ldots, \phi_n$, the following are equivalent for $m \in \mathbb{N}$:*

(i) *S_Φ has approximation order m, for $f \in W_2^m(\mathbb{R}^d)$;*
(ii) *there exists a unique function $\tilde{\psi} \in S_\Phi$ which has the following properties:*

   (a) *$\tilde{\psi}^\wedge$ is of the form (4.10), where*

$$\tilde{\tau}_i \in \mathrm{span}\{e^{-i\beta \cdot \xi} : \beta \in \mathbb{Z}_+^d, \ |\beta| < m\};$$

   (b) *$\tilde{\psi}^\wedge(0) = 1$ and $D^\beta \tilde{\psi}^\wedge(0) = 0$ for $\beta \in \mathbb{Z}_+^d$ $0 < |\beta| < m$;*
   (c) *$S_{\tilde{\psi}}$ has approximation order m, for $f \in W_2^m(\mathbb{R}^d)$.*

The superfunction $\tilde{\psi}$ of this theorem is called the *canonical superfunction*, and the vector $\mathbf{v} := (\tilde{\tau}_1, \ldots, \tilde{\tau}_n)$ is sometimes referred to as the *canonical Φ-vector of order m*.

4.2.4 Notes and extensions

4.2.4.1

Shift-invariant spaces have a long tradition in signal processing and approximation theory. As the most important and widely used examples of generators we mention the famous sinc function (giving rise to expansions of band-limited signals in terms of the Whittaker cardinal series, see Whittaker (1935)), and Schoenberg's cardinal B-splines generating his cardinal B-spline series, Schoenberg (1946). More recently, shift-invariant spaces have been studied quite thoroughly since they appear in Mallat's setup of multiresolution analysis, see Daubechies (1992), Chapter 5. In this section we have tried to give an up-to-date discussion of the approximation orders provided by (scaled versions of) such spaces. The main results are worked out in a form which is perhaps not the most general, but which hopefully can be understood without essential recourse to the original literature.

4.2.4.2

This section is very much influenced by the fundamental work of de Boor, DeVore and Ron on L_2-approximation orders of shift invariant spaces, (de Boor et al. (1994a,1994b,1998)). We did not try to include the general L_p case, for which we would like to refer to Jia's survey (Jia (1995a)) and the references therein. Much of the original interest in approximation orders stems from attempts to understand the approximation orders of box spline spaces, see de Boor et al. (1993), Chapter 3, and the discussion of various notions of approximation orders (like controlled or local approximation orders, see, e.g., de Boor and Jia (1985), Halton and Light (1993), Jia and Lei (1993a), Jia and Lei (1993b), Light (1991), Light and Cheney (1992). Meanwhile, Ron and his coworkers have shown an interesting connection of approximation orders of a PSI space generated by a refinable function to the convergence of the corresponding subdivision scheme (Ron (1999)).

4.2.4.3

It is probably not fruitful to attribute the notion of bracket product to any one author. Periodization techniques have been used for a long time in Fourier analysis, a typical result being Poisson's summation formula which holds true under various assumptions, and with varying interpretations of identities. However, Jia and Micchelli (1991) and de Boor et al. (1994a) were the first to use this notion in a form leading to a useful characterization of orthogonality of PSI spaces, see de Boor et al. (1994a), Lemma 2.8. This same paper contains Lemma 4.2.1 for the PSI case, while the general case is given in de Boor et al. (1994b), Theorem 1.7. A simple proof of this result is due to Jia (1998a), Theorem 2.1, see also Jia (1995a), Theorem 1.1.

4.2.4.4

The construction of the L_2-projector in the stable PSI case is straightforward, and the orthogonalization process (4.3) has been used in the construction of orthonormal spline wavelets (with infinite support, but exponentially decaying at infinity) by Battle and Lemarié (Battle (1987), Lemarié (1988)). Much more involved is the general case of (4.4) and Theorem 4.2.3 (with the convention (4.5) which is again due to de Boor et al. (1994a), Theorems 2.9 and 2.20. In the FSI case the notion of the Gramian was first used by Goodman and Lee (1993) and Goodman, Lee and Tang (1993). They were also aware of the orthogonalization process (4.7) (see Goodman, Lee and Tang (1993), Theorem 3.3). The representation of the L_2-projector in Theorem 4.2.5 and the error formula of Theorem 4.2.6 are again due to de

Boor et al. (1994b), Theorem 3.9. An abstract form of the results in Section 4.2.2.2, and in particular of Theorem 4.2.5, can be found in Aldroubi (1996).

4.2.4.5

Section 4.2.3 closely follows the ideas and methods in de Boor et al. (1998), with some slight modifications. The arguments are easier since we invoke the assumption of stability. Concerning the more general (unstable) case, we refer to the Remark stated in de Boor et al. (1998) after the proof of Theorem 2.2.

4.2.4.6

The compactly supported PSI case as dealt with in Theorem 4.2.14 is well-understood for polynomial spline functions. First results in this direction go back to the seminal paper of Schoenberg (1946). The condition given in statement (ii) of the theorem (together with $\phi^\wedge(0) \neq 0$) is nowadays called the *Strang–Fix conditions*, due to their contribution in Strang and Fix (1973). Assertions (i), (ii) of Theorem 4.2.14 are also equivalent to the statement that algebraic polynomials of degree $\leq m - 1$ can be locally exactly reproduced in S_ϕ (Jia (1998b), Theorem 2.1).

4.2.4.7

In the univariate PSI case $d = 1$ the following assertion can be shown. Let $\phi \in L^2(\mathbb{R}^d)$ be compactly supported. If S_ϕ has approximation order $m \in \mathbb{N}$, for $f \in W_2^m(\mathbb{R}^d)$, then there exists a compactly supported tempered distribution η such that $\psi = N_m * \eta$, with N_m the cardinal B-spline of order m. Moreover, if $\operatorname{supp} \psi \subseteq [a, b]$, then η can be chosen in such a way that $\operatorname{supp} \eta \subseteq [a, b - m]$; see Ron (1990), Proposition 3.6. A generalization of this idea to the FSI case is treated by Plonka and Ron (2000).

4.3 Shift invariant spaces spanned by refinable functions

4.3.1 Dilation matrices and refinement equations

Refinable shift-invariant spaces are defined with recourse to a *dilation matrix* M, i.e., a regular *integer* $(d \times d)$-matrix satisfying

$$\lim_{n \to \infty} M^{-n} = 0 \ .$$

Equivalently, all eigenvalues of M have modulus greater than 1. Given such a matrix, a shift-invariant subspace S of $L_2(\mathbb{R}^d)$ is called M-refinable, if

$$s \in S \quad \Longrightarrow \quad s(M^{-1} \cdot) \in S \ .$$

In the literature, the dilation matrix is often taken as $M = 2I$ (with I the identity matrix), but we allow here the general case.

A shift-invariant space $S(\Phi) \subset L_2(\mathbb{R}^d)$ generated by $\Phi = \{\phi_1, \ldots, \phi_n\} \subset L_2(\mathbb{R}^d)$, is M-refinable if and only if the function vector $\Phi := (\phi_1, \ldots, \phi_n)^t$ satisfies a *matrix refinement equation*

$$\Phi = \sum_{\alpha \in \mathbb{Z}^d} \mathbf{P}_\alpha \Phi(M \cdot -\alpha) \,. \tag{4.11}$$

Here the "coefficients" $\mathbf{P}_\alpha$ are real or complex $(n \times n)$-matrices. The matrix-valued sequence $\mathbf{P} = (\mathbf{P}_\alpha)_{\alpha \in \mathbb{Z}^d}$ is usually called the *refinement mask*. In the case of a single generator, (4.11) takes the scalar form

$$\phi = \sum_{\alpha \in \mathbb{Z}^d} p_\alpha \, \phi(M \cdot -\alpha)$$

with the scalar-valued mask $p = (p_\alpha)_{\alpha \in \mathbb{Z}^d}$.

For simplicity, we shall only consider compactly supported function vectors Φ, and we suppose that the refinement mask $\mathbf{P}$ is finitely supported on $\mathbb{Z}^d$. Then $\Phi \subset L_1(\mathbb{R}^d) \cap L_2(\mathbb{R}^d)$, and in the Fourier domain the refinement equation reads

$$\Phi^\wedge = \mathbf{H}(M^{-t}\cdot) \, \Phi^\wedge(M^{-t}\cdot) \tag{4.12}$$

with $M^{-t} := (M^t)^{-1}$ and

$$\mathbf{H}(\xi) := \frac{1}{|\det M|} \sum_{\alpha \in \mathbb{Z}^d} \mathbf{P}_\alpha \, e^{-i\alpha \cdot \xi} \,, \quad \xi \in \mathbb{R}^d, \tag{4.13}$$

the so-called *refinement mask symbol*. This symbol $\mathbf{H}$ is an $(n \times n)$-matrix of trigonometric polynomials on $\mathbb{R}^d$. Again, in the PSI case we get a scalar-valued symbol which we denote by

$$H(\xi) := \frac{1}{|\det M|} \sum_{\alpha \in \mathbb{Z}^d} p_\alpha \, e^{-i\alpha \cdot \xi} \,, \quad \xi \in \mathbb{R}^d.$$

The goal of this section is a characterization of the approximation order of an M-refinable shift-invariant space S_Φ in terms of the refinement mask $\mathbf{P}$, or its symbol $\mathbf{H}$, or the associated subdivision and transfer operators (see Section 4.3.4). It turns out that the structure of the sublattices $M\mathbb{Z}^d$ and $M^t\mathbb{Z}^d$ of $\mathbb{Z}^d$ is important. We have the partitioning

$$\mathbb{Z}^d = \bigcup_{e \in E} (e + M\mathbb{Z}^d) = \bigcup_{e' \in E'} (e' + M^t\mathbb{Z}^d) \,, \tag{4.14}$$

where E and E' denote any set of representatives of the equivalence classes

$\mathbb{Z}^d/M\mathbb{Z}^d$ and $\mathbb{Z}^d/M^t\mathbb{Z}^d$, respectively. Both E and E' contain $\mu = |\det M|$ representatives, and in a standard form we shall always take

$$E = M([0,1[^d) \cap \mathbb{Z}^d \quad \text{and} \quad E' = M^t([0,1[^d) \cap \mathbb{Z}^d .$$

We also let

$$E_0 := E \setminus \{0\} \quad \text{and} \quad E'_0 := E' \setminus \{0\}.$$

The following fact will be used below. Any $0 \neq \alpha \in \mathbb{Z}^d$ has a unique representation

$$\alpha = (M^t)^\ell (e' + M^t \beta) , \quad \ell \geq 0, \ e' \in E'_0, \ \beta \in \mathbb{Z}^d . \tag{4.15}$$

This can be seen as follows. Since $\lim_{\ell \to \infty} (M^{-t})^\ell \alpha = 0$, there is a unique minimal integer $\ell \geq 0$ such that $\alpha \in (M^t)^\ell \mathbb{Z}^d \setminus (M^t)^{\ell+1} \mathbb{Z}^d$. Using the partitioning (4.14), we find the required unique representation.

4.3.1.1 The PSI case

Here refinability already implies that S_ϕ has some approximation order.

Theorem 4.3.1 *Let $\phi \in L_2(\mathbb{R}^d)$ be compactly supported and M-refinable with finitely supported refinement mask, and assume that $\phi^\wedge(0) \neq 0$. Then ϕ satisfies the Strang–Fix conditions of order one, whence S_ϕ has at least approximation order one for $f \in W_2^1(\mathbb{R}^d)$.*

Proof From the scalar refinement equation

$$\phi^\wedge = H(M^{-t}\cdot) \, \phi^\wedge(M^{-t}\cdot) \tag{4.16}$$

we conclude that $H(0) = 1$, and due to the periodicity of H we have

$$\phi^\wedge(2\pi M^t \alpha) = \phi^\wedge(2\pi \alpha) , \quad \alpha \in \mathbb{Z}^d .$$

From this, by an application of the Riemann–Lebesgue Lemma, $\phi^\wedge(2\pi\alpha) = \lim_{k\to\infty} \phi^\wedge(2\pi(M^t)^k \alpha) = 0$ for $0 \neq \alpha \in \mathbb{Z}^d$. The assertion follows from Theorem 4.2.14. □

4.3.1.2 The FSI case

A generalization of this result depends upon the refinement equation (4.12) at the origin. If $\Phi^\wedge(0) \neq \mathbf{0}$, then $\Phi^\wedge(0)$ is a right eigenvector of $\mathbf{H}(0)$ for the eigenvalue 1. Consider a left eigenvector $\mathbf{v}$, say, for this same eigenvalue, and put $\psi^\wedge := \mathbf{v}\Phi^\wedge$. Then $\psi \in S_\Phi$, and as in the previous proof $\psi^\wedge(2\pi\alpha) = 0$ for $0 \neq \alpha \in \mathbb{Z}^d$. In order to again apply Theorem 4.2.14, we just have to require that $\psi^\wedge(0) \neq 0$. This gives

Theorem 4.3.2 *Let $\Phi \subset L_2(\mathbb{R}^d)$ be compactly supported and M-refinable with finitely supported refinement mask and corresponding mask symbol $\mathbf{H}$. Assume that $\Phi^\wedge(0) \neq \mathbf{0}$, and that $\mathbf{vH}(0) = \mathbf{v}$ for a row vector $\mathbf{v}$ satisfying $\mathbf{v}\Phi^\wedge(0) \neq 0$. Then $\psi \in S_\Phi$ given by $\psi^\wedge := \mathbf{v}\Phi^\wedge$ satisfies the Strang–Fix conditions of order one, whence S_ψ and, a fortiori, S_Φ has at least approximation order one for $f \in W_2^1(\mathbb{R}^d)$.*

4.3.1.3 The spectral condition on $\mathbf{H}(0)$

The additional assumptions made in this theorem will appear later in a more general form, namely when $\mathbf{v} = (\tilde{\tau}_1, \ldots, \tilde{\tau}_n)$ is a row vector of trigonometric polynomials and

$$\psi^\wedge := \mathbf{v}\Phi^\wedge .$$

Then $\psi \in S_\Phi$ is compactly supported (whenever Φ is), and we refer to 4.2.3.3 where we have constructed compactly supported superfunctions $\tilde{\psi}$ in this way. We say that $\mathbf{v}$ *satisfies the spectral condition (of order 1) on $\mathbf{H}$ at the origin*, if

(i) $\psi^\wedge(0) = \mathbf{v}(0)\Phi^\wedge(0) \neq 0$ and
(ii) $\mathbf{v}(0)\mathbf{H}(0) = \mathbf{v}(0)$.

For a stronger version of this condition, see Remark 4.3.6 below.

4.3.2 The zero condition on the mask symbol

The Strang–Fix conditions can be expressed as a zero condition on the mask symbol. This condition is sufficient for the PSI space S_ϕ to have approximation order. The condition is also necessary in the stable case, and even under an assumption weaker than stability.

Theorem 4.3.3 *Let $\phi \in L_2(\mathbb{R}^d)$ be compactly supported and M-refinable with finitely supported refinement mask and corresponding mask symbol H. Assume that $\phi^\wedge(0) \neq 0$. For $m \in \mathbb{N}$ we have:*

(i) *the zero condition of order m on the mask symbol,*

$$D^\mu \{H(M^{-t}\cdot)\}(2\pi e') = 0 \quad \text{for all } e' \in E_0' \text{ and } |\mu| < m, \quad (4.17)$$

implies that S_ϕ has approximation order m for $f \in W_2^m(\mathbb{R}^d)$;
(ii) *conversely, if S_ϕ has approximation order m for $f \in W_2^m(\mathbb{R}^d)$, then (4.17) holds true subject to*

$$[\phi^\wedge | \phi^\wedge](2\pi M^{-t}e') \neq 0 \quad \text{for any } e' \in E_0' . \quad (4.18)$$

Proof First let $m = 1$. In (4.16) we substitute $\xi = 2\pi\alpha$ with

$$\alpha = e' + M^t\beta, \quad e' \in E', \ \beta \in \mathbb{Z}^d, \ (e', \beta) \neq (0,0),$$

to obtain

$$\phi^\wedge(2\pi\alpha) = H(2\pi M^{-t}e') \, \phi^\wedge(2\pi M^{-t}e' + 2\pi\beta).$$

While for $m = 1$ (i) is always satisfied due to Theorem 4.3.1, we see that assertion (ii) follows since $\{\phi^\wedge(2\pi M^{-t}e' + 2\pi\beta)\}_{\beta \in \mathbb{Z}^d}$ cannot be a zero sequence for $e' \in E'_0$ by (4.18).

For $m \geq 1$ we use induction and Leibniz' rule in (4.16),

$$D^\gamma \phi^\wedge = \sum_{\mu \leq \gamma} \binom{\gamma}{\mu} D^\mu \{H(M^{-t}\cdot)\} \, D^{\gamma-\mu}\{\phi^\wedge(M^{-t}\cdot)\}, \quad (4.19)$$

with $|\gamma| < m+1$. In the case of statement (i) we see that the zero condition of order $m + 1$ immediately verifies the Strang–Fix conditions $D^\gamma \phi^\wedge(2\pi\alpha) = 0$ in the following situations:

$$|\gamma| < m \quad \text{and} \quad 0 \neq \alpha \in \mathbb{Z}^d \quad \text{(via Theorem 4.2.14)};$$
$$|\gamma| = m \quad \text{and} \quad \alpha = e' + M^t\beta, \quad e' \in E'_0, \ \beta \in \mathbb{Z}^d.$$

The first situation is the induction assumption. For $|\gamma| = m$ and $0 \neq \alpha \in \mathbb{Z}^d$ as in (4.15) with $\ell > 0$, equation (4.19) reduces to

$$D^\gamma \phi^\wedge\bigl(2\pi\{(M^t)^\ell(e' + M^t\beta)\}\bigr)$$
$$= H(M^{-t}2\pi(M^t)^\ell(e' + M^t\beta)) \, D^\gamma\{\phi^\wedge(M^{-t}\cdot)\}(2\pi(M^t)^\ell(e' + M^t\beta))$$
$$= H(0) \sum_{|\gamma'|=|\gamma|} c_{\gamma',\gamma} \, \{D^{\gamma'}\phi^\wedge\}(2\pi(M^t)^{\ell-1}(e' + M^t\beta))$$

for some constants $c_{\gamma',\gamma}$, and a simple induction argument with respect to ℓ shows that the full Strang–Fix conditions of order $m + 1$ are satisfied. Thus S_ϕ has approximation order $m + 1$. In case of statement (ii), ϕ satisfies the Strang–Fix conditions of order $m + 1$, and due to the induction hypothesis, the symbol satisfies the zero conditions of order m. Hence (4.19) yields, for $|\gamma| = m$, $e' \in E'_0$, and any $\beta \in \mathbb{Z}^d$,

$$0 = D^\gamma \phi^\wedge\bigl(2\pi(e' + M^t\beta)\bigr) = D^\gamma\{H(M^{-t}\cdot)\}(2\pi e') \, \phi^\wedge(2\pi M^{-t}e' + 2\pi\beta).$$

Due to our assumption (4.18), this yields an additional order in the zero condition for the mask symbol. □

Remark 4.3.4 The zero condition on the mask symbol (4.17) is actually equivalent to

$$\{D^\mu H\}(2\pi M^{-t} e') = 0 \quad \text{for all } e' \in E_0' \text{ and } |\mu| < m \ . \quad (4.20)$$

In the univariate case $d = 1$, the refinement equation is of the form

$$\phi = \sum_{\alpha \in \mathbb{Z}} p_\alpha \, \phi(k \cdot - \alpha)$$

for some integer $k \geq 2$. Here this condition simplifies to

$$D^\mu H\left(\frac{2\pi j}{k}\right) = 0 \ , \quad j = 1, \ldots, k-1, \ \mu = 0, \ldots, m-1.$$

Since H is a trigonometric polynomial the zeros can be factored out as

$$H(\xi) = \left(\frac{1}{k} \cdot \frac{1 - e^{-ik\xi}}{1 - e^{-i\xi}}\right)^m G(\xi)$$

where G is another trigonometric polynomial satisfying $G(0) = 1$.

4.3.2.1 Condition (Z_m) and the FSI case

The above Theorem 4.3.3 is a special case of the theorem to follow. We say that *the matrix symbol $\mathbf{H}$ satisfies condition (Z_m)* if there exists a row vector

$$\mathbf{v} = (\tilde{\tau}_1, \ldots, \tilde{\tau}_n)$$

of trigonometric polynomials for which

(a) $\mathbf{v}$ satisfies the spectral condition 4.3.1.3 on $\mathbf{H}$ at the origin and
(b) $D^\mu \{\mathbf{v} \mathbf{H}(M^{-t} \cdot)\}(2\pi e') = 0$ for $|\mu| < m$ and $e' \in E_0'$.

Theorem 4.3.5 *Let $\Phi \subset L_2(\mathbb{R}^d)$ be compactly supported and M-refinable with finitely supported refinement mask and corresponding mask symbol $\mathbf{H}$. Assume that $\Phi^\wedge(0) \neq \mathbf{0}$. Then for $m \in \mathbb{N}$ we have:*

(i) *if $\mathbf{H}$ satisfies condition (Z_m), then S_Φ has approximation order m for $f \in W_2^m(\mathbb{R}^d)$;*
(ii) *conversely, if S_Φ has approximation order m for $f \in W_2^m(\mathbb{R}^d)$, and if the Gramian $\mathbf{G}_\Phi$ is regular at $\xi = 2\pi M^{-t} e'$ for any $e' \in E'$, then $\mathbf{H}$ satisfies condition (Z_m).*

It should be noted that the condition on the regularity of the Gramian is equivalent to the fact that the sequences

$$\{\phi_j(2\pi M^{-t}e' + 2\pi\beta)\}_{\beta\in\mathbb{Z}^d}, \quad j = 1, \ldots, n,$$

are linearly independent.

Proof We can follow the proof of Theorem 4.3.3 with the same notation as in 4.3.1.3. In particular, let $\psi^\wedge := \mathbf{v}\,\Phi^\wedge$. Due to the refinement equation we have

$$\psi^\wedge(2\pi\alpha) = \mathbf{v}(0)\,\mathbf{H}(2\pi M^{-t}e')\,\Phi^\wedge(2\pi M^{-t}e' + 2\pi\beta) \qquad (4.21)$$

for any $\alpha = e' + M^t\beta \in \mathbb{Z}^d$, and the induction procedure will use Leibniz' rule in the form

$$D^\gamma\psi^\wedge = \sum_{\mu\leq\gamma} \binom{\gamma}{\mu} D^\mu\{\mathbf{v}\mathbf{H}(M^{-t}\cdot)\}\,D^{\gamma-\mu}\{\Phi^\wedge(M^{-t}\cdot)\}\,.$$

For (i) the induction is based on Theorem 4.3.2 (case $m = 1$), and the induction step of the above proof extends almost word for word.

For (ii) we first consider the stronger assumption that the Gramian is everywhere regular. Then according to Theorem 4.2.15 we can find $\mathbf{v} = (\tilde{\tau}_1, \ldots, \tilde{\tau}_n)$ such that $\psi^\wedge = \mathbf{v}\Phi^\wedge$ satisfies the Strang–Fix condition of order m. In particular, the Strang–Fix conditions of order 1 already show that $\psi^\wedge(0) \neq 0$ and

$$\mathbf{v}(0)\,\mathbf{H}(0)\,\Phi^\wedge(2\pi\alpha) = \mathbf{v}(0)\,\Phi^\wedge(2\pi\alpha)$$

for all $\alpha \in \mathbb{Z}^d$ by putting $e' = 0$ in (4.21). Hence $\mathbf{v}$ satisfies the spectral condition on $\mathbf{H}(0)$ due to the regularity of $\mathbf{G}_\Phi(0)$. The induction step is now analogous to the earlier one.

In order to see (ii) with the weaker assumptions, it should be emphasized that Remark 4.2.10 holds true in a neighborhood of the origin as long as the Gramian is regular there. In this way, the construction of the superfunction in 4.2.3.3 can also be performed in this neighborhood. □

Remark 4.3.6 In the statement of Theorem 4.3.5, the row vector $\mathbf{v}$ can be chosen so that its components are linear combinations of the exponentials

$$e^{-i\alpha\cdot\xi} \quad \text{with} \quad \alpha \in \mathbb{Z}_+^d \quad \text{and} \quad |\alpha| < m,$$

and moreover such that, besides the zero condition (Z_m), the *spectral condition of order* m,

$$D^\mu\{\mathbf{v}\,\mathbf{H}(M^{-t}\cdot)\}(0) = D^\mu\{\mathbf{v}(M^{-t}\cdot)\}(0) \qquad (4.22)$$

is satisfied for $|\mu| < m$.

Remark 4.3.7 Condition (Z_m) can be simplified in the univariate case $d = 1$. It then implies a matrix factorization of the refinement mask symbol **H**. Here equation (4.11) reads

$$\Phi = \sum_{\alpha \in \mathbb{Z}} \mathbf{P}_\alpha \, \Phi(k \cdot -\alpha)$$

for some integer $k \geq 2$. It can be shown that there exists a trigonometric polynomial matrix **A** such that

$$\mathbf{H}(\xi) = k^{-m} \, \mathbf{A}(k\xi) \, \mathbf{G}(\xi) \, \mathbf{A}(\xi)^{-1},$$

where **G** is another trigonometric polynomial matrix. Moreover, the *factorization matrix* **A** necessarily satisfies the following two conditions:

(a) $\{D^\mu (\det \mathbf{A})\}(0) = 0$ for $|\mu| < m$;
(b) if ψ defined by $\psi^\wedge := \mathbf{v}\, \Phi^\wedge$ is a superfunction (i.e., S_ψ has approximation order m) $\{D^\mu(\mathbf{vA})\}(0) = \mathbf{0}$ for $|\mu| < m$.

Furthermore, setting $\Psi^\wedge(\xi) := (i\xi)^m \, \mathbf{A}(\xi)^{-1} \, \Phi^\wedge(\xi)$, we obtain

$$\Psi^\wedge(\xi) = \mathbf{G}(k^{-1}\xi) \, \Psi^\wedge(k^{-1}\xi)$$

with **G** the trigonometric polynomial matrix occurring in the above factorization of **H**. Thus Ψ is a compactly supported refinable distribution vector with refinement mask symbol **G**.

4.3.3 The sum rules

The zero condition on the mask symbol can be given in an equivalent form. Here the group structure of $E := \mathbb{Z}^d / M\mathbb{Z}^d$ enters the discussion. It is well-known that the dual group is given by $E^\wedge = 2\pi(M^{-t}\mathbb{Z}^d/\mathbb{Z}^d)$, hence *for the Fourier matrix*

$$\mathbf{F}_M := \left(e^{-2\pi i e \cdot (M^{-t} e')} \right)_{e \in E, e' \in E'}$$

we have

$$\frac{1}{|\det M|} \, \mathbf{F}_M \, \mathbf{F}_M^* = \mathbf{I} \,, \tag{4.23}$$

i.e., *the matrix is unitary up to the given factor*. A simple corollary of this is

Lemma 4.3.8 *For a trigonometric polynomial* $h(\xi) = \sum_{e \in E} b_e \, e^{-ie \cdot \xi}$ *the*

following are equivalent:

(i) $h(2\pi M^{-t}e') = 0$ *for all* $e' \in E'_0$;
(ii) $b_e = b_0$ *for all* $e \in E$.

Proof Condition (i) is equivalent to the fact that the vector $(b_e)_{e \in E}$ is orthogonal to all columns of the Fourier matrix except the first one, and this property is equivalent to the vector being a multiple of the first column of the Fourier matrix. □

In the PSI case this can be applied to the scalar-valued mask symbol H in the following way. Given any algebraic polynomial q and corresponding differential operator $q(iD)$, then

$$\{q(iD)\,H\}(\xi) = \frac{1}{|\det M|} \sum_{\alpha \in \mathbb{Z}^d} p_\alpha\, q(\alpha)\, e^{-i\alpha \cdot \xi}.$$

Rearranging the sum in terms of $\alpha = e + M\gamma$ and inserting the dual lattice gives $\alpha \cdot 2\pi M^{-t}e' = e \cdot 2\pi M^{-t}e'$ (modulo 2π), whence

$$\{q(iD)\,H\}(2\pi M^{-t}e') = \frac{1}{|\det M|} \sum_{e \in E} \left(\sum_{\gamma \in \mathbb{Z}^d} p_{e+M\gamma}\, q(e + M\gamma) \right) e^{-ie \cdot 2\pi M^{-t}e'}.$$

Combining this with the above lemma and with (4.17) yields the following *sum rules of order* m:

Theorem 4.3.9 *In the PSI case the zero condition (4.17) of order m on the mask symbol H is equivalent to the fact that the mask p satisfies*

$$\sum_{\gamma \in \mathbb{Z}^d} p_{e+M\gamma}\, q(e + M\gamma) = \sum_{\gamma \in \mathbb{Z}^d} p_{M\gamma}\, q(M\gamma)\,, \quad e \in E, \qquad (4.24)$$

for any algebraic polynomial q of degree less than m.

In particular we observe that for ϕ satisfying $\phi^\wedge(0) \neq 0$ the sum rule of order one,

$$\sum_{\gamma \in \mathbb{Z}^d} p_{e+M\gamma} = \sum_{\gamma \in \mathbb{Z}^d} p_{M\gamma}\,, \quad e \in E,$$

is sufficient for S_ϕ to have approximation order 1. The converse holds true subject to the additional condition in Theorem 4.3.3 (ii).

Turning to the FSI case, we note that Theorem 4.3.9 is a special case of the following statement (when dealing with the scalar case and putting $v_\sigma = \delta_\sigma$):

Theorem 4.3.10 *In the stable FSI case the zero condition (Z_m) on the row vector $\mathbf{vH}(M^{-t}\cdot)$, with $\mathbf{v}(\xi) = \sum_{\alpha \in \mathbb{Z}^d} \mathbf{v}_\alpha e^{-i\alpha\cdot\xi}$ a row vector of trigonometric polynomials, is equivalent to the fact that the mask $\mathbf{P}$ satisfies the sum rules*

$$\sum_{\sigma \in \mathbb{Z}^d} \sum_{\gamma \in \mathbb{Z}^d} \mathbf{v}_{\gamma-\sigma} \mathbf{P}_{e+M\sigma} \, q(e+M\gamma)$$

$$= \sum_{\sigma \in \mathbb{Z}^d} \sum_{\gamma \in \mathbb{Z}^d} \mathbf{v}_{\gamma-\sigma} \mathbf{P}_{M\sigma} \, q(M\gamma) \,, \quad e \in E \,, \quad (4.25)$$

for any algebraic polynomial q of degree less than m.

Proof We have

$$\mathbf{v}(\xi) \, \mathbf{H}(M^{-t}\xi) = \frac{1}{|\det M|} \sum_{\sigma \in \mathbb{Z}^d} \sum_{\alpha \in \mathbb{Z}^d} \mathbf{v}_\sigma \, \mathbf{P}_\alpha \, e^{-i(M^{-1}\alpha + \sigma)\cdot\xi} \,.$$

Thus with $\alpha = e + M\gamma$, $e' \in E'$, and q any algebraic polynomial as before,

$$q(iD)\{\mathbf{v}\,\mathbf{H}(M^{-t}\cdot)\}(2\pi e')$$
$$= \frac{1}{|\det M|} \sum_{e \in E} \left(\sum_{\sigma \in \mathbb{Z}^d} \sum_{\gamma \in \mathbb{Z}^d} \mathbf{v}_\sigma \, \mathbf{P}_{e+M\gamma} \, q(M^{-1}e + \gamma + \sigma) \right) e^{-ie\cdot 2\pi M^{-t}e'}.$$

Applying Lemma 4.3.8, the zero condition on the vector $\mathbf{vH}(M^{-t}\cdot)$, i.e., $(Z_m)(b)$ in 4.3.2.1, is now equivalent to the fact that the expression within the outer brackets does not depend upon e for any polynomial q of degree less than m. Substituting $q(M^{-1}\cdot)$ by q we arrive at the required sum rules. □

In particular, the sum rules of order 1 are satisfied if there exists a row vector $\mathbf{v}_0 \in \mathbb{R}^d$ such that

$$\mathbf{v}_0 \sum_{\sigma \in \mathbb{Z}^d} \mathbf{P}_{e+M\sigma} = \mathbf{v}_0 \sum_{\sigma \in \mathbb{Z}^d} \mathbf{P}_{M\sigma} \quad \text{for all} \quad e \in E \,.$$

4.3.4 The subdivision operator and the transfer operator

In this section we consider two linear operators which come with the refinement mask $\mathbf{P}$ in (4.11). These operators have been shown to be excellent tools for the characterization of refinable function vectors.

For a given (complex) mask $\mathbf{P} = (\mathbf{P}_\alpha)_{\alpha \in \mathbb{Z}^d}$ of $(n \times n)$-matrices, the *subdivision operator* S_P is the linear operator on the sequence space $X = \left(\ell(\mathbb{Z}^d)\right)^n$

Chapter 4 L_2-approximation orders from SI spaces

defined by
$$(S_P \mathbf{b})_\alpha := \sum_{\beta \in \mathbb{Z}^d} \mathbf{P}^*_{\alpha - M\beta} \mathbf{b}_\beta, \quad \alpha \in \mathbb{Z}^d. \tag{4.26}$$

Here, $\mathbf{b} = (\mathbf{b}_\alpha)_{\alpha \in \mathbb{Z}^d} \in \bigl(\ell(\mathbb{Z}^d)\bigr)^n$, i.e., $\mathbf{b}$ is a vector of (complex-valued) sequences indexed by $\mathbb{Z}^d$. Restricting the subdivision operator to the case of sequences which are square summable, i.e., $\|\mathbf{b}\|^2 := \sum_{\alpha \in \mathbb{Z}^d} \mathbf{b}^*_\alpha \mathbf{b}_\alpha < \infty$, or $\mathbf{b} \in (\ell_2(\mathbb{Z}^d))^n$ for short, we have the vector-valued Fourier series
$$\mathbf{b}^\wedge(\xi) := \sum_{\alpha \in \mathbb{Z}^d} \mathbf{b}_\alpha \, e^{i\alpha \cdot \xi},$$
and (4.26) leads to
$$(S_P \mathbf{b})^\wedge(\xi) = |\det M| \, \mathbf{H}(\xi)^* \, \mathbf{b}^\wedge(M^t \xi)$$
with the mask symbol $\mathbf{H}$ as in (4.13).

The *transfer operator* T_P (sometimes also called the *transition operator*) associated with the mask $\mathbf{P}$ is the linear operator operating on $\bigl(\ell_0(\mathbb{Z}^d)\bigr)^n \subset \bigl(\ell(\mathbb{Z}^d)\bigr)^n$, the subspace of compactly supported vector-valued sequences defined by
$$(T_P \mathbf{d})_\alpha := \sum_{\beta \in \mathbb{Z}^d} \mathbf{P}_{M\alpha - \beta} \mathbf{d}_\beta, \quad \alpha \in \mathbb{Z}^d, \quad \mathbf{d} \in \bigl(\ell_0(\mathbb{Z}^d)\bigr)^n.$$

This definition naturally extends to $\bigl(\ell_2(\mathbb{Z}^d)\bigr)^n$ due to the compact support of the matrix mask, and the Fourier series of the image vector sequence is given by
$$(T_P \mathbf{d})^\wedge(\xi) = \sum_{e' \in E'} \mathbf{H}\bigl(M^{-t}(\xi + 2\pi e')\bigr) \, \mathbf{d}^\wedge\bigl(M^{-t}(\xi + 2\pi e')\bigr). \tag{4.27}$$

In the case of a scalar-valued mask $p = (p_\alpha)_{\alpha \in \mathbb{Z}^d}$ the linear operators are simply
$$(S_p c)_\alpha = \sum_{\beta \in \mathbb{Z}^d} \overline{p_{\alpha - M\beta}} \, c_\beta, \quad c \in \ell(\mathbb{Z}^d),$$
$$(T_p d)_\alpha = \sum_{\beta \in \mathbb{Z}^d} p_{M\alpha - \beta} \, d_\beta, \quad d \in \ell_0(\mathbb{Z}^d).$$

For $c, d \in \ell_2(\mathbb{Z}^d)$ we observe that
$$(S_p c)^\wedge(\xi) = |\det M| \, \overline{H(\xi)} \, c^\wedge(M^t \xi) \quad \text{and}$$
$$(T_p d)^\wedge(\xi) = \sum_{e' \in E'} H(M^{-t}(\xi + 2\pi e')) \, d^\wedge(M^{-t}(\xi + 2\pi e')).$$

It is this version of the transfer operator $d^\wedge \mapsto (T_p d)^\wedge$ in the Fourier domain and operating on $(L_2(C))^n$ which frequently appears in the literature.

There is a close connection between the two operators. When operating on $(\ell_2(\mathbb{Z}^d))^n$ considered as a Hilbert space with scalar product

$$\langle \mathbf{b} \mid \mathbf{d} \rangle := \sum_{\alpha \in \mathbb{Z}^d} \mathbf{b}_\alpha^* \mathbf{d}_\alpha, \qquad (4.28)$$

it is easy to see that the adjoint operator S_P^* defined by $\langle S_P \mathbf{b} \mid \mathbf{d} \rangle = \langle \mathbf{b} \mid S_P^* \mathbf{d} \rangle$ is given by

$$S_P^* \mathbf{d} = \overline{(T_P(\mathbf{d}^{\sim}))} \qquad (4.29)$$

with $\mathbf{d}^{\sim}_\alpha := \mathbf{d}_{-\alpha}$ denoting the reflection of a sequence. From this it is clear that, considered as operators on $X = (\ell_2(\mathbb{Z}^d))^n$, the spectra $\sigma_X(S_P)$ and $\sigma_X(T_P)$ are related through complex conjugation, i.e.,

$$\sigma_X(T_P) = \overline{\sigma_X(S_P)} \ .$$

The situation is slightly more involved if we look at eigenvalues of the operators. Here it is opportune to consider (4.28) as a sesquilinear form on the dual pairing

$$X \times X' := (\ell(\mathbb{Z}^d))^n \times (\ell_0(\mathbb{Z}^d))^n$$

(the canonical bilinear form being (4.28) with the superscript "$*$" replaced by "t"). Then the formal extension of (4.29) is still true, and the following theorem can be shown along the lines of Theorem 5.1 in Jia (1998b).

Theorem 4.3.11 *Considered as an operator on $(\ell_0(\mathbb{Z}^d))^n$, T_P has only finitely many nonzero eigenvalues. In particular σ is an eigenvalue of T_P if and only if $\bar{\sigma}$ is an eigenvalue of S_P, the latter being considered as an operator on $(\ell(\mathbb{Z}^d))^n$.*

It is possible to characterize the approximation order of $S(\Phi)$ in terms of properties of invariance of the subdivision and the transfer operator. A linear subspace $Y \subset (\ell_0(\mathbb{Z}^d))^n$ is called T_P-invariant if

$$\mathbf{d} \in Y \quad \Longrightarrow \quad T_P \mathbf{d} \in Y \ ,$$

and S_P-invariant subspaces $Z \subset (\ell(\mathbb{Z}^d))^n$ are defined analogously.

4.3.4.1 The PSI case

For given $m \in \mathbb{N}$ we put

$$V_m := \{ d \in \ell_0(\mathbb{Z}^d) \ ; \ D^\mu d^\wedge(0) = 0 \text{ for all } \mu \in \mathbb{Z}_+^d \text{ with } |\mu| < m \}. \qquad (4.30)$$

Since $d^\wedge$ is a trigonometric polynomial, the derivatives at the origin are well-defined. These zero conditions at the origin are moment conditions for the sequence d. Indeed since $d^\wedge(\xi) = \sum_{\alpha \in \mathbb{Z}^d} d_\alpha e^{i\alpha \cdot \xi}$, we have

$$(-iD)^\mu d^\wedge(0) = 0 \iff \sum_{\alpha \in \mathbb{Z}^d} d_\alpha\, \alpha^\mu = 0\,,$$

whence

Lemma 4.3.12 *The following are equivalent for $m \in \mathbb{N}$ and $d \in \ell_0(\mathbb{Z}^d)$:*

(i) $d \in V_m$;

(ii) $\sum_{\alpha \in \mathbb{Z}^d} d_\alpha\, q(\alpha) = 0$ *for any algebraic polynomial q of degree less than m.*

We are now ready to present the abovementioned invariance properties which imply an approximation order result, in view of Theorem 4.3.3. By a slight misuse of notation we identify the polynomial space P_{m-1} (i.e., polynomials of degree less than m) with the sequence space of polynomials of order m, evaluated at $\mathbb{Z}^d$, namely,

$$\left\{ (q(\alpha))_{\alpha \in \mathbb{Z}^d} \,;\, q \in P_{m-1} \right\} \subset \ell(\mathbb{Z}^d).$$

Theorem 4.3.13 *Let $p \in \ell_0(\mathbb{Z}^d)$ be a finitely supported mask with corresponding scalar mask symbol H. Then the following assertions are equivalent for $m \in \mathbb{N}$:*

(i) $V_m \subset \ell_0(\mathbb{Z}^d)$ *is invariant under the transfer operator T_p;*

(ii) P_{m-1} *is invariant under the subdivision operator S_p;*

(iii) H *satisfies the zero condition (4.17) of order m;*

(iv) p *satisfies the sum rules (4.24) of order m.*

The equivalence of (i) and (iii) is obvious from the Fourier transform expression of the transfer operator,

$$(T_p d)^\wedge(\xi) = \sum_{e' \in E'} H\big(M^{-t}(\xi + 2\pi e')\big)\, d^\wedge\big(M^{-t}(\xi + 2\pi e')\big)\,.$$

The equivalence of (iii) and (iv) was established in Theorem 4.3.9. The implications (i) $\Rightarrow$ (ii) $\Rightarrow$ (iv) can be taken from Jia (1998b), Theorem 5.2.

4.3.4.2 The FSI case

Theorem 4.3.13 is a special case of the following statements. For a given row vector

$$\mathbf{v}(\xi) = \sum_{\alpha \in \mathbb{Z}^d} \mathbf{v}_\alpha e^{-i\alpha \cdot \xi} \tag{4.31}$$

of trigonometric polynomials we extend the definition in (4.30) to

$$V_m(\mathbf{v}) := \{\mathbf{d} \in \left(\ell_0(\mathbb{Z}^d)\right)^n;\ D^\mu\{\mathbf{v}\,\mathbf{d}^\wedge\}(0) = 0 \text{ for all } \mu \in \mathbb{Z}_+^d \text{ with } |\mu| < m\}.$$

As previously in the PSI case, the zero conditions for $\mathbf{v}\,\mathbf{d}^\wedge$ at the origin are moment conditions for the "convolved" series $\mathbf{v} * \tilde{\mathbf{d}} \in \ell_0(\mathbb{Z}^d)$ given by

$$(\mathbf{v} * \tilde{\mathbf{d}})_\alpha := \sum_{\beta \in \mathbb{Z}^d} \mathbf{v}_{\alpha-\beta}\,\tilde{\mathbf{d}}_\beta = \sum_{\beta \in \mathbb{Z}^d} \mathbf{v}_{\alpha+\beta}\,\mathbf{d}_\beta\,,\quad \alpha \in \mathbb{Z}^d.$$

Note that $\mathbf{d}^\wedge(\xi) = \sum_{\alpha \in \mathbb{Z}^d} \mathbf{d}_\alpha e^{i\alpha \cdot \xi}$ is a column vector of trigonometric polynomials. Therefore Lemma 4.3.12 gives

Lemma 4.3.14 *Given the row vector* $\mathbf{v}$ *in (4.31), the following are equivalent for* $m \in \mathbb{N}$ *and* $\mathbf{d} \in \left(\ell_0(\mathbb{Z}^d)\right)^n$:

(i) $\mathbf{d} \in V_m(\mathbf{v})$;
(ii) $\mathbf{v} * \tilde{\mathbf{d}} \in V_m$;
(iii) $\sum_{\gamma \in \mathbb{Z}^d} \left(\sum_{\beta \in \mathbb{Z}^d} \mathbf{v}_{\gamma+\beta}\,\mathbf{d}_\beta\right) q(\gamma) = 0$ *for any algebraic polynomial* q *of degree less than* m.

Theorem 4.3.15 *Let* $\mathbf{P}$ *be a finitely supported* $(n \times n)$-*matrix mask with mask symbol* $\mathbf{H}$. *Let* $\mathbf{v}$ *be a row vector (4.31) of trigonometric polynomials such that the spectral conditions (4.22) of order* $m \in \mathbb{N}$ *are satisfied. Then the following assertions are equivalent:*

(i) $V_m(\mathbf{v})$ *is invariant under the transfer operator* T_P;
(ii) $\mathbf{H}$ *satisfies condition* (Z_m) *with* $\mathbf{v}$.

Proof For $\mathbf{d} \in V_m(\mathbf{v})$ we have, using (4.22),

$$\sum_{\nu \leq \mu} \binom{\mu}{\nu} D^\nu \{\mathbf{v}\,\mathbf{H}(M^{-t}\cdot)\}(0)\,D^{\mu-\nu}\{\mathbf{d}^\wedge(M^{-t}\cdot)\}(0)$$

$$= \sum_{\nu \leq \mu} \binom{\mu}{\nu} D^\nu \{\mathbf{v}(M^{-t}\cdot)\}(0)\,D^{\mu-\nu}\{\mathbf{d}^\wedge(M^{-t}\cdot)\}(0)$$

$$= D^\mu\{(\mathbf{v}\,\mathbf{d}^\wedge)(M^{-t}\cdot)\}(0) = 0\,,\quad |\mu| < m\,.$$

Chapter 4 L_2-approximation orders from SI spaces   101

On the other hand, by (4.27), for any $\mathbf{d} \in (\ell_0(\mathbb{Z}^d))^n$,

$$D^\mu\{\mathbf{v}\,(T_P\mathbf{d})^\wedge\}(0) =$$

$$\sum_{e' \in E'} \sum_{\nu \leq \mu} \binom{\mu}{\nu} D^\nu\{\mathbf{v}\mathbf{H}(M^{-t}(\cdot + 2\pi e'))\}(0)\ D^{\mu-\nu}\{\mathbf{d}^\wedge(M^{-t}(\cdot + 2\pi e'))\}(0).$$

Since $T_P\mathbf{d}$ is finitely supported whenever $\mathbf{d}$ is, statement (i) is equivalent to the fact that

$$\sum_{e' \in E'_0} \sum_{\nu \leq \mu} \binom{\mu}{\nu} D^\nu\{\mathbf{v}\mathbf{H}(M^{-t}(\cdot + 2\pi e'))\}(0)\ D^{\mu-\nu}\{\mathbf{d}^\wedge(M^{-t}(\cdot + 2\pi e'))\}(0)$$
$$= 0 \quad \text{for any } \mathbf{d} \in V_m(\mathbf{v}) \text{ and } |\mu| < m\ .$$

It is not hard to see that this is equivalent to condition (Z_m). □

As a corollary of this theorem we have

Theorem 4.3.16 *In the stable FSI case, any of the statements in Theorem 4.3.15 can be replaced by the sum rules of Theorem 4.3.10. Here the appropriate form of the spectral conditions (4.22) is*

$$|\det M| \sum_{\sigma \in \mathbb{Z}^d} \sum_{\gamma \in \mathbb{Z}^d} \mathbf{v}_{\gamma-\sigma}\,\mathbf{P}_{M\sigma}\,q(M\gamma) = \sum_{\alpha \in \mathbb{Z}^d} \mathbf{v}_\alpha\,q(\alpha) \qquad (4.32)$$

for any polynomial q of degree less than m.

In order to relate the conditions in Theorem 4.3.15 to an invariance property of the subdivision operator we let

$$\Pi(\mathbb{Z}^d) \subset \ell(\mathbb{Z}^d)$$

denote the space of all sequences which increase at most polynomially at infinity. In addition, we put

$$W_m(\mathbf{v}) := \{\mathbf{b} \in (\Pi(\mathbb{Z}^d))^n;\ \langle \mathbf{b} \mid \mathbf{d} \rangle = 0 \text{ for all } \mathbf{d} \in V_m(\mathbf{v})\}.$$

Theorem 4.3.17 *In the stable FSI case, any of the statements in Theorem 4.3.15 can be replaced by either of the following equivalent conditions:*

(i) $W_m(\mathbf{v})$ *is invariant under the subdivision operator S_P;*
(ii) $\mathbf{P}$ *satisfies the sum rules (4.25) with $\mathbf{v}$.*

Proof (ii) $\Longrightarrow$ (i): Using Theorem 4.3.10 and Theorem 4.3.15 we see that (ii) implies the invariance property of T_P. Therefore, by (4.29), $V_m(\mathbf{v})$ is also invariant under the operator S_P^*. Due to the finite support of the mask

P we also have
$$\mathbf{b} \in \left(\Pi(\mathbb{Z}^d)\right)^n \implies S_P \mathbf{b} \in \left(\Pi(\mathbb{Z}^d)\right)^n.$$

Therefore, given $\mathbf{b} \in W_m(\mathbf{v})$ we obtain
$$\langle S_P \mathbf{b} \mid \mathbf{d} \rangle = \langle \mathbf{b} \mid S_P^* \mathbf{d} \rangle = 0,$$
for any $\mathbf{d} \in V_m(\mathbf{v})$, whence $S_P \mathbf{b} \in W_m(\mathbf{v})$.

(ii) $\Longleftarrow$ (i): By Lemma 4.3.14, $\mathbf{d} \in V_m(\mathbf{v})$ if and only if
$$\sum_{\alpha \in \mathbb{Z}^d} \sum_{\delta \in \mathbb{Z}^d} \mathbf{v}_\delta \, q(\delta - \alpha) \, \mathbf{d}_\alpha = 0$$
for all algebraic polynomials q of degree less than m. Setting .5
$$q(\delta - \alpha) = \sum_{|\mu| < m} r_\mu(\delta)(-\alpha)^\mu \quad \text{with} \quad r_\mu := \frac{1}{\mu!} D^\mu q,$$
the spectral conditions (4.32) yield
$$\begin{aligned} 0 &= \sum_{|\mu|<m} \sum_{\alpha \in \mathbb{Z}^d} \left(\sum_{\delta \in \mathbb{Z}^d} \mathbf{v}_\delta \, r_\mu(\delta) \right) (-\alpha)^\mu \, \mathbf{d}_\alpha \\ &= |\det M| \sum_{|\mu|<m} \sum_{\beta \in \mathbb{Z}^d} \left(\sum_{\gamma \in \mathbb{Z}^d} \mathbf{v}_{\gamma - \beta} \, \mathbf{P}_{M\beta} \, r_\mu(M\gamma) \right) \sum_{\alpha \in \mathbb{Z}^d} (-\alpha)^\mu \, \mathbf{d}_\alpha. \end{aligned} \quad (4.33)$$

Now let q be any algebraic polynomial of degree less than m. Then the vector sequence $\mathbf{b} = \mathbf{b}(q)$ given by
$$\mathbf{b}_\alpha := \sum_{\beta \in \mathbb{Z}^d} \mathbf{v}_{\alpha+\beta}^* \, \overline{q(M\beta)}, \quad \alpha \in \mathbb{Z}^d,$$
is an element of $\left(\Pi(\mathbb{Z}^d)\right)^n$, since $(\mathbf{v}_\alpha)_{\alpha \in \mathbb{Z}^d}$ is compactly supported. It is contained in $W_m(\mathbf{v})$ since for all $\mathbf{d} \in V_m(\mathbf{v})$, using Lemma 4.3.14,
$$\begin{aligned} \langle \mathbf{b} \mid \mathbf{d} \rangle &= \sum_{\alpha \in \mathbb{Z}^d} \left(\sum_{\beta \in \mathbb{Z}^d} \mathbf{v}_{\alpha+\beta}^* \, \overline{q(M\beta)} \right)^* \mathbf{d}_\alpha \\ &= \sum_{\beta \in \mathbb{Z}^d} q(M\beta) \sum_{\alpha \in \mathbb{Z}^d} \mathbf{v}_{\alpha+\beta} \, \mathbf{d}_\alpha = 0. \end{aligned}$$

From our assumption (i) we conclude that $S_P \mathbf{b} \in W_m(\mathbf{v})$ as well, i.e., for any $\mathbf{d} \in V_m(\mathbf{v})$:

$$\begin{aligned}
0 &= \langle S_P \mathbf{b} \mid \mathbf{d} \rangle = \sum_{\alpha \in \mathbb{Z}^d} (S_P \mathbf{b})^*_\alpha \, \mathbf{d}_\alpha \\
&= \sum_{\alpha \in \mathbb{Z}^d} \left(\sum_{\beta \in \mathbb{Z}^d} \mathbf{P}^*_{\alpha - M\beta} \mathbf{b}_\beta \right)^* \mathbf{d}_\alpha \\
&= \sum_{\alpha \in \mathbb{Z}^d} \left(\sum_{\beta \in \mathbb{Z}^d} \mathbf{P}^*_{\alpha - M\beta} \left(\sum_{\gamma \in \mathbb{Z}^d} \mathbf{v}^*_{\beta+\gamma} \, \overline{q(M\gamma)} \right) \right)^* \mathbf{d}_\alpha \\
&= \sum_{\alpha \in \mathbb{Z}^d} \left(\sum_{\beta \in \mathbb{Z}^d} \sum_{\gamma \in \mathbb{Z}^d} \mathbf{v}_{\beta+\gamma} \, \mathbf{P}_{\alpha - M\beta} \, q(M\gamma) \right) \mathbf{d}_\alpha.
\end{aligned}$$

Equivalently, putting

$$q(M\gamma) = q(M\gamma + \alpha - \alpha) = \sum_{|\mu| < m} \frac{1}{\mu!} (D^\mu q)(M\gamma + \alpha)\,(-\alpha)^\mu,$$

we have

$$0 = \sum_{|\mu| < m} \sum_{\alpha \in \mathbb{Z}^d} \left(\sum_{\beta \in \mathbb{Z}^d} \sum_{\gamma \in \mathbb{Z}^d} \mathbf{v}_{\gamma - \beta} \, \mathbf{P}_{\alpha + M\beta} \, (D^\mu q)(\alpha + M\gamma) \right) (-\alpha)^\mu \, \mathbf{d}_\alpha$$

for all $\mathbf{d} \in V_m(\mathbf{v})$. Comparing this with (4.33) shows that the expression within the outer brackets must be independent of α, for any polynomial q of degree less than m. Whence the sum rules (4.25) of order m hold true, as we desired. □

4.3.5 Notes and extensions

4.3.5.1

A stable, compactly supported M-refinable function or function vector generates a multiresolution analysis for $L_2(\mathbb{R}^d)$, hence allows for a (now) standard construction of a wavelet basis.

The dilation matrix M in the matrix refinement equation (4.11) is often chosen as $M = 2I$, with I the $d \times d$ unit matrix. For this special case, refinable functions can be obtained from tensor products of refinable univariate functions φ_i, say, as

$$\phi(\mathbf{t}) = \prod_{i=1}^{d} \varphi_i(t_i), \quad \mathbf{t} = (t_1, \ldots, t_d)^t \in \mathbb{R}^d.$$

All results on univariate refinable functions can be simply transfered to the multivariate situation. The corresponding wavelet basis then requires $2^d - 1$ different generating wavelets.

In general, dilation matrices M of smallest possible determinant are of special interest, since the construction of a corresponding wavelet basis consists of $|\det M| - 1$ wavelets (or multiwavelets). If M is not a diagonal matrix, these wavelets are called *non-separable*, see Cohen and Daubechies (1993). An important instant of this is the bivariate construction based on

$$M = \begin{pmatrix} 1 & 1 \\ 1 & -1 \end{pmatrix} \quad \text{or} \quad M = \begin{pmatrix} 1 & -1 \\ 1 & 1 \end{pmatrix}.$$

For example, the so-called *Zwart–Powell element* is refinable with respect to the first matrix. See de Boor et al. (1993), Chapter 7 for more information on approximation orders and subdivision of so-called *box splines*.

4.3.5.2

Considering the refinement equation (4.12) in the Fourier domain, it follows that

(i) $\Phi^\wedge(0)$ is either a right eigenvector of $\mathbf{H}(0)$ with corresponding eigenvalue 1, or

(ii) $\Phi^\wedge(0)$ is the zero vector.

The latter case is not of real interest since then Φ can be considered as a derivative of an M-refinable function vector $\tilde{\Phi}$ with $\tilde{\Phi}^\wedge \neq \mathbf{0}$. Thus the assertion that $\mathbf{H}(0)$ has an eigenvalue 1 is the fundamental condition, and the spectral condition on $\mathbf{H}(0)$ in Section 4.3.1.3 just ensures that $\mathbf{v}(0)$ and $\Phi(0)$ are non-orthogonal left and right eigenvectors of $\mathbf{H}(0)$ to this eigenvalue. The condition of these eigenvectors being non-orthogonal is clearly vacuous if 1 is a simple eigenvalue.

Moreover, if $\Phi \subset L_2(\mathbb{R}^d)$ generates a stable, M-refinable FSI-space, then the spectral radius of $\mathbf{H}(0)$ necessarily equals 1, with 1 being a simple eigenvalue and the only eigenvalue of absolute value 1, see e.g. Dahmen and Micchelli (1997).

4.3.5.3

For a given refinement mask the refinement equation (4.11) can be interpreted as a functional equation for Φ. In the Fourier domain the solution vector can be formally written as

$$\Phi^\wedge(\xi) = \lim_{L \to \infty} \prod_{j=1}^{L} \mathbf{H}((M^{-t})^j \xi) \mathbf{r},$$

where **r** is a right eigenvector of **H**(0) to the eigenvalue 1. In particular it follows that $\Phi^\wedge(0) = \mathbf{r}$.

The convergence of the infinite product (in the sense of uniform convergence on compact sets) is ensured if the spectral radius of **H**(0) is 1, and if there are no further eigenvalues of **H**(0) on the unit circle. In this case a non-degenerate eigenvalue 1 defines a solution vector $\Phi^\wedge$. If the eigenvalue 1 is simple (see 4.3.5.2), then this solution vector Φ is unique.

4.3.5.4

Section 4.3 often refers to work of Jia on approximation properties of multivariate wavelets. In his paper (Jia (1998b)) the PSI case was completely settled for general dilation matrices; see also the following remarks. The proof of Theorem 4.3.1 is a trivial extension of the proof of Jia and Micchelli (1991), Theorem 2.4.

4.3.5.5

In the univariate FSI case, $d = 1$, the zero condition (Z_m) and its consequence for the approximation order of S_Φ have been considered by Heil et al. (1996) and Plonka (1995a,1997) in the Fourier domain, while Lian (1995) has given conditions in the time domain.

In particular, in Plonka (1997) it is shown that the approximation order induces a matrix factorization of the symbol **H**; see Remark 4.3.7. Subsequently Micchelli and Sauer (1997) observed an analogous factorization property for the representing matrix of the subdivision operator. Unfortunately, for $d > 1$ the zero conditions on the mask symbols (4.17) and (4.20) do not lead a priori to a factorization of the symbols.

The multivariate FSI case with arbitrary dilation matrices is, for example, treated in Cabrelli et al. (1998). The observed conditions relate to the sum rules of Theorem 4.3.10.

4.3.5.6

The generalized discrete Fourier transform matrix $\mathbf{F}_M$ and its property in (4.23) of being unitary are well-known, see e.g. Chui and Li (1994). These properties have been used by Jia (1998b) to derive the sum rules as in Theorem 4.3.9.

4.3.5.7

While the notion of subdivision operator S_P (for general dilation matrices) has been coined by Cavaretta et al. (1991), the set-up for the transfer or transition operator is often altered in the literature. We prefer here to say that T_P is (essentially, i.e., modulo reflection) the adjoint of S_P. In this way, Theorem 4.3.13 dealing with the PSI case is identical with Jia (1998b), Theorem 5.2. As far as the FSI case is considered our results are new.

It should be noted that for the PSI case and $M = 2I$, the invariance of P_{m-1} under the subdivision operator S_p is equivalent to the property that polynomials of order less than m can be reproduced from multi-integer translates of ϕ. A result along these lines is already contained in Cavaretta et al. (1991).

4.3.5.8

The *symmetrized form of the transfer operator* (in the Fourier domain) is given by $\tilde{T}_P{}^\wedge$ operating on $(n \times n)$-matrices $\mathbf{C}$ of trigonometric polynomials as follows,

$$(\tilde{T}_P{}^\wedge \mathbf{C})(\xi) :=$$
$$\sum_{e' \in E'} \mathbf{H}(M^{-t}(\xi + 2\pi e')) \, \mathbf{C}(M^{-t}(\xi + 2\pi e')) \, \mathbf{H}(M^{-t}(\xi + 2\pi e'))^*. \quad (4.34)$$

For $M = 2I$, Shen (1998), Theorem 3.8 has shown that the stability of S_Φ is equivalent to the following condition: *the operator $\tilde{T}^\wedge$ has spectral radius 1, with 1 being a simple eigenvalue and all other eigenvalues lying strictly inside the unit circle. Moreover, the eigenmatrix of $\tilde{T}_P{}^\wedge$ corresponding to the eigenvalue 1 is nonsingular on the d-dimensional torus.* We conjecture that this equivalence is also true for arbitrary dilation matrices.

4.3.5.9

When considering the dilation matrix $M = 2I$, the connection between properties of the subdivision operator S_P and the approximation order m provided by the FSI-space S_Φ can be simply stated as follows: *the stable FSI space S_Φ has approximation order m, for $f \in W_2^m(\mathbb{R}^d)$, if and only if there exists a nontrivial vector $\mathbf{q}$ of polynomial sequences $q_1, \ldots, q_n \in P_{m-1}$ such that*

$$S_P \mathbf{q} = 2^{-(m-1)} \, \mathbf{q}.$$

In particular S_P necessarily has the eigenvalues 2^{-k} for $k = 0, \ldots, m-1$.

This result can be generalized to distribution vectors Φ which do not satisfy any conditions of linear independence (see Jia et al. (1997), Theorem 3.1).

References

Aldroubi, A. (1996). Oblique projections in atomic spaces. *Proc. Amer. Math. Soc.*, **124**, 2051–2060.

Battle, G. (1987). A block spin construction of ondelettes. Part I: Lemarié functions. *Comm. Math. Phys.*, **110**, 601–615.

Blu, T. and Unser, M. (1999). Approximation error for quasi-interpolators and (multi-) wavelet expansions. *Appl. Comput. Harmonic Anal.*, **6**, 219–251.

Boor, C. de (1987). The polynomials in the linear span of integer translates of a compactly supported function. *Constr. Approx.*, **3**, 199–208.

Boor, C. de (1990). Quasiinterpolants and approximation order of multivariate splines. In *Computation of Curves and Surfaces*, ed. W. Dahmen, M. Gasca and C.A. Micchelli, pp. 313–345. Kluwer Academic Publ., Dordrecht.

Boor, C. de (1993). Approximation order without quasi-interpolants. In *Approximation Theory VII*, ed. E.W. Cheney, C.K. Chui and L.L. Schumaker, pp. 1–18. Academic Press, New York.

Boor, C. de, DeVore, R.A. and Ron, A. (1994a). Approximation from shift-invariant subspaces of $L_2(\mathbb{R}^d)$. *Trans. Amer. Math. Soc.*, **341**, 787–806.

Boor, C. de, DeVore, R.A. and Ron, A. (1994b). The structure of finitely generated shift-invariant spaces in $L_2(\mathbb{R}^d)$. *J. Funct. Anal.*, **119**, 37–78.

Boor, C. de, DeVore, R.A. and Ron, A. (1998). Approximation orders of FSI spaces in $L_2(\mathbb{R}^d)$. *Constr. Approx.*, **14**, 411–427.

Boor, C. de and Fix, G. (1973). Spline approximation by quasi-interpolants. *J. Approx. Theory*, **8**, 19–45.

Boor, C. de, Höllig, K. and Riemenschneider, S.D. (1993). *Box Splines*. Springer-Verlag, New York.

Boor, C. de and Jia, R.Q. (1985). Controlled approximation and a characterization of local approximation order. *Proc. Amer. Math. Soc.*, **95**, 547–553.

Boor, C. de and Jia, R.Q. (1993). A sharp upper bound on the approximation order of smooth bivariate pp functions. *J. Approx. Theory*, **72**, 24–33.

Boor, C. de and Ron, A. (1992). Fourier analysis of approximation power of principal shift-invariant spaces. *Constr. Approx.*, **8**, 427–462.

Burchard, H.G. and Lei, J.J. (1995). Coordinate order of approximation by functional-based approximation operators. *J. Approx. Theory*, **82**, 240–256.

Cabrelli, C., Heil, C. and Molter, U. (1998). Accuracy of lattice translates of several multidimensional refinable functions. *J. Approx. Theory*, **95**, 5–52.

Cavaretta, A.S., Dahmen, W. and Micchelli, C.A. (1991). *Stationary Subdivision*. Memoirs of the Amer. Math. Soc., Vol. 93, Providence.

Chui, C.K. (1988). *Multivariate Splines*. CBMS-NSF Reg. Conf. Series in Applied Math., Vol. 54, SIAM, Philadelphia.

Chui, C.K. and Diamond, H. (1987). A natural formulation of quasi-interpolation by multivariate splines. *Proc. Amer. Math. Soc.*, **99**, 643–646.

Chui, C.K., Jetter, K. and Stöckler, J. (1994). Wavelets and frames on the four-directional mesh. In *Wavelets: Theory, Algorithms, and Applications*, ed. C.K. Chui, L. Montefusco and L. Puccio, pp. 213–230. Academic Press, New York.

Chui, C.K., Jetter, K. and Ward, J.D. (1987). Cardinal interpolation by multivariate splines. *Math. Comp.*, **48**, 711–724.

Chui, C.K. and Li, C. (1994). A general framework of multivariate wavelets with duals. *Appl. Comput. Harmonic Anal.*, **1**, 368–390.

Cohen, A. and Daubechies, I. (1993). Non-separable bidimensional wavelet bases. *Revista Mathemática Iberoamericana*, **9**, 51–137.

Cohen, A., Daubechies, I. and Feauveau, J. (1992). Biorthogonal basis of compactly supported wavelets. *Comm. Pure and Appl. Math.*, **45**, 485–560.

Dahmen, W. and Micchelli, C.A. (1983). Translates of multivariate splines. *Linear Algebra Appl.*, **52(3)**, 217–234.

Dahmen, W. and Micchelli, C.A. (1984). On the approximation order from certain multivariate splines. *J. Austral. Math. Soc. Ser. B*, **26**, 233–246.

Dahmen, W. and Micchelli, C.A. (1997). Biorthogonal wavelet expansions. *Constr. Approx.*, **13**, 293–328.

Daubechies, I. (1992). *Ten Lectures on Wavelets*. CBMS-NSF Reg. Conf. Series in Applied Math., Vol. 61, SIAM, Philadelphia.

DeVore, R.A. and Lorentz, G.G. (1993). *Constructive Approximation*. Springer-Verlag, Berlin.

Dyn, N. (1989). Interpolation and approximation by radial and related functions. In *Approximation Theory VI*, ed. C.K. Chui, L.L. Schumaker and J.D. Ward, pp. 211–234. Academic Press, New York.

Dyn, N., Jackson, I.R.H., Levin, D. and Ron, A. (1992). On multivariate approximation by the integer translates of a basis function. *Israel J. Math.*, **78**, 95–130.

Goodman, T. and Lee, S. (1993). Wavelets of multiplicity r. *Trans. Amer. Math. Soc.*, **342**, 307–324.

Goodman, T., Lee, S. and Tang, W.S. (1993). Wavelets in wandering subspaces. *Trans. Amer. Math. Soc.*, **338**, 639–654.

Goodman, T., Micchelli, C.A. and Ward, J.D. (1994). Spectral radius formulas for subdivision operators. In *Recent Advances in Wavelet Analysis*, ed. L.L. Schumaker and G. Webb, pp. 335–360. Academic Press, New York.

Halton, E.J. and Light, W.A. (1993). On local and controlled approximation order. *J. Approx. Theory*, **72**, 268–277.

Han, B. and Jia, R.Q. (1998). Multivariate refinement equations and convergence of subdivision schemes. *SIAM J. Math. Anal.*, **29**, 1177–1199.

Heil, C., Strang, G. and Strela, V. (1996). Approximation by translates of refinable functions. *Numer. Math.*, **73**, 75–94.

Jetter, K. (1993). Multivariate approximation from the cardinal interpolation point of view. In *Approximation Theory VII*, ed. E.W. Cheney, C.K. Chui and L.L. Schumaker, pp. 131–161. Academic Press, New York.

Jetter, K. and Zhou, D.X. (1995a). Order of linear approximation from shift-invariant spaces. *Constr. Approx.*, **11**, 423–438.

Jetter, K. and Zhou, D.X. (1995b). Seminorm and full norm order of linear approximation from shift-invariant spaces. *Rendiconti del Seminario Matematico e Fisico di Milano*, **LXV**, 277–302.

Jetter, K. and Zhou, D.X. (2000). Approximation order of linear operators onto finitely generated shift-invariant spaces. Unpublished manuscript, 28 pages.

Jia, R.Q. (1991). A characterization of the approximation order of translation invariant spaces. *Proc. Amer. Math. Soc.*, **111**, 61–70.

Jia, R.Q. (1994). Multiresolution of L_p spaces. *J. Math. Anal. Appl.*, **184**, 620–639.

Jia, R.Q. (1995a). Refinable shift-invariant spaces: From splines to wavelets. In *Approximation Theory VIII, vol. 2, Wavelets and Multilevel Approximation*, ed. C.K. Chui and L.L. Schumaker, pp. 179–208. World Scientific, Singapore.

Jia, R.Q. (1995b). Subdivision schemes in L_p spaces. *Advances Comp. Math.*, **3**, 309–341.

Jia, R.Q. (1996). The subdivision and transition operators associated with a refinement equation. In *Advanced Topics in Multivariate Approximation*, ed. F. Fontanella, K. Jetter and P.-J. Laurent, pp. 139–154. World Scientific, Singapore.

Jia, R.Q. (1997). Shift-invariant spaces on the real line. *Proc. Amer. Math. Soc.*, **125**, 785–793.

Jia, R.Q. (1998a). Shift-invariant spaces and linear operator equations. *Israel J. Math.*, **103**, 259–288.

Jia, R.Q. (1998b). Approximation properties of multivariate wavelets. *Math. Comp.*, **67**, 647–665.

Jia, R.Q. (1998c). Convergence of vector subdivision schemes and construction of biorthogonal multiple wavelets. In *Advances in Wavelets*, ed. K.S. Lau, pp. 189–216. Springer-Verlag, New York.

Jia, R.Q. (2000). Stability of the shifts of a finite number of functions. *J. Approximation Theory*, to appear.

Jia, R.Q., Lee, S.L. and Sharma, A. (1998). Spectral properties of continuous refinement operators. *Proc. Amer. Math. Soc.*, **126**, 729–737.

Jia, R.Q. and Lei, J.J. (1993a). Approximation by multiinteger translates of functions having global support. *J. Approx. Theory*, **72**, 2–23.

Jia, R.Q. and Lei, J.J. (1993b). A new version of the Strang–Fix conditions. *J. Approx. Theory*, **74**, 221–225.

Jia, R.Q. and Micchelli, C.A. (1991). Using the refinement equation for the construction of pre-wavelets II: Power of two. In *Curves and Surfaces*, ed. P.-J. Laurent, A. Le Méhauté, and L.L. Schumaker, pp. 209–246. Academic Press, New York.

Jia, R.Q., Riemenschneider, S.D. and Zhou, D.X. (1997). Approximation by multiple refinable functions. *Canad. J. Math.*, **49**, 944–962.

Jia, R.Q., Riemenschneider, S.D. and Zhou, D.X. (1998). Vector subdivision schemes and multiple wavelets. *Math. Comp.*, **67**, 1533–1563.

Jia, R.Q., Riemenschneider, S.D. and Zhou, D.X. (2000). Smoothness of multiple refinable functions and multiple wavelets. *SIAM J. Matrix Anal. Appl.*, to appear.

Jia, R.Q. and Shen, Z. (1994). Multiresolution and wavelets. *Proc. Edinburgh Math. Soc.*, **37**, 271–300.

Jia, R.Q. and Wang, J.Z. (1993). Stability and linear independence associated with wavelet decompositions. *Proc. Amer. Math. Soc.*, **117**, 1115–1224.

Johnson, M.J. (1997a). An upper bound on the approximation power of principal shift-invariant spaces. *Constr. Approx.*, **13**, 155–176.

Johnson, M.J. (1997b). On the approximation power of principal shift-invariant subspaces of $L_p(\mathbb{R}^d)$. *J. Approx. Theory*, **91**, 279–319.

Kyriazis, G.C. (1995). Approximation from shift-invariant spaces. *Constr. Approx.*, **11**, 141–164.

Kyriazis, G.C. (1996a). Approximation orders of principal shift-invariant spaces generated by box splines. *J. Approx. Theory*, **85**, 218–232.

Kyriazis, G.C. (1996b). Approximation of distribution spaces by means of kernel operators. *J. Fourier Anal. Appl.*, **2**, 261–286.

Kyriazis, G.C. (1997). Wavelet-type decompositions and approximations from shift-invariant spaces. *J. Approx. Theory*, **88**, 257–271.

Lei, J.J. (1994a). On approximation by translates of globally supported functions. *J. Approx. Theory*, **77**, 123–138.

Lei, J.J. (1994b). $L_p(\mathbb{R}^d)$-approximation by certain projection operators. *J. Math. Anal. Appl.*, **185**, 1–14.

Lei, J.J., Jia, R.Q. and Cheney, E.W. (1997). Approximation from shift-invariant spaces by integral operators. *SIAM J. Math. Anal.*, **28**, 481–498.

Lemarié, P.G. (1988). Une nouvelle base d'ondelettes de $L^2(\mathbb{R}^n)$. *J. de Math. Pures et Appl.*, **67**, 227–236.

Lian, J.A. (1995). Characterization of the order of polynomial reproduction for multiscaling functions. In *Approximation Theory VIII, vol. 2, Wavelets and Multilevel Approximation*, ed. C. K. Chui and L. L. Schumaker, pp. 251–258. World Scientific, Singapore.

Light, W.A. (1991). Recent developments in the Strang–Fix theory for approximation orders. In *Curves and Surfaces*, ed. P.-J. Laurent, A. Le Méhauté and L. L. Schumaker, pp. 285–292. Academic Press, New York.

Light, W.A. and Cheney, E.W. (1992). Quasi-interpolation with translates of a function having non-compact support. *Constr. Approx.*, **8**, 35–48.

Madych, W.R and Nelson, S.A. (1990). Polyharmonic cardinal splines. *J. Approx. Theory*, **40**, 141–156.

Micchelli, C.A. and Sauer, T. (1997). Regularity of multiwavelets. *Advances in Comp. Math.*, **7**, 455–545.

Micchelli, C.A. and Sauer, T. (1998). On vector subdivision. *Math. Z.*, **229**, 621–674.

Plonka, G. (1995a). Approximation properties of multi-scaling functions: A Fourier approach. *Rostock Math. Kolloq.*, **49**, 115–126.

Plonka, G. (1995b). Factorization of refinement masks of function vectors. In *Approximation Theory VIII, vol. 2, Wavelets and Multilevel Approximation*, ed. C.K. Chui and L.L. Schumaker, pp. 317–324. World Scientific, Singapore.

Plonka, G. (1997). Approximation order provided by refinable function vectors. *Constr. Approx.*, **13**, 221–244.

Plonka, G. and Ron, A. (2000). A new factorization technique of the matrix mask of univariate refinable functions. *Numer. Math.*, to appear.

Plonka, G. and Strela, V. (1998a). Construction of multiscaling functions with approximation and symmetry. *SIAM J. Math. Anal*, **29**, 481–510.

Plonka, G. and Strela, V. (1998b). From wavelets to multiwavelets. In *Mathematical Methods for Curves and Surfaces II*, ed. M. Daehlen, T. Lyche, and L.L. Schumaker, pp. 1–25. Vanderbilt University Press, Nashville.

Riemenschneider, S.D. (1989). Multivariate cardinal interpolation. In *Approximation Theory VI*, ed. C.K. Chui, L.L. Schumaker and J.D. Ward, pp. 561–580. Academic Press, New York.

Riemenschneider, S.D. and Shen, Z. (1991). Box splines, cardinal series and wavelets. In *Approximation Theory and Functional Analysis*, ed. C.K. Chui, pp. 133–149. Academic Press, New York.

Ron, A. (1990). Factorization theorems for univariate splines on regular grids. *Israel J. Math.*, **70**, 48–68.

Ron, A. (1991). A characterization of the approximation order of multivariate spline spaces. *Studia Math.*, **98**, 73–90.

Ron, A. (1992). The L_2-approximation orders of principal shift-invariant spaces generated by a radial basis function. In *Numerical Methods in Approximation*

Theory, vol. 9, ed. D. Braess and L.L. Schumaker, pp. 245–268. ISNM, vol. 105, Birkhäuser-Verlag, Basel.

Ron, A. (1995). Approximation orders of and approximation maps from local principal shift-invariant spaces. *J. Approx. Theory*, **81**, 38–65.

Ron, A. (1997). Smooth refinable functions provide good approximation orders. *SIAM J. Math. Anal.*, **28**, 731–748.

Ron, A. (1999). Wavelets and their associated operators. In *Approximation Theory IX*, ed. C.K. Chui and L.L. Schumaker. Vanderbilt University Press, Nashville.

Ron A. and Sivakumar, N. (1993). The approximation order of box spline spaces. *Proc. Amer. Math. Soc.*, **117**, 473–482.

Schoenberg, I.J. (1946). Contributions to the problem of approximation of equidistant data by analytic functions. *Quart. Appl. Math.*, **4**, 45–99 and 112–141.

Shen, Z. (1998). Refinable function vectors. *SIAM J. Math. Anal.*, **29**, 235–250.

Strang, G. and Fix, G. (1973). A Fourier analysis of the finite-element variational approach. In *Constructive Aspects of Functional Analysis*, ed. G. Geymonat, pp. 793–840. C.I.M.E., Erice.

Strang, G. and Strela, V. (1994). Orthogonal multiwavelets with vanishing moments. *J. Optical Eng.*, **33**, 2104–2107.

Unser, M. and Blu, T. (2000). Fractional splines and wavelets. *SIAM Review*, **42**, 43–67.

Whittaker, J.M. (1935). *Interpolatory Function Theory*. Cambridge University Press, Cambridge.

Zhao, K. (1992). Approximation order achieved by semi-discrete convolution, unpublished manuscript.

Zhao, K. (1994). Density of dilates of a shift-invariant subspace. *J. Math. Anal. Appl.*, **184**, 517–532.

Zhao, K. (1995). Simultaneous approximation from PSI spaces. *J. Approx. Theory*, **81**, 166–184.

Zhao, K. (1996). Approximation from locally finite-dimensional shift-invariant spaces. *Proc. Amer. Math. Soc.*, **124**, 1857–1867.

5
Introduction to shift-invariant spaces. Linear independence

A. RON

Abstract

Shift-invariant spaces play an increasingly important role in various areas of mathematical analysis and its applications. They appear either implicitly or explicitly in studies of wavelets, splines, radial basis function approximation, regular sampling, Gabor systems, uniform subdivision schemes, and perhaps in some other areas. One must keep in mind, however, that the shift-invariant system explored in one of the above-mentioned areas might be very different from those investigated in another. For example, in *splines* the shift-invariant system is generated by elements of compact support, while in the area of *sampling* the shift-invariant system is generated by band-limited elements, i.e., elements whose Fourier transform is compactly supported. The *theory of shift-invariant spaces* attempts to provide a uniform platform for all these different investigations of shift-invariant spaces. The two main pillars of the theory are the study of the *approximation properties* of such spaces, and the study of *generating sets* for these spaces. The chapter by Kurt Jetter and Gerlind Plonka provides excellent up-to-date information about the first topic. The present chapter is devoted to the latter topic. My goal is to provide the reader with an easy and friendly introduction to the basic principles of that topic. The core of the presentation is devoted to the study of *local principal shift-invariant* spaces, while the more general cases are treated as extensions of that basic setup.

5.1 Introduction

A shift-invariant (SI) space is a linear space S, consisting of functions (or distributions) defined on $\mathbb{R}^d$ ($d \geq 1$), that is invariant under lattice

translations:
$$f \in S \Longrightarrow E^j f \in S, \quad j \in \mathcal{L}, \tag{5.1}$$
where E^j is the **shift operator**
$$(E^j f)(x) := f(x - j). \tag{5.2}$$
The most common choice for $\mathcal{L}$ is the integer lattice $\mathcal{L} = \mathbb{Z}^d$. Here and hereafter we use the notion of a **shift** as a synonym for *integer translation* and/or *integer translate*. Given a set Φ of functions defined on $\mathbb{R}^d$, we say that Φ **generates** the SI space S, if the collection
$$E(\Phi) := (E^j \phi : \phi \in \Phi, \, j \in \mathbb{Z}^d) \tag{5.3}$$
of shifts of Φ is **fundamental** in S, i.e., the span of $E(\Phi)$ is dense in S. Of course, the definition just given assumes that S is endowed with some topology, so we will give a more precise definition of this notion later.

Shift-invariant spaces are usually defined in terms of their generating set Φ, and they are classified according to its properties. For example, a **principal shift-invariant (PSI)** space is generated by a single function, i.e., $\Phi = \{\phi\}$, and a **finitely generated shift-invariant (FSI)** space is generated by a finite Φ. In some sense, the PSI space is the simplest type of SI space. Another possible classification is according to smoothness or decay properties of the generating set Φ. For example, an SI space is **local** if it is generated by a *compactly supported* Φ. Local PSI spaces are, probably, the bread and butter of shift-invariant spaces. At the other end of this classification are the **band-limited** SI spaces; their generators have their Fourier transforms supported in some given compact domain.

Studies in several areas of analysis employ, explicitly or implicitly, SI spaces, and the *Theory of Shift-Invariant Spaces* attempts to provide a uniform platform for all these studies. In certain areas, the SI space appears as an *approximation space*. More precisely, in *Spline Approximation*, local PSI and local FSI spaces are employed, the most notable examples being the box splines and the exponential box spline spaces (see de Boor et al. (1993), de Boor and Ron (1992) and the references therein). In contrast, in *radial basis function approximation*, PSI spaces generated by functions of global support are typical; e.g., fundamental solutions of elliptic operators are known to be useful generators there (see Buhmann (1993), Dyn and Ron (1995) and the references therein). In *Uniform Sampling*, band-limited SI spaces are the rule (see e.g., Marks (1993)). *Uniform Subdivision* (see Dyn (1992), Cavaretta et al. (1991)) is an example where SI spaces appear in an implicit way: the SI spaces appear there in the analysis, not in the setup. The SI spaces in this area are usually local PSI/FSI, and possess the additional important property of *refinability* (that we define and discuss in the body of this article).

In other areas, the shift-invariant space is the "building block" of a larger system, or, to put it differently, a multitude of SI spaces is employed simultaneously. In the area of *Weyl–Heisenberg* (WH, also known as *Gabor*) *systems* (see Feichtinger and Strohmer (1997)), the SI space S is PSI/FSI and is either local or "near-local" (e.g., generated by functions that decay exponentially fast at ∞; the generators are sometime referred to as "windows"). The complete system is then of the form $(S_i)_{i \in I}$, with each S_i a *modulation* of S, i.e., the multiplication product of S by a suitable exponential function. Finally, in the area of *Wavelets* (see Meyer (1992), Daubechies (1992), Ron (1998)), SI spaces appear in two different ways. First, the *wavelet system* is of the form $(S_i)_{i \in I}$ where all the S_i spaces obtained from a single SI space (which, again, is a PSI/FSI space and is usually local), but this time *dilation* replaces the modulation from the WH case. Second, refinable PSI/FSI spaces are crucial in the construction of wavelet systems via the vehicle of *Multiresolution Analysis*.

There are two foci in the study of shift-invariant spaces. The first is the study of their approximation properties (see de Boor and Ron (1992), de Boor et al. (1994a), de Boor et al. (1994b), de Boor et al. (1998)). The second is the study of the shift-invariant system $E(\Phi)$ as a basis for the SI space it spans. The present article discusses the basics of that latter topic. In view of the prevalence of local PSI spaces in the relevant application areas, we first develop in Section 5.2 the theory for that case, and then discuss various extensions of the basic theory. Most of the theory presented in the present article was developed in the early 90s, but, to the best of my knowledge, has not been summarized before in a self-contained manner.

The rest of this article is laid out as follows:

5.2 Bases for PSI spaces

    5.2.1 The analysis and synthesis operators

    5.2.2 Basic theory: Linear independence in local PSI spaces

    5.2.3 Univariate local PSI spaces

    5.2.4 The space $S_2(\phi)$

    5.2.5 Basic theory: Stability and frames in PSI spaces

5.3 Beyond local PSI spaces

    5.3.1 L_p-stability in PSI spaces

    5.3.2 Local FSI spaces: Resolving linear dependence, injectability

5.3.3 Local FSI spaces: Linear independence
5.3.4 L_2-stability and frames in FSI spaces, fiberization
5.4 Refinable shift-invariant spaces
   5.4.1 Local linear independence in univariate refinable PSI spaces
   5.4.2 The simplest application of SI theory to refinable functions

5.2 Bases for PSI spaces

Let ϕ be a compactly supported function in $L_1(\mathbb{R}^d)$, or, more generally, a compactly supported distribution in $\mathcal{D}'(\mathbb{R}^d)$. We analyse in detail the "basis" properties of the set of shifts $E(\phi)$ (cf. (5.3)). The compact support assumption simplifies some technical details in the analysis, and, more importantly, allows the introduction and analysis of a fine scale of possible "basis" properties.

5.2.1 The analysis and synthesis operators

The basic operators associated with a shift-invariant system are the *analysis operator* and the *synthesis operator*. There are several different variants of these operators, due to different choices of the domain of the corresponding map. It is then important to stress right from the beginning that these differences are *very significant*. We illustrate this point later.

Let
$$\mathcal{Q} \tag{5.4}$$
be the space of all complex valued functions defined on $\mathbb{Z}^d$. More generally, let Φ be some set of functions/distributions; letting the elements in Φ index themselves, we set
$$\mathcal{Q}(\Phi) := \mathcal{Q} \times \Phi. \tag{5.5}$$

The space $\mathcal{Q}(\Phi)$ is equipped with the topology of *pointwise convergence* (which makes it into a Fréchet space). For the lion's share of the study below, however, it suffices to treat $\mathcal{Q}$ merely as a linear space.

Given a finite set Φ of compactly supported distributions, the **synthesis operator** $\mathcal{T}_\Phi$ is defined by
$$\mathcal{T}_\Phi : \mathcal{Q}(\Phi) \to S_\star(\Phi) : c \mapsto \sum_{\phi \in \Phi} \sum_{j \in \mathbb{Z}^d} c(j, \phi)\, E^j \phi.$$

The notation
$$S_\star(\Phi)$$

that we have just used stands, by definition, for the range of $\mathcal{T}_\Phi$. In this section, we focus on the PSI case, i.e., the case when Φ is a singleton $\{\phi\}$. Thus,
$$\mathcal{T}_\phi : \mathcal{Q} \to S_\star(\phi) : c \mapsto \sum_{j \in \mathbb{Z}^d} c(j)\, E^j \phi.$$

Example If ϕ is the support function of the interval $[0,1]$ (in one dimension), then $S_\star(\phi)$ is the space of all piecewise-constants with (possible) discontinuities at $\mathbb{Z}$.

Note that, thanks to the compact support assumption on ϕ, the operator $\mathcal{T}_\phi$ is well-defined on the entire space $\mathcal{Q}$. In what follows, we either consider the operator $\mathcal{T}_\phi$ as above, or inspect its restriction to some subspace $C \subset \mathcal{Q}$ (and usually equip that subspace with a stronger topology). The compact support assumption on ϕ "buys" the largest possible domain, viz., $\mathcal{Q}$, hence the full range of subdomains to inspect. The properties of $\mathcal{T}_\phi$ (or a restriction of it) that are of immediate interest are the *injectivity* of the operator, its *continuity*, and its *invertibility*. Of course, the two latter properties make sense only if we rigorously define the target space, and equip the domain and the target space with appropriate topologies.

Discussion As mentioned before, the choice of the domain of $\mathcal{T}_\phi$ is crucial. Consider for example the injectivity property of $\mathcal{T}_\phi$. This property, known as the linear independence of $E(\phi)$, is one of the most fundamental properties of the shift-invariant system. On the other hand, the restriction of $\mathcal{T}_\phi$ to, say, $\ell_2(\mathbb{Z}^d)$ is always injective (recall that ϕ is assumed to be compactly supported), hence that restricted type of injectivity is void of any value. Consequently, keeping the domain of $\mathcal{T}_\phi$ "large" allows us to get meaningful definitions, hence is important for the development of the theory.

An important alternative to the above is to study the formal adjoint $\mathcal{T}_\phi^*$ of $\mathcal{T}_\phi$, known as the **analysis operator** and defined by
$$\mathcal{T}_\phi^* : f \mapsto (\langle f, E^j \phi \rangle)_{j \in \mathbb{Z}^d}.$$
(Here and elsewhere, we write
$$\langle f, \lambda \rangle$$
for the action of the linear functional λ on the function f.) We intentionally avoided the task of defining the domain of this adjoint: it is defined on the largest domain that can make sense! For example, if $\phi \in L_2(\mathbb{R}^d)$ (and is of compact support), $\mathcal{T}_\phi^*$ is naturally defined on $L_{2,\mathrm{loc}}(\mathbb{R}^d)$. On the other hand,

if ϕ is merely a compactly supported distribution, f should be assumed to be a $C^\infty(\mathbb{R}^d)$-function. In any event, and unless the surjectivity of T_ϕ^* is the goal, the target space here can be taken to be $\mathcal{Q}$.

The study of $E(\phi)$ via the analysis operator is done by considering pre-images of certain sequences in the target space. For example, a non-empty pre-image $T_\phi^{*-1}\delta$ of the δ-sequence indicates that the shifts of ϕ can, at least to some extent, be separated. A much finer analysis is obtained by studying properties of the functions in $T_\phi^{*-1}\delta$; first and foremost *decaying* properties of such functions. The desire is to find in $T_\phi^{*-1}\delta$ a function that decays as fast as possible, ideally a compactly supported function. Note that $f \in T_\phi^{*-1}\delta$ if and only if $E(f)$ forms a dual basis to the shifts of ϕ in sense that

$$\langle E^k f, E^j \phi \rangle = \delta_{j,k}.$$

One of the advantages in this complementary approach is that certain functions in $T_\phi^{*-1}\delta$ can be represented explicitly in terms of ϕ, hence their decay properties can be examined directly.

5.2.2 Basic theory: Linear independence in local PSI spaces

We say that the shifts of the compactly supported distribution ϕ are **(globally) linearly independent (=:gli)** if T_ϕ is injective, i.e., if the condition

$$(T_\phi c = 0, \quad \text{for some } c \in \mathcal{Q}) \Longrightarrow c = 0$$

holds. The discussion concerning this basic, important, property is two-fold. First, we discuss characterizations of the linear independence property that are useful for checking its validity. Then, we discuss other properties that are either equivalent to linear independence or are implied by it, and that are useful for the construction of approximation maps into $S_\star(\phi)$.

We start with the first task, characterizing linear independence in terms of more verifiable conditions. To this end, we recall that, since ϕ is compactly supported, its Fourier transform extends to an entire function. We still denote that extension by $\widehat{\phi}$. The following fairly immediate observation is crucial.

Observation $\ker T_\phi$ is a *closed shift-invariant* subspace of $\mathcal{Q}$.

In Lefranc (1958), closed SI subspaces of $\mathcal{Q}$ are studied. It is proved there that $\mathcal{Q}$ admits *spectral analysis*, and, moreover, admits *spectral synthesis*. The latter property means that every closed SI subspace of $\mathcal{Q}$ contains a dense exponential subspace (here "an exponential" is a linear combination of the restriction to $\mathbb{Z}^d$ of products of exponential functions by polynomials). More details on Lefranc's synthesis result, as well as complete details of its proof, can be found in de Boor and Ron (1989). We need here only the much weaker *analysis* part of Lefranc's theorem, which says the following.

Theorem 5.2.1 *Every (nontrivial) closed SI subspace of $\mathcal{Q}$ contains an exponential sequence*

$$e_\theta : j \mapsto e^{\theta \cdot j}, \quad \theta \in \mathbb{C}^d.$$

We sketch the proof below, and refer to Ron (1989) for the complete one.

Proof (sketch). The continuous dual space of $\mathcal{Q}$ is the space $\mathcal{Q}_0$ of all finitely supported sequences, with $\langle \lambda, c \rangle := \sum_{j \in \mathbb{Z}^d} \lambda(j) c(j)$. Given a closed non-zero SI subspace $C \subset \mathcal{Q}$, its annihilator $C^\perp$ in $\mathcal{Q}_0$ is a proper SI subspace. The sequences $C_+^\perp$ in $C^\perp$ that are entirely supported on $\mathbb{Z}_+^d$ can be viewed as polynomials via the association

$$Z : c \mapsto \sum_{j \in \mathbb{Z}_+^d} c(j) X^j,$$

with X^j the standard monomial. Since $C^\perp$ is SI, $Z(C_+^\perp)$ is an ideal in the space of all d-variate polynomials. Since $C^\perp$ is proper, $Z(C_+^\perp)$ cannot contain any monomial. Hilbert's *(Weak) Nullstellensatz* then implies that the polynomials in $Z(C_+^\perp)$ all vanish at some point $e^\theta := (e^{\theta_1}, \ldots, e^{\theta_d}) \in (\mathbb{C} \backslash 0)^d$. One then concludes that the sequence

$$e_\theta : j \mapsto e^{\theta \cdot j}$$

vanishes on $C^\perp$, hence, by Hahn-Banach, lies in C. □

Unaware of Lefranc's work, Dahmen and Micchelli (1983) proved that, assuming the compactly supported ϕ to be a continuous function, $\ker \mathcal{T}_\phi$, if non-trivial, must contain an exponential e_θ. Their argument is essentially the same as Lefranc's, save some simplifications that are available due to their additional assumptions on ϕ.

The following characterization of linear independence appears first in Ron (1989).

Theorem 5.2.2 *The shifts of the compactly supported distribution ϕ are linearly independent if and only if $\widehat{\phi}$ does not have any 2π-periodic zero (in $\mathbb{C}^d$).*

Proof (sketch) Poisson's summation formula implies that, for any $\theta \in \mathbb{C}^d$,

$$\mathcal{T}_\phi e_\theta = 0 \quad \Longleftrightarrow \quad (\widehat{\phi} \text{ vanishes on } -\mathrm{i}\theta + 2\pi\mathbb{Z}^d). \tag{5.6}$$

Therefore, $\ker \mathcal{T}_\phi$ contains an exponential e_θ iff $\widehat{\phi}$ has a 2π-periodic zero. Since $\ker \mathcal{T}_\phi$ is SI and closed, Theorem 5.2.1 completes the proof. □

Example Let ϕ be a univariate exponential B-spline (Dyn and Ron (1990)). The Fourier transform of such a spline is of the form

$$\omega \mapsto \prod_{j=1}^{n} \int_0^1 e^{(\mu_j - \mathrm{i}\omega)t} \, \mathrm{d}t,$$

with $(\mu_j)_j \subset \mathbb{C}$. One observes that $\widehat{\phi}$ vanishes exactly on the set

$$\bigcup_j (-\mathrm{i}\mu_j + 2\pi(\mathbb{Z}\backslash 0)).$$

From Theorem 5.2.2 it then follows that $E(\phi)$ are linearly dependent iff there exist j and k such that $\mu_j - \mu_k \in 2\pi(\mathbb{Z}\backslash 0)$.

Example Let ϕ be the k-fold convolution of a compactly supported distribution ϕ_0. It is fairly obvious that $\mathcal{T}_\phi$ cannot be injective when $\mathcal{T}_{\phi_0}$ is not. Since the zero sets $\widehat{\phi}$ and $\widehat{\phi_0}$ are identical, Theorem 5.2.2 proves that the converse is valid, too, linear independence of $E(\phi_0)$ implies that of $E(\phi)$!

Among the many applications of Theorem 5.2.2, we mention Ron (1989) where the theorem is applied to exponential box splines, Jia and Sivakumar (1990), Sivakumar (1991) and Ron (1992a) where the theorem is used for the study of box splines with rational directions, Jia (1993) where discrete box splines are studied, and Chui and Ron (1991) where convolution products of box splines and compactly supported distributions are considered.

We now turn our attention to the second subject, useful properties of $E(\phi)$ that are implied or equivalent to linear independence. We work initially in a slightly more general setup. Instead of studying $E(\phi)$, we treat any countable set F of distributions with **locally finite** supports. Given any bounded set Ω, the supports of almost all the elements of F are disjoint from Ω. For convenience, we index F by $\mathbb{Z}^d$: $F = (f_j)_{j \in \mathbb{Z}^d}$. The relevant synthesis operator is

$$\mathcal{T} : \mathcal{Q} \to \mathcal{D}'(\mathbb{R}^d) \; : \; c \mapsto \sum_{j \in \mathbb{Z}^d} c(j) f_j,$$

and is well-defined, thanks to the local finiteness assumption. The notion of *linear independence* remains unchanged; it is the injectivity of the map $\mathcal{T}$.

We start with the following result of Ben-Artzi and Ron (1990). The proof given follows Zhao (1992).

Theorem 5.2.3 *Let $F = (f_j)_{j \in \mathbb{Z}^d}$ be a collection of compactly supported distributions with locally finite supports. Then the following conditions are equivalent:*

(a) *F is linearly independent;*
(b) *there exists a dual basis to F in $\mathcal{D}(\mathbb{R}^d)$, i.e., a sequence $G := (g_j)_{j \in \mathbb{Z}^d} \subset \mathcal{D}(\mathbb{R}^d)$ such that*
$$\langle g_k, f_j \rangle = \delta_{j,k};$$
(c) *for every $j \in \mathbb{Z}^d$, there exists a bounded $A_j \subset \mathbb{R}^d$ such that, if $\mathcal{T}c$ vanishes on A_j, we must have $c(j) = 0$.*

Proof (sketch) Condition (c) clearly implies (a). Also, (b) implies (c). Indeed, with $G \subset \mathcal{D}(\mathbb{R}^d)$ the basis dual to F, we have that $\langle g_j, \mathcal{T}c \rangle = c(j)$, hence we may take A_j to be any bounded open set that contains supp g_j.

(a)$\Longrightarrow$(b): Recall that $\mathcal{Q}_0$ is the collection of finitely supported sequences in $\mathcal{Q}$. It is equipped with the inductive-limit topology, a discrete analog of the $\mathcal{D}(\mathbb{R}^d)$-topology. The only facts required on this topological space $\mathcal{Q}_0$ are that (a) $\mathcal{Q}$ and $\mathcal{Q}_0$ are each the continuous dual space of the other, and (b) $\mathcal{Q}_0$ does not contain proper dense subspaces (see Trèves (1967) for details). From that and the definition of $\mathcal{T}$, one concludes that $\mathcal{T}$ is continuous, that its adjoint is the operator
$$\mathcal{T}^* : \mathcal{D}(\mathbb{R}^d) \to \mathcal{Q}_0 \ : \ g \mapsto (\langle g, f_j \rangle_{j \in \mathbb{Z}^d}),$$
and that $\mathcal{T}^{**} = \mathcal{T}$. Thus, if $\mathcal{T}$ is injective, then $\mathcal{T}^*$ has a dense range, hence must be surjective. This surjectivity implies, in particular, that all sequences supported at one point are in the range of $\mathcal{T}^*$, and (b) follows. □

For the choice $F = E(\phi)$, the sets $(A_j)_j$ are obtained by shifting A_0, and a dual basis G can be chosen to have the form $E(g_0)$. Moreover, if $F = E(\Phi)$, Φ finite, the sets $(A_j)_j$ are still obtained as the shifts of some finitely many compact sets (viz., with $E(G)$ the dual basis of $E(\Phi)$ which is guaranteed to exist by Theorem 5.2.3, the finitely many compact sets are the supports of the functions in G). Thus, for this case the sets $(A_j)_j$ are locally finite. With this in hand, Jia and Micchelli (1991) concluded the following from Theorem 5.2.3.

Corollary 5.2.4 *Let Φ be a finite set of compactly supported distributions. If $E(\Phi)$, the set of shifts of Φ, is linearly independent, then every compactly supported $f \in S_\star(\Phi)$ is a finite linear combination of $E(\Phi)$.*

Proof Let $E(G) \subset \mathcal{D}(\mathbb{R}^d)$ be the basis dual to $E(\Phi)$ from Theorem 5.2.3. Given a compactly supported f, we have for every $g \in G$, and for almost every $j \in \mathbb{Z}^d$, that the function $E^j g$ has its support disjoint from supp f. This finishes the proof since, if $f = \mathcal{T}c = \sum_{\phi \in \Phi} \sum_{j \in \mathbb{Z}^d} c(j, \phi) E^j \phi$, then $\langle f, E^j g \rangle = c(j, \phi)$. □

Theorem 5.2.3 allows us, in the presence of linear independence, to construct projectors into $S_\star(\phi)$ of the form $\mathcal{T}_\phi \mathcal{T}_g^*$ that are based on compactly supported functions. In fact, the theorem also shows that nothing less than linear independence suffices for such construction.

Corollary 5.2.5 *Let $\phi \in L_2(\mathbb{R}^d)$ be compactly supported, and assume $E(\phi)$ to be orthonormal. Then $E(\varphi)$ is linearly independent.*

The next theorem summarizes some of the observations made above, and adds a few more.

Theorem 5.2.6 *Let ϕ be a compactly supported distribution. Consider the following conditions:*

- *(gli) global linear independence: $\mathcal{T}_\phi$ is injective;*
- *(ldb) local dual basis: $E(\phi)$ has a dual basis $E(g)$, $g \in \mathcal{D}(\mathbb{R}^d)$;*
- *(ls) local spanning: all compactly supported elements of $S_\star(\phi)$ are finitely spanned by $E(\phi)$;*
- *(ms) minimal support: for every compactly supported $f \in S_\star(\phi)$, there exists some $j \in \mathbb{Z}^d$ such that supp ϕ lies in the convex hull of supp $E^j f$. Equality can happen only if $f = cE^j \phi$, for some constant c, and some $j \in \mathbb{Z}^d$.*

Then (gli) $\iff$ (ldb) $\implies$ (ls) $\implies$ (ms). Also, if $S_\star(\phi)$ is known to contain a linearly independent SI basis $E(\phi_0)$, then all the above conditions are equivalent. In particular, a linearly independent generator of $S_\star(\phi)$ is unique, up to shifts and multiplication by constants.

Proof (sketch) The fact that (ls) $\implies$ (ms) follows from basic geometric observations. In view of previous results, it remains only to show that, if $S_\star(\phi)$ contains a linear independent $E(\phi_0)$, then (ms) implies (gli). That, however, is simple: since $E(\phi_0)$ is linearly independent, ϕ_0 has minimal support among all compactly supported elements of $S_\star(\phi)$. On the other hand, if ϕ has that minimal support property, then, since it is finitely spanned by $E(\phi_0)$, it must be a constant multiple of a shift of ϕ_0, hence $E(\phi_0)$ is linearly independent, too.

The uniqueness of the linearly independent generator follows now from the analogous uniqueness property of the minimally supported generator. □

Until now, we have retained the compact support assumption on ϕ. This allowed us to strive for the superior property of linear independence. However, PSI spaces that contain no non-trivial compactly supported functions are of interest, too. How do we analyse the set $E(\phi)$ if ϕ is not of compact support? One way, the customary one, is to apply a cruder analysis. One should define the synthesis operator on whatever domain that operator may make sense, and study then this restricted operator. A major effort in this direction is presented in the next subsection, where the notions of stability and frames are introduced and studied. There, only mild decay assumptions on ϕ are imposed, e.g., that $\phi \in L_2(\mathbb{R}^d)$. On the other hand, if ϕ decays at ∞ in a more substantial way, more can be said. For example, if ϕ decays rapidly (i.e., faster than any fixed rational polynomial) then the following is true (Ron (1989)). Recall that a sequence c is said to have **polynomial growth** if there exists a polynomial p such that $|c(j)| \leq |p(j)|$ for all $j \in \mathbb{Z}^d$.

Proposition 5.2.7 *Assume that ϕ decays rapidly, and let $\mathcal{T}_{\phi,T}$ be the restriction of the synthesis operator to sequences of (at most) polynomial growth. Then the following conditions are equivalent:*

(a) $\mathcal{T}_{\phi,T}$ *is injective;*
(b) *the restriction of $\mathcal{T}_{\phi,T}$ to $\ell_\infty(\mathbb{Z}^d)$ is injective;*
(c) $\widehat{\phi}$ *(that is defined now on $\mathbb{R}^d$ only) does not have a (real) 2π-periodic zero;*
(d) *there is a basis $E(g)$ dual to $E(\phi)$, with g a rapidly decaying, $C^\infty(\mathbb{R}^d)$-function.*

Proof (sketch) (a) trivially implies (b). If $\widehat{\phi}$ vanishes identically on $\theta + 2\pi\mathbb{Z}^d$, θ real, then, by Poisson's summation formula, $e_{i\theta} \in \ker \mathcal{T}_{\phi,T}$ (with $e_\theta : j \mapsto e^{\theta \cdot j}$). This exponential is bounded (since θ is real), hence (b) implies (c). The fact that (d) implies (a) follows from the relation $\langle \mathcal{T}_{\phi,T}c, E^j g\rangle = c(j)$, a relation that holds for any polynomially growing sequence, thanks to the rapid decay assumptions on ϕ and g.

We prove the missing implication "(c) implies (d)" in the next section (after the proof of Theorem 5.3.4), under the assumption that ϕ is a function. If one likes to stick with a *distribution* ϕ, then, instead of proving the missing implication directly, (a) can be proved from (c) as in Ron (1989), (see also the proof of the implication (c)$\Longrightarrow$(b) of Theorem 5.3.2), and the equivalence of (a) and (d) can be proved by an argument similar to that used in Theorem 5.2.3. □

Thus, if ϕ is not of compact support, we settle for notions weaker than linear independence. This approach is not entirely satisfactory, as illustrated in the following example.

Example Assume that ϕ is compactly supported and univariate. Then, by Theorem 5.2.8 below, there is, up to shifting and multiplication by scalars, exactly one linearly independent generator for $S_\star(\phi)$, and one may wish to select this, and only this generator. In contrast, there are many other compactly supported generators of $S_\star(\phi)$ that satisfy all the properties of Proposition 5.2.7. Unfortunately, the supports of these seemingly "good" generators may be as large as one wishes.

The example indicates that properties weaker than linear independence may fail to distinguish between "good" generators and "better" generators. Therefore, an alternative approach to the above (i.e., to the idea of restricting the synthesis operator to a smaller domain) is desired, extending the notion of "linear independence" beyond *local* PSI spaces. For that, we note first that both Theorem 5.2.3 and Proposition 5.2.7 characterize injectivity properties of $\mathcal{T}_\phi$ in terms of surjectivity properties of $\mathcal{T}_\phi^*$, or, more precisely, in terms of the existence of the "nicely decaying" dual basis. Thus, I suggest the following extension of the linear independence notion.

Definition (linear independence) Let ϕ be any distribution. We say that $E(\phi)$ is *linearly independent* if there exists $g \in \mathcal{D}(\mathbb{R}^d)$ whose set of shifts $E(g)$ is linearly independent and is a dual basis to $E(\phi)$.

In view of Theorem 5.2.3, this definition is, indeed, an extension of the previous linear independence definition.

Open Problem Find an effective characterization of the above general notion of linear independence, similar to Theorem 5.2.2.

Proposition 5.2.7 deals with the synthesis operator of a rapidly decaying $E(\phi)$. There are cases where ϕ decays faster than rapidly, for example at a certain exponential rate $\rho \in \mathbb{R}_+^d$. Roughly speaking, it means that

$$|\phi(x)| \leq c e^{\rho(x)}, \quad \rho(x) := \sum_j \rho_j |x_j|.$$

In this case, the synthesis operator is well-defined on all sequences that grow (at most) at that exponential rate. The injectivity of the synthesis operator on this extended domain can be shown, once again, to be equivalent to the lack of 2π-periodic zeros of $\widehat{\phi}$, with $\widehat{\phi}$ the analytic extension of the Fourier transform. This analytic extension is well-defined in the multistrip

$$\{z \in \mathbb{C}^d : |\Im z_j| \le \rho_j, \ j = 1, \ldots, d\}.$$

The above strip can be shown to be the spectrum of a suitably chosen commutative Banach algebra (with the Gelfand transform being the Fourier transform), and those basic observations yield the above-mentioned injectivity result; see Ron (1993).

5.2.3 Univariate local PSI spaces

There are three properties of local PSI spaces that, while not valid in general, are always valid in case the spatial dimension is 1. The first, and possibly the most important one, is the existence of a "canonical" generator (Theorem 5.2.8). The second is the equivalence of the linear independence property to another, seemingly stronger, notion of linear independence, termed here "weak local linear independence". A third fact is discussed in Section 5.4.1.

The following result can be found in Ron (1990) (it was already stated without proof in Strang and Fix (1973)). The result is invalid in two variables. The space generated by the shifts of the characteristic function of the square with vertices $(0,0), (1,1), (0,2), (-1,1)$ is a simple counter-example.

Theorem 5.2.8 *Let S be a univariate local PSI space. Then S contains a generator ϕ whose shifts $E(\phi)$ are linearly independent.*

Proof Let ϕ_0 be a generator of S with support in $[a, b]$. If $E(\phi_0)$ is linearly independent, we are done. Otherwise, $\ker \mathcal{T}_{\phi_0}$ is non-zero, hence, by Theorem 5.2.1, contains an exponential $e_\theta : j \mapsto e^{\theta j}$. We define two distributions

$$\phi_1 := \sum_{j=0}^{\infty} e_\theta(j) \, E^j \phi_0,$$

and

$$-\sum_{j=-\infty}^{-1} e_\theta(j) \, E^j \phi_0.$$

Obviously, ϕ_1 is supported on $[a, \infty)$, and ϕ_2 is supported on $(-\infty, b-1]$. However, since $e_\theta \in \ker \mathcal{T}_{\phi_0}$, the two distributions are equal, and hence ϕ_1 is supported on $[a, b-1]$. Also, ϕ_0 is spanned by $\{\phi_1, E^1\phi_1\}$, and one concludes that $S_\star(\phi_0) = S_\star(\phi_1)$. Since $[a,b]$ is of finite length, we may proceed by induction until arriving at the desired linearly independent ϕ. □

Combining Theorem 5.2.8 and Theorem 5.2.6, we conclude that the properties (gli), (ldb), (ls), and (ms) that appear in Theorem 5.2.6 are all equivalent for *univariate* local PSI spaces. In fact, there is another, seemingly stronger, property that is equivalent here to linear independence.

Definition (local linear independence) Let F be a countable, locally finite, family of distributions/functions. Let $A \subset \mathbb{R}^d$ be an open set. We say that F is *locally linearly independent on A* if, for every $c \in \mathcal{Q}$ (and with $\mathcal{T}$ the synthesis operator of F) the condition

$$\mathcal{T}c = 0 \quad \text{on } A$$

implies that $c(f)f = 0$ on A, for every $f \in F$. The set F is *weakly locally linearly independent =: (wlli)* if F is locally linearly independent on some open, bounded set A. These distributions are *strongly locally linearly independent :=(slli)* is they are locally linearly independent on any open set A.

Note that in the case $\operatorname{supp} f \cap A = \emptyset$, we trivially obtain that $c(f)f = 0$ on A. So, the non-trivial part of the definition is that $c(f) = 0$ whenever $\operatorname{supp} f$ intersects A.

Trivially, weak local linear independence implies linear independence. The converse fails to hold already in the PSI space setup. In Ben-Artzi and Ron (1990), there is an example of a bivariate ϕ whose shifts are globally linearly independent, but are not weakly locally independent. However, as claimed in Ben-Artzi and Ron (1990), these two different notions of independence do coincide in the univariate case. The proof provided here is taken from Ron (1991). Another proof is given in Sun (1993).

Proposition 5.2.9 *Let ϕ be a univariate distribution supported in $[0, N]$ with $E(\phi)$ globally linearly independent. Then $E(\phi)$ is weakly locally independent. More precisely, $E(\phi)$ is locally linearly independent on any interval A whose length is greater than $N - 1$.*

Proof To avoid technical "end-point" problems, we assume herein that ϕ is a *function* supported in $[0, N]$, and prove that $E(\phi)$ is locally linearly independent over $[0, N-1]$. For that, we assume that some sequence $c \in \mathcal{Q}\backslash 0$ satisfies $\mathcal{T}_\phi c = 0$ on $[0, N-1]$. We will show that this implies the existence of a non-zero sequence, say b, such that $\mathcal{T}_\phi b = 0$ a.e.

For $j = 1, ..., N$, let f_j be the periodic extension of $\phi_{|[j-1,j]}$. Note that

$$(\mathcal{T}_\phi b)_{|[j,j+1]} = 0 \iff \sum_{i=1}^{N} b(j-N+i) f_{N-i+1} = 0. \tag{5.7}$$

Suppose that we are given some $b \in \mathcal{Q}$, and would like to check whether $\mathcal{T}_\phi b = 0$. To that end, let B be the matrix indexed by $\{1,\ldots,N\} \times \mathbb{Z}$ whose (i,j)th entry is $b(j-N+i)$. Then, as (5.7) shows, the condition $\mathcal{T}_\phi b = 0$ is equivalent to $FB = 0$, with F the row vector $[f_N, f_{N-1}, \ldots, f_1]$. Our aim is to construct such B. Initially, we select $b := c$, and then know that $FB(\cdot,j) = 0$, $j = 0, 1, \ldots, N-2$, since $\mathcal{T}_\phi c = 0$ on $[0, N-1]$.

Let $C = B(1{:}N{-}1, 0{:}N{-}1)$ be the submatrix of B made up from the first $N-1$ rows and from columns $0, \ldots, N-1$ of B. Since C has more columns than rows, there is a first column that is in the span of the columns to its left. In other words, $C(\cdot, r) = \sum_{j=0}^{r-1} a(j) C(\cdot, j)$ for some $r > 0$ and some $a(0), \ldots, a(r-1)$. In other words, the sequence $b(-N+1), \ldots, b(r-1)$ satisfies the constant coefficient difference equation

$$b(i) = \sum_{j=0}^{r-1} a(j) b(i-r+j), \qquad i = r+1-N, \ldots, r-1.$$

Now use this very equation to define $b(i)$ inductively for $i = r, r+1, \ldots$. Then the corresponding columns of our matrix B satisfy the equation

$$B(\cdot, i) = \sum_{j=0}^{r-1} a(j) B(\cdot, i-r+j), \qquad i = r, r+1, \ldots,$$

and, since $FB(\cdot, j) = 0$ for $j = 0, \ldots, r-1$, this now also holds for $j = r, r+1, \ldots$. In other words, $\mathcal{T}_\phi b = 0$ on $[0, \infty)$. The corresponding further modification of b to also achieve $\mathcal{T}_\phi b = 0$ on $(\infty, 0]$ is now obvious. $\square$

5.2.4 The space $S_2(\phi)$

The notion of the "linear independence of $E(\phi)$" is the "right one" for local PSI spaces. While we were able to extend this notion to PSI (and other) spaces that are not necessarily local, there do not exist at present effective methods for checking this more general notion.

This means that, in the case the generator ϕ of the PSI space $S(\phi)$ is *not* compactly supported; we need other, weaker, notions of "independence". We have already described some possible notions of this type that apply to generators ϕ that decay exponentially fast or at least decay rapidly.

However, generators ϕ that decay at ∞ at slower rates are also of interest. Two pertinent univariate examples of this type are the *sinc function*

$$\text{sinc} : x \mapsto \frac{\sin(\pi x)}{\pi x},$$

and the *inverse multiquadric*

$$x \mapsto \frac{1}{\sqrt{1+x^2}}.$$

While these functions decay very slowly at ∞, they both still lie in L_2, as well as in any L_p, $p > 1$. It is thus natural to seek a theory that will only assume ϕ to lie in L_2, or more generally, in some L_p space. The two basic notions in that development are the notions of *stability* and *a frame*. For $p = 2$, the notion of stability is also known as the *Riesz basis* property.

In what follows, we assume ϕ to lie in $L_2(\mathbb{R}^d)$. Under this assumption, the PSI space $S_\star(\phi)$ is no longer well-defined, nor is there any hope of meaningfully defining the synthesis operator $\mathcal{T}_\phi$. We replace $S_\star(\phi)$ by the PSI space variant

$$S_2(\phi),$$

which is defined as the L_2-closure of the finite span of $E(\phi)$. We also replace the domain $\mathcal{Q}$ of $\mathcal{T}_\phi$ by $\ell_2(\mathbb{Z}^d)$, and denote this restriction by

$$\mathcal{T}_{\phi,2}.$$

Note that, since we are only assuming here that $\phi \in L_2(\mathbb{R}^d)$, we do not know a priori that $\mathcal{T}_{\phi,2}$ is well-defined.

Very useful in this context is the **bracket product**. Given $f, g \in L_2(\mathbb{R}^d)$, the bracket product $[f, g]$ is defined as follows.

$$[f,g] := \sum_{\alpha \in 2\pi\mathbb{Z}^d} E^\alpha f \, \overline{E^\alpha g}.$$

It is easy to see that $[f, g] \in L_1(\mathbb{T}^d)$. We assign also a special notation for the square root of $[\widehat{f}, \widehat{f}]$:

$$\widetilde{f} := \sqrt{[\widehat{f}, \widehat{f}]}. \tag{5.8}$$

Note that the map $f \mapsto \widetilde{f}$ is a unitary map from $L_2(\mathbb{R}^d)$ into $L_2(\mathbb{T}^d)$.

We collect below a few of the basic facts about the space $S_2(\phi)$, which are taken from de Boor et al. (1994a).

Theorem 5.2.10 *Let $\phi \in L_2(\mathbb{R}^d)$. Then:*

(a) *the orthogonal projection Pf of $f \in L_2(\mathbb{R}^d)$ onto $S_2(\phi)$ is given by*

$$\widehat{Pf} = \frac{[\widehat{f}, \widehat{\phi}]}{[\widehat{\phi}, \widehat{\phi}]} \widehat{\phi};$$

(b) *a function $f \in L_2(\mathbb{R}^d)$ lies in the PSI space $S_2(\phi)$ if and only if $\widehat{f} = \tau \widehat{\phi}$ for some measurable 2π-periodic τ;*

(c) *the set $\operatorname{supp} \widetilde{\phi} \subset \mathbb{T}^d$ is independent of the choice of ϕ, i.e., if $S_2(\phi) = S_2(\psi)$, then, up to a null set, $\operatorname{supp} \widetilde{\phi} = \operatorname{supp} \widetilde{\psi}$.*

We call $\operatorname{supp} \widetilde{\phi}$ the **spectrum** of the PSI space $S_2 := S_2(\phi)$, and denote it by

$$\sigma(S_2).$$

Note that the spectrum is defined up to a null set. A PSI space S_2 is **regular** if $\sigma(S_2) = \mathbb{T}^d$ (up to a null set). Note that a *local* PSI space $S_2(\phi)$ is always regular.

The bracket product was introduced in Jia and Micchelli (1991) (in a slightly different form), and in de Boor et al. (1994a) in the present form. A key fact concerning the bracket product is the following identity, which is valid for any $\phi, \psi \in L_2(\mathbb{R}^d)$, and every $c \in \ell_2(\mathbb{Z}^d)$, provided, say, that the operators $\mathcal{T}_{\phi,2}$ and $\mathcal{T}_{\psi,2}$ are bounded.

$$(\mathcal{T}_\phi^* \mathcal{T}_\psi c)\widehat{} = [\widehat{\psi}, \widehat{\phi}]\, \widehat{c}. \tag{5.9}$$

Lemma 5.2.11 *Let $\phi, \psi \in L_2(\mathbb{R}^d)$, and assume that the operators $\mathcal{T}_{\phi,2}$ and $\mathcal{T}_{\psi,2}$ are bounded.*

(a) *The kernel $\ker \mathcal{T}_{\phi,2}^* \mathcal{T}_{\psi,2} \subset \ell_2(\mathbb{Z}^d)$ is the space*

$$K_{\phi,\psi} := \left\{ c \in \ell_2(\mathbb{Z}^d) : \operatorname{supp} \widehat{c} \subset \mathbb{T}^d \backslash \left(\operatorname{supp}[\widehat{\psi}, \widehat{\phi}] \right) \right\}.$$

(b) *The operator $\mathcal{T}_{\phi,2}^* \mathcal{T}_{\psi,2}$ is a projector if and only if $[\widehat{\psi}, \widehat{\phi}] = 1$ on its support.*

Proof (sketch) (a) follows from (5.9), since the latter implies that $\mathcal{T}_{\phi,2}^* \mathcal{T}_{\psi,2} c = 0$ if and only if $\operatorname{supp} \widehat{c}$ is disjoint from

$$\sigma := \operatorname{supp}[\widehat{\psi}, \widehat{\phi}].$$

For (b), note that (5.9) implies that the range of $\mathcal{T}_{\phi,2}^* \mathcal{T}_{\psi,2}$ is $K_{\phi,\psi}^\perp = \{c \in \ell_2(\mathbb{Z}^d) : \operatorname{supp} \widehat{c} \subset \sigma\}$. Thus $\mathcal{T}_{\phi,2}^* \mathcal{T}_{\psi,2}$ is a projector if and only if it is the identity on $(K_{\phi,\psi})^\perp$. The result now easily follows from (5.9). □

5.2.5 Basic theory: Stability and frames in PSI spaces

We need here to make our setup a bit more general. Thus, we assume F to be any countable subset of $L_2(\mathbb{R}^d)$, and define a corresponding *synthesis map* $T_{F,2}$ as follows:

$$T_{F,2} : \ell_2(F) \to L_2(\mathbb{R}^d) : c \mapsto \sum_{f \in F} c(f) f.$$

The choice $F := E(\phi)$ is of immediate interest here, but other choices will be considered in what follows.

Definition (Bessel systems, stable bases and frames) Let $F \subset L_2(\mathbb{R}^d)$ be countable.

(a) We say that F forms a *Bessel system* if $T_{F,2}$ is a well-defined bounded map.
(b) A Bessel system F is a *frame* if the range of $T_{F,2}$ is closed (in $L_2(\mathbb{R}^d)$).
(c) A frame F is a *stable basis* if $T_{F,2}$ is injective.

Discussion The notion of stability effectively says that $T_{F,2}$ is a continuous injective open, hence invertible, map, i.e., that there exist constants $C_1, C_2 > 0$ such that

$$C_1 \|c\|_{\ell_2(F)} \le \|T_{F,2}c\|_{L_2(\mathbb{R}^d)} \le C_2 \|c\|_{\ell_2(F)}, \quad \forall c \in \ell_2(F), \tag{5.10}$$

for every finitely supported c defined on F (hence for every $c \in \ell_2(F)$).

The frame condition is weaker. It does not assume that (5.10) holds for *all* $c \in \ell_2(F)$, but only for c in the orthogonal complement of $\ker T_{F,2}$. In general, it is hard to compute that orthogonal complement, hence it is non-trivial to implement the definition of a frame via the synthesis operator. However, for the case of interest here, i.e., the PSI system $F = E(\phi)$, computing $\ker T_{F,2}$ is quite simple (see Lemma 5.2.11).

There is an alternative definition of the frame property, which is more common in the literature. Assume that F is a Bessel system, and let $T_{F,2}^*$ be its analysis operator,

$$T_{F,2}^* : L_2(\mathbb{R}^d) \to \ell_2(F) : g \mapsto (\langle g, f \rangle)_{f \in F}.$$

The equivalent definition of a frame with the aid of this operator is analogous. There exist constants $C_1, C_2 > 0$ such that

$$C_1 \|f\|_{L_2(\mathbb{R}^d)} \le \|T_{F,2}^* f\|_{\ell_2(F)} \le C_2 \|f\|_{L_2(\mathbb{R}^d)} \tag{5.11}$$

for every f in the orthogonal complement of $\ker T_{F,2}^*$. While it seems that we have gained nothing by switching operators, it is usually easier to identify the above-mentioned orthogonal complement. It is simply the closure in $L_2(\mathbb{R}^d)$ of the finite span of F.

The constant C_1 (C_2, respectively) is sometimes referred to as the **lower** (**upper**, respectively) **stability/frame bound**.

A third, and possibly the most effective, definition of stable bases/frames goes via a **dual system**. Let R be some assignment

$$\mathrm{R}: F \to L_2(\mathbb{R}^d),$$

and assume that F as well as $\mathrm{R}F$ are Bessel systems. We then say that $\mathrm{R}F$ is a dual system for F if the operator

$$T_{F,2} T^*_{\mathrm{R}F,2} : g \mapsto \sum_{f \in F} \langle g, \mathrm{R}f \rangle f$$

is a *projector*, i.e., it is the identity on the closure of span F. The roles of F and $\mathrm{R}F$ in the above definition are interchangeable. We have the following simple lemma.

Lemma 5.2.12 *Let $F \subset L_2(\mathbb{R}^d)$ be countable, and assume that F is a Bessel system. Then:*

(a) *F is a frame if and only if there exists an assignment $\mathrm{R}: F \to L_2(\mathbb{R}^d)$ such that $\mathrm{R}F$ is Bessel, and is a dual system of F;*

(b) *F is a stable basis if and only if there exists an assignment $\mathrm{R}: F \to L_2(\mathbb{R}^d)$ such that $\mathrm{R}F$ is Bessel, and is a dual system for F in the stronger biorthogonal sense, i.e., for $f, g \in F$,*

$$\langle f, \mathrm{R}g \rangle = \begin{cases} 1, & f = g, \\ 0, & f \neq g \end{cases}$$

(i.e., $T^*_{F,2} T_{\mathrm{R}F,2}$ is the identity operator).

The next result is due to de Boor et al. (1994b) and Ron and Shen (1995). It was also established independently by Benedetto and Li (see Benedetto and Walnut (1994)). We use below the convention that

$$0/0 := 0.$$

Theorem 5.2.13 *Let $\phi \in L_2(\mathbb{R}^d)$ be given. Then:*

(a) *$E(\phi)$ is a Bessel system if and only if $\widetilde{\phi} \in L_\infty(\mathbb{R}^d)$. Moreover, $\|\mathcal{T}_{\phi,2}\| = \|\widetilde{\phi}\|_{L_\infty(\mathbb{R}^d)}$;*

(b) assume $E(\phi)$ is a Bessel system. Then $E(\phi)$ is a frame if and only if $1/\widetilde{\phi} \in L_\infty(\sigma(S_2(\phi)))$. Moreover, $\|\mathcal{T}_{\phi,2}{}^{-1}\| = \|1/\widetilde{\phi}\|_{L_\infty(\mathbb{R}^d)}$ (with $\mathcal{T}_{\phi,2}{}^{-1}$ the pseudo-inverse of $\mathcal{T}_{\phi,2}$);

(c) assume $E(\phi)$ is a frame. Then it is also a stable basis if and only if $\widetilde{\phi}$ vanishes almost nowhere, i.e., if and only if $S_2(\phi)$ is regular.

Proof (sketch) Choosing $\psi := \phi$ in (5.9), we obtain that, for $c \in \ell_2(\mathbb{Z}^d)$,

$$\|\mathcal{T}_\phi^* \mathcal{T}_\phi c\|_{\ell_2(\mathbb{Z}^d)} = (2\pi)^{-d/2} \|\widetilde{\phi}^2 \widehat{c}\|_{L_2(\mathbb{T}^d)}.$$

This yields $\|\mathcal{T}_\phi^* \mathcal{T}_\phi\| = \|\widetilde{\phi}^2\|_{L_\infty}$, and (a) follows.

For (b), assume that $1/\widetilde{\phi}$ is bounded on its support $\sigma(S_2(\phi))$, and define ψ by

$$\widehat{\psi} := \frac{\widehat{\phi}}{\widetilde{\phi}^2} = \frac{\widehat{\phi}}{[\widehat{\phi}, \widehat{\phi}]}.$$

Then ψ lies in $L_2(\mathbb{R}^d)$. Moreover, if $c \in \ell_2(\mathbb{Z}^d)$ and $\widehat{c}$ is supported in $\operatorname{supp} \widetilde{\phi}$, then

$$[\widehat{\psi}, \widehat{\phi}]\widehat{c} = \frac{[\widehat{\phi}, \widehat{\phi}]}{[\widehat{\phi}, \widehat{\phi}]}\widehat{c} = \widehat{c}.$$

In view of (5.9), this implies that $\widehat{\mathcal{T}_\phi^* \mathcal{T}_\psi}$ is a projector (whose range consists of all the periodic functions that are supported in $\sigma(S_2(\phi))$), hence that $E(\psi)$ is a system dual to $E(\phi)$. Also, $E(\phi)$ is a Bessel system by assumption, while for ψ we have that

$$[\widehat{\psi}, \widehat{\psi}] = \frac{[\widehat{\phi}, \widehat{\phi}]}{[\widehat{\phi}, \widehat{\phi}]^2} = \frac{1}{[\widehat{\phi}, \widehat{\phi}]} \in L_\infty.$$

Hence, by (a), $E(\psi)$ is a Bessel system, too. We conclude from (a) of Lemma 5.2.12 that $E(\phi)$ is a frame.

For the converse implication in (b), let $E(\psi)$ be a Bessel system that is dual to $E(\phi)$. Then, by (b) of Lemma 5.2.11, $[\widehat{\psi}, \widehat{\phi}] = 1$ on its support. On the other hand, by Cauchy–Schwarz,

$$[\widehat{\phi}, \widehat{\psi}]^2 \leq [\widehat{\phi}, \widehat{\phi}][\widehat{\psi}, \widehat{\psi}].$$

This shows that, on $\operatorname{supp}[\widehat{\psi}, \widehat{\phi}]$,

$$[\widehat{\phi}, \widehat{\phi}]^{-1} \leq [\widehat{\psi}, \widehat{\psi}].$$

The conclusion now follows from (a) and the fact that $E(\psi)$ is assumed to be Bessel.

As for (c), once $E(\phi)$ is known to be a frame, it is also a stable basis if and only if $\mathcal{T}_{\phi,2}$ is injective. In view of (a) of Lemma 5.2.11 (take $\psi := \phi$ there), and the definition of the spectrum of $S_2(\phi)$, that injectivity is equivalent to the non-vanishing a.e. of $\widetilde{\phi}$. □

Note that the next corollary applies, in particular, to any compactly supported $\phi \in L_2(\mathbb{R}^d)$.

Corollary 5.2.14 *Let $\phi \in L_2(\mathbb{R}^d)$. If $\widetilde{\phi}$ is continuous, then $E(\phi)$ is a frame (if and) only if it is a stable basis.*

Proof Since $\widetilde{\phi}$ is continuous and 2π-periodic, the function $1/\widetilde{\phi}$ can be bounded on its support only if it is bounded everywhere. Now apply Theorem 5.2.13. □

5.3 Beyond local PSI spaces

"Extending the theory of local PSI spaces" might be interpreted as one of the following two attempts. One direction is to extend the setup that is studied; another direction is to extend the tools that were developed. These two directions are clearly interrelated, but not necessarily identical.

When discussing more general setups, there are, again, several different, and quite complementary, generalizations. Once such extension concerns the application of the stability notion to p-norms, $p \neq 2$. This is the subject of Section 5.3.1.

Another extension is the study of the linear independence and the related notions in FSI spaces. This is the subject of Sections 5.3.2 and 5.3.3.

A third is the extension of the notions of stability and frames to FSI spaces. We will introduce in that context (Section 5.3.4) the general L_2-tools and briefly discuss the general approach in that direction. Starting with the bracket product, we will be led to the theory of *fiberization*, a theory that goes beyond FSI spaces, and goes even beyond general SI spaces.

5.3.1 L_p-stability in PSI spaces

We denote by
$$\mathcal{T}_{\phi,p}$$
the restriction of the synthesis operator $\mathcal{T}_\phi$ to $\ell_p(\mathbb{Z}^d)$.

Definition (*p*-Bessel systems and *p*-stable bases) Given $1 \leq p \leq \infty$, and $\phi \in L_p(\mathbb{R}^d)$, we say that $E(\phi)$ forms

(a) a *p-Bessel system*, if $\mathcal{T}_{\phi,p}$ is a well-defined bounded map into $L_p(\mathbb{R}^d)$,
(b) a *p-stable basis*, if $\mathcal{T}_{\phi,p}$ is bounded, injective and its range is closed in $L_p(\mathbb{R}^d)$.

We first discuss, in the next result, the p-Bessel property. That property is implied by mild decay conditions on ϕ. We provide *characterizations* for the 1- and ∞-Bessel properties, and a *sufficient condition* for the other cases. As to the proofs, the proof of the 1-case is straightforward, and that of the ∞-case involves routine arguments. The sufficient condition for the general p-Bessel property can be obtained easily from the discrete convolution inequality $\|a * b\|_{\ell_p} \leq \|a\|_{\ell_1} \|b\|_{\ell_p}$ and can also be obtained by interpolation between the $p = 1$ and the $p = \infty$ cases; it is due to Jia and Micchelli (1991). Note that for the case $p = 2$ the sufficient condition listed below is *not* equivalent to the characterization in Theorem 5.2.13.

The following spaces, which were introduced in Jia and Micchelli (1991), are useful here and later.

$$\mathcal{L}_p(\mathbb{R}^d) := \left\{ f \in L_p(\mathbb{R}^d) : \|\phi\|_{\mathcal{L}_p(\mathbb{R}^d)} := \left\| \sum_{j \in \mathbb{Z}^d} |\phi(\cdot + j)| \right\|_{L_p([0,1]^d)} < \infty \right\}. \tag{5.12}$$

Proposition 5.3.1

(a) $E(\phi)$ is 1-Bessel iff $\phi \in L_1(\mathbb{R}^d)$. Moreover, $\|\mathcal{T}_{\phi,1}\| = \|\phi\|_{L_1(\mathbb{R}^d)}$;
(b) $E(\phi)$ is ∞-Bessel iff $\phi \in \mathcal{L}_\infty(\mathbb{R}^d)$. Moreover, $\|\mathcal{T}_{\phi,\infty}\| = \|\phi\|_{\mathcal{L}_\infty(\mathbb{R}^d)}$;
(c) if $\phi \in \mathcal{L}_p(\mathbb{R}^d)$, $1 < p < \infty$, then $E(\phi)$ is p-Bessel and $\|\mathcal{T}_{\phi,p}\| \leq \|\phi\|_{\mathcal{L}_p(\mathbb{R}^d)}$.

We now discuss the p-stability property. A complete characterization of this stability property is known again for the case $p = \infty$ (in addition to the case $p = 2$ that was analysed in Theorem 5.2.13). We start with this case, which is due to Jia and Micchelli (1991).

Theorem 5.3.2 *Let $\phi \in \mathcal{L}_\infty(\mathbb{R}^d)$. Then the following conditions are equivalent:*

(a) $E(\phi)$ is ∞-stable;
(b) $\mathcal{T}_{\phi,\infty}$ is injective;
(c) $\widehat{\phi}$ does not have a real 2π-periodic zero.

Proof (sketch) The implication (a)$\Longrightarrow$(b) is trivial, while the proof of (b)$\Longrightarrow$(c) has already been outlined in Proposition 5.2.7.

(b)$\Longrightarrow$(a) (Ron (1991)). Assuming (a) is violated, we find sequences $(a_n)_n \subset \ell_\infty(\mathbb{Z}^d)$ such that $a_n(0) = 1 = \|a_n\|_{\ell_\infty(\mathbb{Z}^d)}$, all n, and such that $T_\phi a_n$ tends to 0 in $L_\infty(\mathbb{R}^d)$. Without loss, $(a_n)_n$ converges pointwise to a sequence a; necessarily $a \neq 0$. Since $(a_n)_n$ is bounded in $\ell_\infty(\mathbb{Z}^d)$, and since $\phi \in \mathcal{L}_\infty(\mathbb{R}^d)$, it follows that $T_\phi a_n$ converges pointwise a.e. to $T_\phi a$. Hence $T_\phi a = 0$, in contradiction to (b).

(c)$\Longrightarrow$(b): If (b) is violated, say, $T_\phi a = 0$, then $\widehat{a}\widehat{\phi} = 0$. Since $\widehat{a}$ is a pseudo-measure, and $\phi \in \mathcal{L}_\infty(\mathbb{R}^d) \subset L_1(\mathbb{R}^d)$, we conclude that $\widehat{a}$ is supported in the zero set of $\widehat{\phi}$. However, $\widehat{a}$ is periodic and non-zero, hence $\widehat{\phi}$ must have a periodic zero. $\square$

The reader should be warned that the above reduction of stability to injectivity is very much an L_∞-result. For example, for a univariate compactly supported bounded ϕ, T_ϕ is injective on $\ell_p(\mathbb{Z})$ for all $p < \infty$ (Ron (1989)), while certainly $E(\phi)$ may be unstable (in any chosen norm). On the other hand (as follows from some results in the sequel), assuming that ϕ lies in $\mathcal{L}_\infty(\mathbb{R}^d)$, the injectivity of $T_{\phi,\infty}$ characterizes the p-stability for all $1 \leq p \leq \infty$!

In order to investigate the stability property for other norms, we follow the approach of Jia and Micchelli (1991), assume that $\widetilde{\phi}^2 = [\widehat{\phi}, \widehat{\phi}]$ is bounded away from 0 (cf. Theorem 5.2.13), and consider the function g defined by its Fourier transform as

$$\widehat{g} := \frac{\widehat{\phi}}{\widetilde{\phi}^2}.$$

We have the following

Proposition 5.3.3 *Let $\phi, g \in \mathcal{L}_p(\mathbb{R}^d) \cap \mathcal{L}_{p'}(\mathbb{R}^d)$, $1 \leq p \leq \infty$, and p' be its conjugate. Assume that*

$$[\widehat{\phi}, \widehat{g}] = 1. \tag{5.13}$$

Then $E(\phi)$ is p-stable.

Proof (sketch) From Proposition 5.3.1 and the assumptions here, we conclude that $E(\phi)$ as well as $E(g)$ are p-, as well as p'-Bessel systems. Also Poisson's summation formula can be invoked to infer (from (5.13)) that

$$\langle \phi, E^j g \rangle = \delta_{j,0}, \quad j \in \mathbb{Z}^d,$$

i.e., that the shifts of g are biorthogonal to the shifts of ϕ.

Consider the operator

$$\mathcal{T}_{g,p}^* : L_p(\mathbb{R}^d) \to \ell_p(\mathbb{Z}^d) : f \mapsto (\langle f, E^j g \rangle)_{j \in \mathbb{Z}^d}.$$

Then, $\mathcal{T}_{g,p}^*$ is the adjoint of the operator $\mathcal{T}_{g,p'}$ (for $p = 1$ it is the restriction of the adjoint to $L_1(\mathbb{R}^d)$). Since $E(g)$ is p'-Bessel, it follows that $\mathcal{T}_{g,p}^*$ is bounded. On the other hand, $\mathcal{T}_{g,p}^* \mathcal{T}_{\phi,p}$ is the identity, hence $\mathcal{T}_{\phi,p}$ is boundedly invertible, i.e., $E(\phi)$ is p-stable. □

The following result is due to Jia and Micchelli (1991). We use in that result, for $1 \leq p < \infty$, the notation

$$S_p(\phi)$$

for the $L_p(\mathbb{R}^d)$-closure of the finite span of $E(\phi)$ (for $\phi \in L_p(\mathbb{R}^d)$).

Theorem 5.3.4 *Let $1 \leq p \leq \infty$ be given, let p' be its conjugate exponent, and assume that $\phi \in \mathcal{L} := \mathcal{L}_p(\mathbb{R}^d) \cap \mathcal{L}_{p'}(\mathbb{R}^d)$. Then:*

(a) *if $\widetilde{\phi}$ does not have (real) zeros, then $E(\phi)$ is p- and p'- stable. In this case, a generator g of a basis dual to $E(\phi)$ lies in $\mathcal{L}$, and, for $p < \infty$, the dual space $S_p(\phi)^*$ is isomorphic to $S_{p'}(\phi)$;*
(b) *if $\widetilde{\phi}$ has a 2π-periodic zero, then $E(\phi)$ is not p-stable.*

Proof (sketch) (a) Poisson's summation yields that the Fourier coefficients of $\widetilde{\phi}^2$ are the inner products $(\langle \phi, E^j \phi \rangle)_{j \in \mathbb{Z}^d}$. The assumption $\phi \in \mathcal{L}$ implies that $\phi \in \mathcal{L}_2(\mathbb{R}^d)$, and that latter condition implies that the above Fourier coefficients are summable, i.e., $\widetilde{\phi}^2$ lies in the Wiener algebra $\mathcal{A}(\mathbb{T}^d)$.

Now, if the continuous function $\widetilde{\phi}$ does not vanish, then, since $\phi \in L_2(\mathbb{R}^d)$, g defined by $\widehat{g} = \widehat{\phi}/\widetilde{\phi}^2$ is also in $L_2(\mathbb{R}^d)$, and we have $[\widehat{\phi}, \widehat{g}] = 1$. However, by Wiener's Lemma, $1/\widetilde{\phi}^2 \in \mathcal{A}(\mathbb{T}^d)$, too. This means that $g = \mathcal{T}_\phi a$ for some $a \in \ell_1(\mathbb{Z}^d)$. From that it follows that $g \in \mathcal{L}$ and is bounded, hence, by Proposition 5.3.3, $E(\phi)$ is p-stable. The p'-stability is obtained by symmetry, which directly implies, for $p < \infty$, that $S_p(\phi)^* = S_{p'}(\phi)$. Incidentally, we have proved that a dual basis of $E(\phi)$ lies, indeed, in $\mathcal{L}$.

We refer to Jia and Micchelli (1991) for the proof of (b). □

Proof of the implication (c) $\Longrightarrow$ (d) *in Proposition 5.2.7.* If ϕ decays rapidly, $\widehat{\phi}$ is infinitely differentiable. Furthermore, it vanishes nowhere in case $\widetilde{\phi}$ does not have a 2π-periodic zero. Thus, the Fourier coefficients a of $\widetilde{\phi}^{-2}$ are rapidly decaying, hence the function g defined by $\widehat{g} = \widehat{\phi}/\widetilde{\phi}^2$ is rapidly decaying, too. □

5.3.2 Local FSI spaces: Resolving linear dependence, injectability

An FSI space S is almost always given in terms of a generating set Φ for it. In many cases, the generating set has unfavorable properties. For example, $E(\Phi)$ might be linearly dependent in the sense that $\mathcal{T}_\Phi c = 0$, for some non-zero $c \in \mathcal{Q}(\Phi)$, i.e.,

$$\sum_{j \in \mathbb{Z}^d \times \Phi} c(j, \phi) \, E^j \phi = 0. \tag{5.14}$$

Theorem 5.2.8 provides a remedy to this situation for *univariate local* PSI spaces. If the compactly supported generator ϕ of S has linearly dependent shifts, we can replace it by another compactly supported generator, whose shifts are linearly *independent*.

The argument extends to univariate local FSI spaces, and that extension is presented later on. The essence of these techniques extend to spaces that are *not* local (see e.g., Plonka and Ron (2000). I should warn the reader that it may not be trivial to see the connection between the factorization techniques of Plonka and Ron (2000) and those that are discussed here; nonetheless, a solid connection does exist), but definitely not to shift-invariant spaces in *several* dimensions. The attempt to find an alternative method that is applicable in several dimensions will lead us, as is discussed near the end of this section, to the notion of *injectability*.

We start with the following result from de Boor et al. (1994b).

Lemma 5.3.5 *Let $S_2(\Phi)$ and $S_2(\Psi)$ be two local FSI spaces. Then the orthogonal projection of $S_2(\Psi)$ into $S_2(\Phi)$, as well as the orthogonal complement of this projection, are each local (FSI) spaces, i.e., each is generated by compactly supported functions.*

Proof (sketch) The key for the proof is the observation (see de Boor et al. (1994b)) that, given any compactly supported $f \in L_2(\mathbb{R}^d)$, and any FSI space S that is generated by a compactly supported vector $\Phi \subset L_2(\mathbb{R}^d)$, there exist trigonometric polynomials τ_f, $(\tau_\phi)_{\phi \in \Phi}$ such that

$$\tau_f \widehat{Pf} = \sum_{\phi \in \Phi} \tau_\phi \widehat{\phi}. \tag{5.15}$$

Here, Pf is the orthogonal projection of f on $S_2(\Phi)$.

Now let g be the inverse transform of $\tau_f \widehat{Pf}$. From (5.15) (and the fact that Φ is compactly supported) we get that g too is of compact support. From (b) of Theorem 5.2.10, it follows that $S_2(g) = S_2(Pf)$. Thus, $S_2(Pf)$ is a *local* PSI space. Varying f over Ψ, we get the result concerning projection.

When proving the claim concerning complements, we may assume without loss (in view of the first part of the proof) that $S_2(\Psi) \subset S_2(\Phi)$. Let now P denote the orthogonal projector onto $S_2(\Psi)$. Then, by (5.15), given ϕ in Φ, there exists a compactly supported g, and a trigonometric polynomial τ such that $\widehat{g} = \tau \widehat{P\phi}$. Thus also $\tau(\widehat{\phi} - \widehat{P\phi})$ is the transform of a compactly supported function g_1 (since ϕ is compactly supported by assumption). By (b) of Theorem 5.2.10, $S_2(g_1) = S_2(\phi - P\phi)$, and the desired result follows. □

The first part of the next result is taken from de Boor et al. (1994b). The second part is due to R.-Q. Jia.

Corollary 5.3.6

(a) *Every local FSI space $S_2(\Phi)$ is the orthogonal sum of local PSI spaces.*
(b) *Every local univariate shift-invariant space $S_\star(\Phi)$ is generated by a compactly supported Ψ whose shift set $E(\Psi)$ is linearly independent.*

Proof (sketch) Note that in part (a) we tacitly assume that $\Phi \subset L_2(\mathbb{R}^d)$. Let $f \in \Phi$. By Lemma 5.3.5, the orthogonal complement of $S_2(f)$ in $S_2(\Phi)$ is a *local* PSI space. Iterating with this argument, we obtain the result.

Part (b) now follows for $\Phi \subset L_2(\mathbb{R})$. We simply write then $S_2(\Phi)$ as an orthogonal sum of local PSI spaces, apply Theorem 5.2.8, and use the fact orthonormality implies linear independence. If Φ are merely distributions (still with compact support), then we can reduce this case to the former one by convolving Φ with a suitable smooth compactly supported mollifier. □

In more than one dimension, it is usually impossible to resolve the dependence relations of the shifts of Φ. This is already true in the case of a single generator (cf. the discussion in Section 5.2.3). One of the possible alternative approaches (which I learned from the work of Jia, see e.g., Jia (1998); the basic idea can already be found in the proof of Lemma 3.1 of Jia and Micchelli (1992)), is to embed the given SI space in a larger SI space that has generators with "better" properties.

Definition 5.3.7 Let Φ be a finite collection of compactly supported distributions. We say that the local FSI space $S_\star(\Phi)$ is *injectable* if there exists another finite set Φ_0 of compactly supported distributions so that

(a) $S_\star(\Phi) \subset S_\star(\Phi_0)$, and
(b) Φ_0 has linearly independent shifts.

The injectability assumption is quite mild. For example, if Φ consists of (compactly supported) *functions* then $S_\star(\Phi)$ is injectable. One can take Φ_0 to be any basis for the finite-dimensional space span$\{\chi E^j \phi : \phi \in \Phi, j \in \mathbb{Z}^d\}$, where χ is the support function of the unit cube.

At the time this chapter is being written, I am not entirely convinced that the injectability notion is the right one for the general studies of local FSI spaces, and for several reasons. First, if some of the entries of the compactly supported Φ are merely distributions, it is not clear how to inject $S_\star(\Phi)$ into a better space. Moreover, the above-mentioned canonical injection of a *function* Φ into $S_\star(\Phi_0)$ is not smoothness-preserving, i.e., while the entries of Φ may be smooth, the entries of Φ_0 are not expected to be so. We do not know of a general technique for smoothness-preserving injection.

My last comment in this context is about the actual notion of "linear independence". The analysis in Sections 5.2.2 and 5.2.3 provides ample evidence that this notion is the right one in the context of local *principal* shift-invariant spaces. The same cannot be said about local *finitely-generated* SI spaces, as the following example indicates.

Example Let $\Phi := \{\phi_1, \phi_2\}$ be a set of two compactly supported functions, and assume that $E(\Phi)$ is linear independent. Then, with f any finite linear combination of $E(\phi_1)$, the set $\{\phi_1, \phi_2 + f\}$ also has linearly independent shifts. However, we can select f to have very large support, hence to enforce large support on $\phi_2 + f$. Consequently, the generators of a linearly independent $E(\Phi)$ may have supports as large as one wishes, in stark contrast with the PSI space counterpart (Theorem 5.2.6).

Thus we need, in the context of FSI space theory, a notion that is somewhat stronger than linear independence, and that takes into account the support size of the various elements, as well as an effective characterization of this property, as effective perhaps as that of linear independence that appears in the next section.

5.3.3 Local FSI spaces: Linear independence

Despite of the reservations discussed in the previous section, linear independence is still a basic notion in the theory of local FSI spaces, and the injectability assumption provides one at times with a very effective tool. The current section is devoted to the study of the linear independence property via the injectability tool. The basic reference on this matter is Jia and Micchelli (1992). The results here are derived under the assumption that the FSI space $S_\star(\Phi)$ is injectable into the FSI space $S_\star(\Phi_0)$ (whose generators have linearly independent shifts). Recall from the last section that every *univariate* local FSI space is injectable (into itself), and that, in higher dimensions, local FSI spaces that are generated by compactly supported *functions* are injectable as well. It is very safe to conjecture that the results here are valid for spaces generated by compactly supported distributions, and it would be nice to find a neat way to close this small gap.

Thus, we are given a local FSI space $S_\star(\Phi)$, and assume that the space is injectable, i.e., it is a subspace of the local FSI space $S_\star(\Phi_0)$, and that Φ_0 has linear independent shifts. In view of Corollary 5.2.4, we conclude that there exists, for every $\phi \in \Phi$, a *finitely supported* $c_\phi \in \mathcal{Q}(\Phi_0)$, such that

$$\mathcal{T}_{\Phi_0} c_\phi = \phi.$$

We then create a matrix Γ, whose columns are the vectors c_ϕ, $\phi \in \Phi$ (thus the columns of Γ are indexed by Φ, the rows are indexed by Φ_0, and all the entries are finitely supported sequences defined on $\mathbb{Z}^d$).

It is useful to consider each of the above sequences as a (Laurent) polynomial, and to write the possible dependence relations among $E(\Phi)$ as formal power series. Let, thus, A be the space of formal power series in d variables, i.e., $a \in A$ has the form

$$a = \sum_{j \in \mathbb{Z}^d} a(j) X^j,$$

with X^j the formal monomial. Recall that

$$\mathbb{Z}^d \supset \operatorname{supp} a := \{j \in \mathbb{Z}^d : a(j) \neq 0\}.$$

Let

$$A_0$$

be the ring of all finitely supported d-variate power series (i.e., Laurent polynomials). Given a finite set Φ, let

$$A(\Phi)$$

be the free A_0-module consisting of $\#\Phi$ copies of A. We recall that the z-transform is the linear bijection

$$Z : \mathcal{Q} \to A : c \mapsto \sum_{j \in \mathbb{Z}^d} c(j) X^j.$$

Applying the z-transform (entry by entry) to our matrix Γ above, we obtain a matrix

$$M,$$

whose entries are in A_0. We consider this matrix M an A_0-homomorphism between the module $A(\Phi)$ and the module $A(\Phi_0)$, i.e.,

$$M \in \operatorname{Hom}_{A_0}(A(\Phi), A(\Phi_0)).$$

It is relatively easy then to conclude the following

Lemma 5.3.8 *In the above notation,*

(a) *if $E(\Phi)$ is linearly independent, then M is injective;*
(b) *the converse is true, too, provided that $E(\Phi_0)$ is linearly independent.*

Thus, the characterization of linear independence in (injectable) local FSI spaces is reduced to characterizing injectivity in $\operatorname{Hom}_{A_0}(A(\Phi), A(\Phi_0))$. We provide below the relevant result, which can be viewed either as a spectral analysis result in $\operatorname{Hom}_{A_0}(A(\Phi), A(\Phi_0))$, or as an extension of the *Nullstellensatz* to modules. The following result is due to Jia and Micchelli (1992). The Jia–Micchelli proof reduces the statement in the theorem below to the case studied in Theorem 5.2.1 by a tricky Gauss elimination argument. The proof provided here is somewhat different, and employs the Quillen–Suslin Theorem, Quillen (1976), Suslin (1976).

Discussion: The Quillen–Suslin Theorem We briefly explain the relevance of this theorem to our present setup. The Quillen–Suslin Theorem affirms a famous conjecture of J.P Serre (see Lam (1978)) that every projective module over a polynomial ring is free. The extension of that result to Laurent polynomial rings is mentioned in Suslin's paper, and was proved by Swan. A simple consequence of that theorem is that every row $(w_1, \ldots, w_m)$ of Laurent polynomials that do not have a common zero in $(\mathbb{C}\backslash 0)^d$, can be extended to a square A_0-valued matrix W which is non-singular everywhere, i.e., $W(\xi)$ is non-singular for every $\xi \in (\mathbb{C}\backslash 0)^d$. A very nice discussion of the above, together with a few more references, can be found in Jia and Micchelli (1991). $\square$

The Nullstellensatz for Free Modules *Let A be the space of formal power series in d-variables. Let*

$$A_0$$

be the ring of all finitely supported d-variate power series. Given a positive integer n, let

$$A^n$$

be the free A_0-module consisting of n copies of A. Let

$$M_{m \times n} \in \operatorname{Hom}_{A_0}(A^n, A^m)$$

be an A_0-valued matrix. Then M is injective if and only if there does not exist $\xi \in (\mathbb{C}\backslash 0)^d$ for which $\operatorname{rank} M(\xi) < n$.

Here, $M(\xi)$ is the constant-coefficient matrix obtained by evaluating each entry of M at ξ.

Proof (sketch) The "only if" follows immediately from the fact that, for every ξ as above, there exists a_ξ in A such that $a_0 a_\xi = a_0(\xi)$, for every $a_0 \in A_0$. Indeed, if $M(\xi)$ is rank-deficient, we can find a vector in $c \in \mathbb{C}^n \backslash 0$ such that $M(\xi) c = 0$, and we get that $a_\xi c \in \ker M$.

We prove the converse by induction on n. For $n = 1$, we let I be the ideal in A_0 generated by the entries of the (single) column of M. If $a \in \ker M \subset A^1 = A$, then $a_0 a = 0$ for every $a_0 \in I$. By an argument identical to that used in the proof of Theorem 5.2.1, we conclude that, if $\ker M \neq 0$, then all the polynomials in I must vanish at a point $\xi \in (\mathbb{C}\backslash 0)^d$, hence $M(\xi) = 0$.

So assume that $n > 1$, and that $a \in \ker M \backslash 0$. We may assume without loss that the entries of the first column of M do not have a common zero $\xi \in (\mathbb{C}\backslash 0)^d$ (otherwise, we obviously have that $\operatorname{rank} M(\xi) < n$). Thus, by the classical *Nullstellensatz*, we can form a combination of the rows of M, with coefficients w_i in A_0, so that the resulting row u has the constant 1 in its first entry. Then the entries w_i cannot have a common zero in $(\mathbb{C}\backslash 0)^d$, and therefore (see Discussion 5.3.3) the row vector $w := (w_i)$ can be extended to an $m \times m$ A_0-valued matrix W that is non-singular at every $\xi \in (\mathbb{C}\backslash 0)^d$. Set $M_1 := WM$.

Since the $(1,1)$th entry of M_1 is the constant 1, we can and do use Gauss elimination to eliminate all the entries in the first column (while preserving, for every $\xi \in (\mathbb{C}\backslash 0)^d$, the rank of $M_1(\xi)$). From the resulting matrix, remove its first row and its first column, and denote the matrix so obtained by M_2, which has $n - 1$ columns. Also, since $\ker M \subset \ker M_1$, we conclude that $\ker M_2 \neq \{0\}$ (since otherwise $\ker M_1$ contains an element whose only non-zero entry is the first one, which is absurd, since the $(1,1)$th entry of M_1 is 1). Thus, by the induction hypothesis, there exists $\xi \in (\mathbb{C}\backslash 0)^d$ such that $\operatorname{rank} M_2(\xi) < n - 1$. It then easily follows that $\operatorname{rank} M_1(\xi) < n$, and since $W(\xi)$ is non-singular, it must be that $\operatorname{rank} M(\xi) < n$, as claimed. □

By converting back the above result to the language of shift-invariant spaces, we get the following result, Jia and Micchelli (1992). Note that the result is not a complete extension of Theorem 5.2.2, due to the injectability assumption here.

Theorem 5.3.9 *Let Φ be a finite set of compactly supported distributions,* and assume that $S_\star(\Phi)$ is injectable. Then $E(\Phi)$ is linearly dependent if and only if there exists a linear combination $\phi_\star$ of Φ for which $E(\phi_\star)$ is linearly dependent, too.

5.3.4 L_2-stability and frames in FSI spaces, fiberization

One of the main results in the theory of local SI spaces is the characterization of linear independence. A seemingly inefficient way of stating the PSI case (Theorem 5.2.2) of this result is as follows. Let ϕ be a compactly supported distribution, $\widehat{\phi}$ its Fourier transform. Given $\omega \in \mathbb{C}^d$, let C_ω be the one-dimensional subspace of $\mathcal{Q}$ spanned by the sequence

$$\phi_\omega : 2\pi\mathbb{Z}^d \to \mathbb{C} : \alpha \mapsto \widehat{\phi}(\omega + \alpha). \tag{5.16}$$

Let G_ω be the map

$$G_\omega : \mathbb{C} \to C_\omega : c \mapsto c\phi_\omega.$$

Then $\mathcal{T}_\phi$ is injective (i.e., $E(\phi)$ is linearly independent) if and only if each of the maps G_ω is injective'.

Armed with this new perspective of the linear independence characterization, we can now find with ease a similar form for the characterization of the linear independence in the FSI space setup. We just need to change the nature of the "fiber" spaces C_ω. Given a finite vector of compactly supported functions Φ, and given $\omega \in \mathbb{C}^d$, we define (cf. (5.16))

$$C_\omega := \operatorname{span}\{\phi_\omega : \phi \in \Phi\},$$

and

$$G_\omega : \mathbb{C}^\Phi \to C_\omega : c \mapsto \sum_{\phi \in \Phi} c(\phi)\phi_\omega. \tag{5.17}$$

Then, the characterization of linear independence for local FSI spaces, Theorem 5.3.9 (when combined with Theorem 5.2.2), says that $\mathcal{T}_\Phi$ is injective if and only if each G_ω, $\omega \in \mathbb{C}^d$, is injective.

The discussion above represents a general principle that turned out to be a powerful tool in the context of shift-invariant spaces: *fiberization*. Here, one is given an operator and is interested in a certain property of the operator T (e.g., its injectivity or its boundedness). The goal of fiberization is to associate the operator with a large collection of much simpler operators G_ω (=: fibers), to associate each one of them with an analogous property P_ω, and to prove that T satisfies P iff each G_ω satisfies the property P_ω (sometimes in some uniform way).

The idea of fiberization appears implicitly in many papers on SI spaces from the early 90s (e.g., Jia and Micchelli (1991), Jia and Micchelli (1992), de Boor et al. (1994b)). It was formalized first in Ron and Shen (1995), and was applied in Ron and Shen (1997a) to Weyl–Heisenberg systems, and in Ron and Shen (1997b) to wavelet systems. We refer to Ron (1998) for more details and references.

In principle, the fiberization techniques of Ron and Shen (1995) apply to the operator $T_\Phi T_\Phi^*$, for some (finite or countable) $\Phi \subset L_2(\mathbb{R}^d)$, as well as to the operator $T_\Phi^* T_\Phi$. The first approach is *dual Gramian analysis*, while the second is *Gramian analysis*. We provide in this section a brief introduction to the latter, by describing its roots in the context of the FSI space $S_2(\Phi)$.

We start our discussion with the **Gramian matrix** $G := G_\Phi$ which is the analog of the function $\widetilde{\phi}^2$ (cf. (5.8)). The Gramian is an $L_2(\mathbb{T}^d)$-valued matrix indexed by $\Phi \times \Phi$, and its (φ, ϕ)th entry is

$$[\widehat{\phi}, \widehat{\varphi}](\omega) = \sum_{\alpha \in 2\pi \mathbb{Z}^d} \widehat{\phi}(\omega + \alpha) \overline{\widehat{\varphi}(\omega + \alpha)}.$$

In analogy to the PSI case (cf. the proof of Theorem 5.2.13), a dual basis for Φ may be given by the functions whose Fourier transforms are

$$G^{-1} \widehat{\Phi},$$

provided that the above expression represents well-defined functions.

In case $\Phi \subset \mathcal{L}_2(\mathbb{R}^d)$, the entries of G, hence the determinant $\det G$, all lie in the Wiener algebra $\mathcal{A}(\mathbb{R}^d)$. If $\det G$ vanishes nowhere, one finds that the functions whose Fourier transforms are given by $G^{-1}\widehat{\Phi}$ can each be written as $T_\Phi c$, for some $c \in \ell_1(\mathbb{Z}^d) \times \Phi$. The functions obtained in this way thus lie in $L_2(\mathbb{R}^d)$, and in this way, Jia and Micchelli (1991) extend Theorem 5.3.4 to FSI spaces. The crucial PSI condition (that $\widehat{\phi}$ vanishes nowhere) is replaced by the condition that the Gramian is non-singular everywhere. That non-singularity is equivalent to the injectivity of the map G_ω (cf. (5.17)) for every *real* $\omega \in \mathbb{R}^d$.

We wish to discuss in more detail the notions of L_2-stability and frames for general Φ in $L_2(\mathbb{R}^d)$. In that case the entries of G, hence its determinant, may not be continuous. The extension of the L_2-results to FSI spaces cannot make use of the mere non-singularity of G (on $\mathbb{R}^d$). Instead, one inspects the norms of the operators

$$G_\omega : \ell_2(\Phi) \to \ell_2(\Phi) : v \mapsto G(\omega)v.$$

We also recall the notion of a **pseudo-inverse**. For a linear operator L on a finite-dimensional (inner product) space, the pseudo-inverse L^{-1} of L is the unique linear map for which $L^{-1}L$ is the orthogonal projector with kernel ker L. If L is non-negative Hermitian (as is G_ω), then $\|L^{-1}\| = 1/\lambda_+$, with λ_+ the smallest non-zero eigenvalue of L.

The result stated below was established in de Boor et al. (1994b) (stability characterization) and in Ron and Shen (1995) (frame characterization). The reference de Boor et al. (1994b) contains a characterization of the so-called *quasi-stability* which is a slightly stronger notion than that of a frame (and which coincides with the frame notion in the PSI case).

In the statement of the result below, we use the norm functions

$$\mathcal{G} : \omega \mapsto \|G_\omega\|,$$

and

$$\mathcal{G}^{-1} : \omega \mapsto \|G_\omega{}^{-1}\|.$$

Recall that $G_\omega{}^{-1}$ is a pseudo-inverse, hence is always well-defined.

Theorem 5.3.10 *Let Φ be a finite vector of $L_2(\mathbb{R}^d)$-functions. Then*

(a) *$E(\Phi)$ is a Bessel system (i.e., $\mathcal{T}_{\Phi,2}$ is bounded) iff $\mathcal{G} \in L_\infty(\mathbb{R}^d)$. Moreover, $\|\mathcal{T}_{\Phi,2}\|^2 = \|\mathcal{G}\|_{L_\infty}$;*
(b) *assume that $E(\Phi)$ is a Bessel system. Then $E(\Phi)$ is a frame iff $\mathcal{G}^{-1} \in L_\infty(\mathbb{R}^d)$. Moreover, the square of the lower frame bound (cf. (5.10)) is then $1/\|\mathcal{G}^{-1}\|_{L_\infty}$;*
(c) *$E(\Phi)$ is stable iff it is a frame and in addition $S_2(\Phi)$ is regular, i.e., G_ω is non-singular a.e.*

5.4 Refinable shift-invariant spaces

Refinable shift-invariant spaces are used in the construction of wavelet systems via the vehicle of *multiresolution analysis*. It is beyond the scope of this article to review, to any extent, the rich connections between shift-invariant space theory on the one hand and refinable spaces (and wavelets) on the other hand. I refer to Ron (1998) for some discussion of these connections.

Definition 5.4.1 *Let N be a positive integer. A compactly supported distribution $\phi \in \mathcal{D}'(\mathbb{R}^d)$ is called N-refinable if $\phi(\cdot/N) \in S_\star(\phi)$.*

Another possible definition of refinability is given on the Fourier domain. An $L_2(\mathbb{R}^d)$-function ϕ is **refinable** if there exists a bounded 2π-periodic τ such that, a.e. on $\mathbb{R}^d$,

$$\widehat{\phi}(N\omega) = \tau(\omega)\widehat{\phi}(\omega).$$

The definitions are not equivalent, but are closely related and are both used in the literature. We will be primarily interested in the case of a univariate compactly supported $L_2(\mathbb{R})$-function ϕ with globally linearly independent shifts. For such a case, the above two definitions coincide, and the **mask** function τ is a trigonometric polynomial.

Our discussion here is divided into two parts. In Section 5.4.1, we present a remarkable property of 2-refinable univariate local PSI spaces. For such spaces, the basic property of global linear independence (which can always be achieved by a suitable selection of the generator of the space, cf. Theorem 5.2.8) implies the much stronger property of *local* linear independence. Unfortunately, this result does not extend to *any* more general setup.

A major problem in the context of refinable functions is the identification of their properties by a mere inspection of the mask function. While we do not attempt to address that topic (this article is devoted exclusively to the intrinsic properties of SI spaces), we give, in Section 5.4.2, a single example that shows how the basic tools and results about SI spaces help in the study of that problem.

5.4.1 Local linear independence in univariate refinable PSI spaces

A strong independence relation is that of *local linear independence* (cf. Section 5.2.3). It is well known that univariate polynomial B-splines satisfy this property. On the other hand, the support function of the interval $[0, 1.5]$ is an example of a function whose shifts are (gli), hence (wlli), but are not (slli), and thus, local linear independence is properly stronger than its global counterpart, even in the univariate context. It is then remarkable to note that, for a 2-*refinable* univariate compactly supported ϕ, global independence and local independence are equivalent.

The theorem below is due to Lemarie-Rieusset and Malgouyres (1991). For a function ϕ with *orthonormal* shifts, it was proved before by Meyer (1991).

Theorem 5.4.2 *Let ϕ be a univariate refinable function whose shifts satisfy the local spanning property (ls) (cf. Theorem 5.2.6). Then, $E(\phi)$ is locally linearly independent.*

Proof By Theorem 5.2.8, the local spanning property is equivalent to global linear independence, and we will use that latter property in the proof.

It will be convenient during the proof to use an alternative notation for the synthesis operator $\mathcal{T}_\phi$. Thus, we set

$$\phi*' : \mathcal{Q} \to S_\star(\phi) : c \mapsto \mathcal{T}_\phi c,$$

i.e., $\phi*' := \mathcal{T}_\phi$.

We assume that ϕ is supported in $[0, N]$ and that $\psi = \phi *' a$, with the sequence a supported in $\{0, 1, 2, ..., N\}$ (and, thus, ψ is supported in $[0, 2N]$); we also assume that the shifts of ϕ are *locally* linearly independent over some interval $[0, k]$, and that the even shifts of ψ are (globally) linearly independent. Under all these assumptions, we prove that the even shifts of ψ are locally linearly independent over $[0, k]$, as well.

The theorem will follow from the above. For $\psi := \phi(\cdot/2)$, the global linear independence of $E(\phi)$ is equivalent to the global linear independence of the even shifts of ψ. Thus, assuming the shifts of ϕ are locally independent over $[0, k]$, the above claim (once proved) would imply that the even shifts of ψ are locally independent over that set also, and this amounts to the local independence of the shifts of ϕ over $[0, k/2]$. Starting with $k := N - 1$ (i.e., invoking Proposition 5.2.9), we can then proceed until the interval is as small as we wish.

Let $f = \sum_{j \in \mathbb{Z}} b'(2j)\psi(\cdot - 2j)$. Assuming f to vanish on $[0, k]$, we want to show that $b'(2j) = 0$, $-N < j < k/2$. Since $\psi \in S_\star(\phi)$, $f \in S_\star(\phi)$. Thus, $f = \phi *' c$. The local linear independence of $E(\phi)$ over $[0, k]$ implies that $c(-N+1) = c(-N+2) = \ldots = c(k-1) = 0$. We define

$$f_1 := \sum_{j \leq -N} c(j)\phi(\cdot - j) =: \phi *' c_1, \quad f_2 := \sum_{j \geq k} c(j)\phi(\cdot - j) =: \phi *' c_2.$$

Since c_2 vanishes on $j < k$, (and assuming without loss that $a(0) \neq 0$), we can find a sequence b_2 supported also on $j \geq k$ such that $c_2 = a * b_2$. Then,

$$f_2 = \phi *' c_2 = \phi *' (a * b_2) = (\phi *' a) *' b_2 = \psi *' b_2.$$

By the same argument (and assuming $a(N) \neq 0$), we can find a sequence b_1 supported on $\{-2N, -2N-1, ...\}$ such that $c_1 = a * b_1$; hence, as before,

$$f_1 = \psi *' b_1.$$

Thus we have found that $f = \psi *' (b_1 + b_2)$, with $b := b_1 + b_2$ vanishing on $-2N < j < k$. Since also $f = \psi *' b'$, we conclude that $\lambda := b - b'$ lies in $\ker \psi*'$. This leads to

$$0 = \psi *' \lambda = (\phi *' a) *' \lambda = \phi *' (a * \lambda).$$

Since $E(\phi)$ is linearly independent, $a * \lambda = 0$. Further, λ vanishes at all odd integers in the interval $(-2N, k)$. If λ vanishes also at all even integers in that interval, so does b' and we are done, since this is exactly what we sought to prove. Otherwise, since $\dim \ker a* = N$, and the interval $(-2N, k)$ contains at least N consecutive odd integers we must have (cf. Lemma 5.4.3 below) $\theta \in \mathbb{C}\backslash 0$ such that $\pm\theta \in \operatorname{spec}(a*)$, i.e., such that the two exponential sequences

$$\mu_1 : j \mapsto \theta^j, \quad \mu_2 : j \mapsto (-\theta)^j$$

lie in $\ker(a*)$. But, then, $\mu := \mu_1 + \mu_2 \in \ker(a*)$, and is supported only on the even integers. Since $\psi *' \mu = \phi *' (a * \mu) = 0$, this contradicts the global linear independence of the 2-shifts of ψ. □

Lemma 5.4.3 *Let $a : \mathbb{Z} \to \mathbb{C}$ be a sequence supported on $[0, N]$. If $0 \neq \lambda \in \ker(a*)$, and λ vanishes at N consecutive even (odd) integers, then there exists $\theta \in \mathbb{C}\backslash 0$ such that the sequences $j \mapsto \theta^j$, and $j \mapsto (-\theta)^j$ both lie in $\ker(a*)$, i.e., $\{\pm\theta\} \subset \operatorname{spec}(a*)$.*

Proof (sketch) We know $\dim \ker(a*) \leq N$. Let $\theta \in \operatorname{spec}(a*)$, and assume that $-\theta$ is not there. Then there exists a difference operator T supported on $N-1$ consecutive even points that maps $\ker(a*)$ onto the one-dimensional span of $j \mapsto \theta^j$. Since $T\lambda$ vanishes at least at one point, it must be that λ lies in the span of the other exponentials in $\ker(a*)$. □

Thus, for a univariate 2-refinable compactly supported function, we have the following remarkable result (compare with Theorem 5.2.6).

Corollary 5.4.4 *Let ϕ be a univariate 2-refinable compactly supported function. Then the properties (slli), (gli), (ldb), (ls) and (ms) are all equivalent for this ϕ.*

Theorem 5.4.2 does not extend to generators that are refinable by a dilation factor $N \neq 2$. To see that, consider, for any integer $N \geq 2$, the refinable function ϕ_N defined as

$$\widehat{\phi}_N(N\omega) = \tau_N(\omega)\widehat{\phi}_N(\omega),$$

with the Fourier coefficients $t_N(k)$ of the 2π-periodic trigonometric polynomial τ_N defined by

$$t_N(k) := \begin{cases} 1/2, & k \in \{0, \ldots, 2(N-1)\}\backslash\{N-1\}, \\ 1, & k = N-1, \\ 0, & \text{otherwise.} \end{cases}$$

The resulting refinable ϕ is supported in $[0,2]$, has globally linearly independent shifts, and has linearly *dependent* shifts on the interval $(\frac{1}{N}, \frac{N-1}{N})$, which is non-empty for every $N \geq 3$. The case $N = 3$ appears in Dai et al. (1996).

5.4.2 The simplest application of SI theory to refinable functions

We close this chapter with an example that shows how general SI theory may be applied in the study of refinable spaces. The example is taken from Ron (1992b).

Suppose that ϕ is a univariate, compactly supported, N-refinable distribution with trigonometric polynomial mask τ. Suppose that we would like, by inspecting τ only, to determine whether the shifts of ϕ are linearly independent. We can invoke to this end Theorem 5.2.8. By this theorem, there exists $\phi_0 \in S_\star(\phi)$, such that (i) $\phi = \mathcal{T}_{\phi_0} c$, for some finitely supported c (defined on $\mathbb{Z}$), and (ii) $E(\phi_0)$ is linearly independent. One then easily conclude that (i) ϕ_0 is also refinable, with a trigonometric polynomial mask t, and (ii)

$$\tau = t \frac{\widehat{c}(N\cdot)}{\widehat{c}}. \tag{5.18}$$

This leads to a characterization of the linear independence property of the univariate $E(\phi)$ in terms of the non-factorability of τ in the form (5.18), Ron (1992b). That characterization leads then easily to the characterization of the linear independence property in terms of the distribution of the zeros of τ (Jia and Wang (1993), Ron (1992b), Zhou (1996)).

Acknowledgments

I am indebted to Carl de Boor for his critical reading of this article, which yielded many improvements including a shorter proof for Proposition 5.2.9. This work was supported by the National Science Foundation under Grants DMS-9626319 and DMS-9872890, by the US Army Research Office under Contract DAAG55-98-1-0443, and by the National Institute of Health.

References

Ben-Artzi, A. and Ron, A. (1990). On the integer translates of a compactly supported function: dual bases and linear projectors. *SIAM J. Math. Anal.*, **21**, 1550–1562.

Benedetto, J.J. and Walnut, D.D. (1994). Gabor frames for L^2 and related spaces. In *Wavelets: Mathematics and Applications*, ed. J. Benedetto and M. Frazier, pp. 97–162. CRC Press, Boca Raton, Florida.

Boor, C. de, DeVore, R. and Ron, A. (1994a). Approximation from shift-invariant subspaces of $L_2(\mathbb{R}^d)$. *Trans. Amer. Math. Soc.*, **341**, 787–806.

Boor, C. de, DeVore, R. and Ron, A. (1994b). The structure of finitely generated shift-invariant spaces in $L_2(\mathbb{R}^d)$. *J. Funct. Anal.*, **119**, 37–78.

Boor, C. de, DeVore, R. and Ron, A. (1998). Approximation orders of FSI spaces in $L_2(\mathbb{R}^d)$. *Constr. Approx.*, **14**, 631–652.

Boor, C. de, Höllig, K. and Riemenschneider, S.D. (1993). *Box Splines*, xvii + 200p. Springer-Verlag, New York.

Boor, C. de and Ron, A. (1989). Polynomial ideals and multivariate splines. In *Multivariate Approximation Theory IV, ISNM 90*, ed. C. Chui, W. Schempp and K. Zeller, pp. 31–40. Birkhäuser Verlag, Basel.

Boor, C. de and Ron, A. (1992). The exponentials in the span of the multi-integer translates of a compactly supported function. *J. London. Math. Soc.*, **45**, 519–535.

Buhmann, M.D. (1993). New developments in the theory of radial basis function interpolation. In *Multivariate Approximation: From CAGD to Wavelets*, ed. K. Jetter and F. Utreras, pp. 35–75. World Scientific Publishing, Singapore.

Cavaretta, A.S., Dahmen, W. and Micchelli, C.A. (1991). *Stationary Subdivision*, Mem. Amer. Math. Soc. **93**, No. 453.

Chui, C.K. and Ron, A. (1991). On the convolution of a box spline with a compactly supported distribution: linear independence for the integer translates. *Canad. J. Math.*, **43**, 19–33.

Dahmen, W. and Micchelli, C.A. (1983). Translates of multivariate splines. *Linear Algebra Appl.*, **52**, 217–234.

Dai, X., Huang, D. and Sun, Q. (1996). Some properties of five-coefficient refinement equation. *Arch. Math.*, **66**, 299–309.

Daubechies, I. (1992). *Ten Lectures on Wavelets*. CBMS Conf. Series in Appl. Math., vol. 61, SIAM, Philadelphia.

Dyn, N. (1992). Subdivision schemes in CAGD. In *Advances in Numerical Analysis Vol. II: Wavelets, Subdivision Algorithms and Radial Basis Functions*, ed. W.A. Light, pp. 36–104. Oxford University Press, Oxford, 1992.

Dyn, N. and Ron, A. (1990). Cardinal translation invariant Tchebycheffian B-splines. *Approx. Theory Appl.*, **6**, 1–12.

Dyn, N. and Ron, A. (1995). Radial basis function approximation: from gridded centers to scattered centers. *Proc. London Math. Soc.*, **71**, 76–108.

Feichtinger, H.G. and Strohmer, T. (1997). *Gabor Analysis and Algorithms: Theory and Applications*, 500p. Birkhauser, Boston.

Jia, R.-Q. (1993). Multivariate discrete splines and linear diophantine equations. *Trans. Amer. Math. Soc.*, **340**, 179–198.

Jia, R.-Q. (1998). Shift-invariant spaces and linear operator equations. *Israel J. Math.*, **103**, 259–288.

Jia, R.-Q. and Micchelli, C.A. (1991). Using the refinement equations for the construction of pre-wavelets II: powers of two. In *Curves and Surfaces*, ed. P.-J. Laurent, A. LeMéhauté, and L.L. Schumaker, pp. 209–246. Academic Press, New York.

Jia, R.-Q. and Micchelli, C.A. (1992). On linear independence for integer translates of a finite number of functions. *Proc. Edinburgh Math. Soc.*, **36**, 69–85.

Jia, R.-Q. and Sivakumar, N. (1990). On the linear independence of integer translates of box splines with rational directions. *Linear Algebra Appl.*, **135**, 19–31.

Jia, R.-Q. and Wang, J. (1993). Stability and linear independence associated with wavelet decompositions. *Proc. Amer. Math. Soc.*, **117**, 1115–1124.

Lam, T.-Y. (1978). *Serre's Conjecture*. Lecture Notes in Math., Vol. 635, Springer-Verlag, Berlin.

Lefranc, M. (1958). Analyse spectrale sur Z_n. *C. R. Acad. Sci. Paris*, **246**, 1951–1953.

Lemarie-Rieusset, P.-G. and Malgouyres, G. (1991). Support des fonctions de base dans une analyse multiresolution. *C. R. Acad. Sci. Paris*, **313**, 377–380.

Marks, R.J. (1993). *Advanced Topics in Shannon Sampling and Interpolation Theory*, 360p. Springer Texts in Electrical Engineering, Springer-Verlag, New York.

Meyer, Y. (1991). The restrictions to $[0,1]$ of the $\phi(x-k)$ are linearly independent. *Rev. Mat. Iberoamericana*, **7**, 115–133.

Meyer, Y. (1992). *Wavelets and Operators*. Cambridge University Press, Cambridge.

Plonka, G. and Ron, A. (2000). A new factorization technique of the matrix mask of univariate refinable functions. *Numer. Math.*, to appear.

Quillen, D. (1976). Projective modules over polynomial rings. *Invent. Math.*, **36**, 167–171.

Ron, A. (1989). A necessary and sufficient condition for the linear independence of the integer translates of a compactly supported distribution. *Constr. Approx.*, **5**, 297–308.

Ron, A. (1990). Factorization theorems of univariate splines on regular grids. *Israel J. Math.*, **70**, 48–68.

Ron, A. (1991). *Lecture Notes on Shift-Invariant Spaces*. Math. 887, U. Wisconsin, Madison.

Ron, A. (1992a). Remarks on the linear independence of integer translates of exponential box splines. *J. Approx. Theory*, **71**, 61–66.

Ron, A. (1992b). Characterizations of linear independence and stability of the shifts of a univariate refinable function in terms of its refinement mask. CMS TSR #93-3, U. Wisconsin, Madison.

Ron, A. (1993). Shift-invariant spaces generated by an exponentially decaying function: linear independence. Manuscript.

Ron, A. (1998). Wavelets and their associated operators. In *Approximation Theory IX, Vol. 2: Computational Aspects*, ed. C.K. Chui and L.L. Schumaker, pp. 283–317. Vanderbilt University Press, Nashville TN.

Ron, A. and Shen, Z. (1995). Frames and stable bases for shift-invariant subspaces of $L_2(\mathbb{R}^d)$. *Canad. J. Math.*, **47**, 1051–1094.

Ron, A. and Shen, Z. (1997a). Weyl–Heisenberg frames and Riesz bases in $L_2(\mathbb{R}^d)$. *Duke Math. J.*, **89**, 237–282.

Ron, A. and Shen, Z. (1997b). Affine systems in $L_2(\mathbb{R}^d)$: the analysis of the analysis operator. *J. Funct. Anal.*, **148**, 408–447.

Sivakumar, N. (1991). Concerning the linear dependence of integer translates of exponential box splines. *J. Approx. Theory*, **64**, 95–118.

Strang, G. and Fix, G. (1973). A Fourier analysis of the finite element variational method. In *Constructive Aspects of Functional Analysis*, ed. G. Geymonat, pp. 793–840. CIME II Ciclo 1971.

Sun, Q. (1993). A note on the integer translates of a compactly supported distribution on $\mathbb{R}$. *Arch. Math.*, **60**, 359–363.

Suslin, A.A. (1976). Projective modules over polynomial rings are free. *Soviet Math. Dokl.*, **17**, 1160–1164.

Trèves, F. (1967). *Topological Vector Spaces, Distributions and Kernels*. Academic Press, New York.

Zhao, K. (1992). Global linear independence and finitely supported dual basis. *SIAM J. Math. Anal.*, **23**, 1352–1355.

Zhou, D.-X. (1996). Stability of refinable functions, multiresolution analysis, and Haar bases. *SIAM J. Math. Anal.*, **27**, 891–904.

6

Theory and algorithms for nonuniform spline wavelets

T. LYCHE, K. MØRKEN and E. QUAK

Abstract

We investigate mutually orthogonal spline wavelet spaces on nonuniform partitions of a bounded interval, addressing the existence, uniqueness and construction of bases of minimally supported spline wavelets. The relevant algorithms for decomposition and reconstruction are considered as well as some stability-related questions. In addition, we briefly review the bivariate case for tensor products and arbitrary triangulations. We conclude this chapter with a discussion of some special cases.

6.1 Introduction

Splines have become the standard mathematical tool for representing smooth shapes in computer graphics and geometric modeling. Wavelets have been introduced more recently, but are by now well established both in mathematics and in applied sciences like signal processing and numerical analysis. The two concepts are closely related as splines provide some of the most important examples of wavelets. Although there is an extensive literature on cardinal spline wavelets (spline wavelets with uniform knot spacing), see Chui (1992), relatively little has been published about spline wavelets on arbitrary, nonuniform knots, which form the subject of this chapter. These kinds of wavelets, however, are needed for performing operations like decomposition, reconstruction and thresholding on splines given on a nonuniform knot vector, which typically occur in practical applications.

The flexibility of splines in modeling is due to good approximation properties, useful geometric interpretations of the B-spline coefficients, and simple algorithms for adding and removing knots. Full advantage of these capabilities can only be taken on general nonuniform knot vectors, where also multiple knots are allowed. In fact, the spline algorithms for general knots are hardly more complicated than the special ones for uniform knots. We will see in this chapter that the same is true for B-wavelets, which are spline wavelets of minimal support. Unlike other types of wavelet functions, spline B-wavelets are given as explicit expressions, namely as rational functions of the knots.

The classical construction of wavelets takes place on the whole real line, with an infinite, uniform grid, but many applications require wavelets on a bounded interval. In the early days of wavelet theory this problem was solved either by making the given data periodic or by extending the data to the whole real line, setting the data to be zero outside the interval or using reflection at the boundaries, see Daubechies (1992). None of these solutions were satisfactory since they introduced various kinds of aliasing near the boundaries. Meyer proposed a better solution for the Daubechies functions in Meyer (1991), namely to restrict the scaling functions and wavelets to the interval in question and then adjust the functions close to the boundary appropriately in order to preserve most of the classical properties of a multiresolution analysis. Meyer's construction suffered from certain shortcomings such as numerical instability, and was improved upon in Cohen et al. (1993).

The construction of general wavelets on nonuniform grids is complicated, especially in higher dimensions. The lifting scheme is a promising general framework for accomplishing this task, see Sweldens and Schröder (1996) and also Carnicer et al. (1996) for the related concept of stable completions. A brief overview of the nonuniform constructions obtained with the lifting scheme can be found in Daubechies et al. (1999).

A spline multiresolution analysis on an interval with mutually orthogonal wavelet spaces can be constructed analogously to Meyer's approach, namely by retaining cardinal B-splines in the interior of the interval and by introducing special boundary splines. These boundary functions, however, are typically not obtained by restricting B-splines with uniform knots, but by introducing B-splines with multiple knots at the ends of the interval. The construction of spline wavelets in this framework was carried out in Chui and Quak (1992). In the more general setting, where direct sum decompositions are considered instead of orthogonal ones, biorthogonal spline wavelets on the interval were investigated in Dahmen et al. (1999). In this chapter, we will restrict our attention to mutually orthogonal spline wavelet spaces, but for general nonuniform knot sequences. Spline wavelets on certain types of nonuniform knots were studied in Buhmann and Micchelli (1992), and spline wavelets on general knots were first constructed in Lyche and Mørken (1992). This construction was generalized to non-polynomial splines in Lyche and

Schumaker (1994). Chui and De Villiers (1998) discuss computations involving spline wavelets on nonuniform, but simple, knots. For an early discussion of orthogonality in spline spaces, see Ahlberg and Nilson (1965).

This chapter is organized as follows. First we summarize some necessary background material in Section 6.2. Then we describe in Section 6.3 the construction of the spline wavelets from Lyche and Mørken (1992). Using the new concept of a minimal interval, we establish that these B-wavelets form the unique basis of minimal support (up to scaling) for the given wavelet space. In Section 6.4 we discuss algorithms for implementing the (fast) wavelet transforms and their inverses for spline wavelets on nonuniform knots. Section 6.5 is devoted to the bivariate setting, describing the straightforward tensor product approach and providing a brief review of what little seems to be known for spline wavelets on arbitrary triangulations. We conclude in Section 6.6 with a discussion of some examples.

6.2 Background material

This section is devoted to a brief review of some basic properties of wavelets and splines we will need later. We will be working with real-valued functions defined on an interval $[a, b]$ and will use the standard L^2-norm defined by

$$||f|| = ||f||_2 = \left(\int_a^b f^2 \right)^{1/2}$$

induced by the inner product

$$\langle f, g \rangle = \int_a^b fg.$$

Vectors will be denoted by bold type, and for a vector $\mathbf{c} = (c_1, \ldots, c_n)^T$ we will use the ℓ^2-norm

$$||\mathbf{c}|| = ||\mathbf{c}||_2 = \left(\sum_{i=1}^n c_i^2 \right)^{1/2}.$$

We will often use vector notation for linear combinations as in

$$\sum_{i=1}^n c_i \phi_i = \boldsymbol{\phi}^T \mathbf{c},$$

where the functions $\{\phi_i\}_{i=1}^n$ have been gathered in the vector $\boldsymbol{\phi} = (\phi_1, \ldots, \phi_n)^T$.

6.2.1 Multiresolution and wavelets on an interval

The starting point for wavelets on an interval $[a, b]$ is usually a *multiresolution analysis*. This is a nested sequence of finite dimensional subspaces of $L^2[a,b]$

$$V_0 \subset V_1 \subset \cdots$$

with the property that

$$\overline{\bigcup_{j=0}^{\infty} V_j} = L^2[a,b], \tag{6.1}$$

and each space V_j is assumed to be spanned by a *Riesz basis* $\{\phi_{i,j}\}_{i=1}^{n_j}$ (the integer n_j giving the dimension of V_j), i.e., the basis satisfies the inequalities

$$K_1 \|\mathbf{c}_j\|_2 \leq \left\|\sum_i c_{i,j} \phi_{i,j}\right\|_2 \leq K_2 \|\mathbf{c}_j\|_2,$$

for any choice of coefficients $\mathbf{c}_j = (c_{1,j}, \ldots, c_{n_j,j})^T$, where the two constants K_1 and K_2 are independent of j. The basis functions $\{\phi_{i,j}\}_{i=1}^{n_j}$ are often referred to as *scaling functions*.

In classical wavelet theory the space V_j is related to V_{j-1} by dilation, i.e., if f is a function in V_{j-1} then $f(2\cdot)$ is a function in V_j. In our more general setting it is only assumed that each scaling function in V_{j-1} is a linear combination of the scaling functions in V_j. If we collect the scaling functions in V_j in the vector $\boldsymbol{\phi}_j = (\phi_{i,j})_i$ then this relation can be written

$$\boldsymbol{\phi}_{j-1}^T = \boldsymbol{\phi}_j^T \mathbf{P}_j,$$

for some matrix $\mathbf{P}_j$ of dimension $n_j \times n_{j-1}$. If $f = \boldsymbol{\phi}_{j-1}^T \mathbf{c}_{j-1}$ is a function in V_{j-1} it can then be represented in V_j as $f = \boldsymbol{\phi}_j^T \mathbf{c}_j$, where $\mathbf{c}_j = \mathbf{P}_j \mathbf{c}_{j-1}$.

The space V_j can be written as the direct sum $V_j = V_{j-1} \oplus W_{j-1}$, where W_{j-1} is the L^2-orthogonal complement of V_{j-1} in V_j. In other words W_{j-1} consists of all the functions in V_j that are orthogonal to V_{j-1},

$$W_{j-1} = \{f_j \in V_j : \langle f_j, f_{j-1} \rangle = 0 \text{ for all } f_{j-1} \in V_{j-1}\},$$

and the dimension of W_{j-1} is given by $\dim W_{j-1} = \dim V_j - \dim V_{j-1} = n_j - n_{j-1}$. By combining this decomposition repeatedly with (6.1), we obtain the direct sum decomposition

$$L^2[a,b] = V_0 \oplus W_0 \oplus W_1 \oplus \cdots . \tag{6.2}$$

We refer to the functions in the spaces $\{W_j\}_j$ as *wavelets* and the spaces as *wavelet spaces*.

For computations we need a suitable basis for each wavelet space, and we shall denote such a basis for W_{j-1} by $\boldsymbol{\psi}_{j-1} = (\psi_{i,j-1})_{i=1}^{n_j-n_{j-1}}$. Because of (6.2) we then in principle have a basis for all of $L^2[a,b]$,

$$L^2[a,b] = \text{span}\{\boldsymbol{\phi}_0, \boldsymbol{\psi}_0, \boldsymbol{\psi}_1, \dots\},$$

at least as long as the combined basis $(\boldsymbol{\phi}_0, \boldsymbol{\psi}_0, \boldsymbol{\psi}_1, \dots)$ forms a Riesz basis, i.e.,

$$K_1 \|(\mathbf{c}_0^T, \mathbf{w}_0^T, \mathbf{w}_1^T, \dots)\|_2 \leq \left\| \boldsymbol{\phi}_0^T \mathbf{c}_0 + \sum_{j=0}^{\infty} \boldsymbol{\psi}_j^T \mathbf{w}_j \right\|_2$$
$$\leq K_2 \|(\mathbf{c}_0^T, \mathbf{w}_0^T, \mathbf{w}_1^T, \dots)\|_2, \qquad (6.3)$$

for constants K_1 and K_2.

Since $W_{j-1} \subset V_j$, the wavelet basis $\boldsymbol{\psi}_{j-1}$ is linked to $\boldsymbol{\phi}_j$ via a matrix relation $\boldsymbol{\psi}_{j-1}^T = \boldsymbol{\phi}_j^T \mathbf{Q}_j$. A function $g = \boldsymbol{\psi}_{j-1}^T \mathbf{w}_{j-1}$ can then be lifted to V_j as $g = \boldsymbol{\phi}_j^T \mathbf{d}_j$, where $\mathbf{d}_j = \mathbf{Q}_j \mathbf{w}_{j-1}$.

The relation $V_j = V_{j-1} \oplus W_{j-1}$ implies that the two bases $\boldsymbol{\phi}_{j-1}$ and $\boldsymbol{\psi}_{j-1}$ together provide an alternative basis for V_j; we have the basis transformation

$$(\boldsymbol{\phi}_{j-1}^T, \boldsymbol{\psi}_{j-1}^T) = \boldsymbol{\phi}_j^T (\mathbf{P}_j, \mathbf{Q}_j) = \boldsymbol{\phi}_j^T \mathbf{M}_j.$$

From the inverse relation

$$\boldsymbol{\phi}_j^T = (\boldsymbol{\phi}_{j-1}^T, \boldsymbol{\psi}_{j-1}^T) \mathbf{M}_j^{-1} = (\boldsymbol{\phi}_{j-1}^T, \boldsymbol{\psi}_{j-1}^T) \begin{pmatrix} \mathbf{A}_j \\ \mathbf{B}_j \end{pmatrix},$$

we obtain formulas for decomposing a function $f = \boldsymbol{\phi}_j^T \mathbf{c}_j$ in V_j into two components $f_{j-1} = \boldsymbol{\phi}_{j-1}^T \mathbf{c}_{j-1}$ and $g_{j-1} = \boldsymbol{\psi}_{j-1}^T \mathbf{w}_{j-1}$ in V_{j-1} and W_{j-1}, namely

$$\mathbf{c}_{j-1} = \mathbf{A}_j \mathbf{c}_j, \qquad \mathbf{w}_{j-1} = \mathbf{B}_j \mathbf{c}_j.$$

By iterating this decomposition we can rewrite a function f in V_N, say, as a coarse approximation f_0 in V_0 and a collection $\{g_j\}_{j=0}^{N-1}$ of detail functions ("wavelets") with $g_j \in W_j$ such that

$$f = f_0 + g_0 + \cdots + g_{N-1}.$$

This corresponds to a change of basis from $\boldsymbol{\phi}_N$ to $(\boldsymbol{\phi}_0, \boldsymbol{\psi}_0, \dots, \boldsymbol{\psi}_{N-1})$ and is referred to as the (fast) *wavelet transform*. In applications the detail functions $\{g_j\}_{j=0}^{N-1}$ of f are typically filtered in some way before applying the inverse transform

$$\widehat{\mathbf{c}}_j = \mathbf{P}_j \widehat{\mathbf{c}}_{j-1} + \mathbf{Q}_j \widehat{\mathbf{w}}_{j-1}, \quad \text{for } j = 1, \dots, N,$$

where $(\widehat{\mathbf{w}}_j)_{j=0}^{N-1}$ are the filtered wavelet coefficients and $\widehat{\mathbf{c}}_0 = \mathbf{c}_0$.

In our setting here, the spaces $\{V_j\}_j$ will be polynomial spline spaces with B-splines as scaling functions, see below. The matrices $\{\mathbf{P}_j\}_j$ are then given and the main challenge is to construct the matrices $\{\mathbf{Q}_j\}_j$ corresponding to suitable wavelets ($\psi_{i,j}$).

Our rudimentary introduction to wavelets has been restricted to orthogonal decompositions of the spaces $\{V_j\}_j$. More generally, one may consider biorthogonal or direct sum decompositions, but for our present purposes this is not necessary.

6.2.2 Splines and B-splines

We will represent splines as linear combinations of B-splines. On a given nondecreasing sequence $\mathbf{t} = (t_i)_{i=1}^{n+d+1}$ of knots we consider for $i = 1, \ldots, n$ the B-splines $B_{i,d} = B_{i,d,\mathbf{t}}$, of degree $d \geq 0$ with support $[t_i, t_{i+d+1}]$. By convention the B-splines are right continuous and are normalized to sum to one, i.e.,

$$\sum_{i=1}^{n} B_{i,d}(x) = 1, \quad x \in [t_{d+1}, t_{n+1}).$$

We assume that $t_i < t_{i+d+1}$ for $i = 1, \ldots, n$ so that the B-splines are linearly independent on $[t_{d+1}, t_{n+1})$. The spline space $\mathcal{S}_{d,\mathbf{t}}$ is the linear space spanned by these B-splines,

$$\mathcal{S}_{d,\mathbf{t}} = \left\{ \sum_{i=1}^{n} c_i B_{i,d} : c_i \in \mathbb{R} \right\}.$$

An important property of B-splines is the stability estimate (see de Boor (1976))

$$D_d^{-1} \|\mathbf{c}\|_p \leq \left\| \sum_i c_i \tilde{B}_{i,d} \right\|_p \leq \|\mathbf{c}\|_p \tag{6.4}$$

for $1 \leq p \leq \infty$, where the scaled B-spline $\tilde{B}_{i,d}$ is defined by

$$\tilde{B}_{i,d} = \left(\frac{d+1}{t_{i+d+1} - t_i} \right)^{1/p} B_{i,d}. \tag{6.5}$$

The important point here is that the constant D_d (in addition to being independent of p) is independent of the knots $\mathbf{t}$. The norms $\|f\|_p$ and $\|\mathbf{c}\|_p$ are the standard L^p- and ℓ^p-norms for functions and vectors.

The inequalities (6.4) raise the question of what scaling to use for the B-splines in a wavelet setting. When working in L^2 the natural scaling is given by (6.5) with $p = 2$, while in most spline applications the partition of unity scaling corresponding to (6.5) with $p = \infty$ is used. We remark that this scaling can also be used in the wavelet setting provided that for highly nonuniform knot vectors we scale the coefficient matrices in the relevant linear systems so that they become well-conditioned. This will be discussed in more detail in Section 6.4.5.

The support of the B-spline $B_{i,d}$ is $[t_i, t_{i+d+1}]$, so the support of the spline function

$$f = \sum_{i=\ell}^{r} c_i B_{i,d}, \qquad (6.6)$$

with nonzero coefficients **c**, is the interval $[t_\ell, t_{r+d+1}]$ if $r \geq \ell$. For our purposes the notion of *index support* (often shortened to support) will prove useful. The index support of a spline gives the index of the first and last B-spline involved in its representation. If c_ℓ and c_r are nonzero, the first and last active B-splines of the spline in (6.6) are ℓ and r, respectively, and its index support is then denoted as $[\ell : r]$.

We will need some notation related to the multiplicities of knots. For each real number z, we let $m_{\mathbf{t}}(z)$ denote the multiplicity of z in $\mathbf{t}$, i.e., the number of times z occurs in $\mathbf{t}$. The left multiplicity $\lambda_{\mathbf{t}}(i)$ (right multiplicity $\rho_{\mathbf{t}}(i)$) of a knot t_i gives the number of knots in $\mathbf{t}$ equal to t_i, but with index less or equal (greater or equal) than i,

$$\lambda_{\mathbf{t}}(i) = \max\{j : t_{i-j+1} = t_i\}, \qquad \rho_{\mathbf{t}}(i) = \max\{j : t_{i+j-1} = t_i\}.$$

Two sufficiently smooth functions f and g are said to *agree* on $\mathbf{t}$ if

$$D^{\lambda_{\mathbf{t}}(i)-1} f(t_i) = D^{\lambda_{\mathbf{t}}(i)-1} g(t_i)$$

for $i = 1, \ldots, n+d+1$. This corresponds to the standard interpretation of Hermite or osculatory interpolation.

Collocation matrices are a central ingredient of our constructions. If $\mathbf{x} = (x_i)_{i=1}^m$ are m real numbers in the interval $[a, b]$ with $a \leq x_1 \leq \cdots \leq x_m \leq b$ and $f_1, \ldots, f_n$ are n sufficiently smooth functions on $[a, b]$, we denote the $m \times n$ collocation matrix with (i, j)-entry $D^{\lambda_{\mathbf{x}}(i)-1} f_j(x_i)$ by

$$\begin{pmatrix} x_1, \ldots, x_m \\ f_1, \ldots, f_n \end{pmatrix}.$$

If $m = n$ we denote the determinant of the collocation matrix by

$$\det \begin{pmatrix} x_1, \ldots, x_n \\ f_1, \ldots, f_n \end{pmatrix}. \qquad (6.7)$$

The Schoenberg–Whitney theorem tells us when a square collocation matrix for splines is non-singular, see Schumaker (1981).

Theorem 6.2.1 *With $f_j = B_{j,d,t}(x)$ for $j = 1, \ldots, n$, the determinant of the corresponding B-spline collocation matrix (6.7) is always nonnegative and it is positive if and only if*

$$x_i \in (t_i, t_{i+d+1}) \cup \{x : D^{\lambda_t(i)-1} B_{i,d,t}(x) \neq 0\}, \quad i = 1, \ldots, n. \quad (6.8)$$

If $x_1 < x_2 < \cdots < x_n$ are distinct then condition (6.8) holds if and only if all the diagonal elements in the collocation matrix are positive.

The following lemma gives a simple criterion which ensures that a collection of spline functions are linearly independent. It is part of the folklore, and a relatively simple consequence of the local support and continuity properties of B-splines. More specifically, it is based on the fact that if $t_i = \cdots = t_{i+q} < t_{i+q+1}$ then $D^{d-q}B_{i,d}(t_i) \neq 0$ (remember that limits are taken from the right at knots).

Lemma 6.2.2 *Let $\{f_i\}_{i=1}^p$ be a collection of spline functions in $\mathcal{S}_{d,t}$ with supports $\{[\ell_i : r_i]\}_{i=1}^p$ that satisfy $\ell_1 < \ell_2 < \cdots < \ell_p$ or $r_1 < r_2 < \cdots < r_p$. Then these functions are linearly independent on $[t_{\ell_1}, t_{r_p+d+1}]$.*

A basic tool for splines is knot insertion or refinement. If $\boldsymbol{\tau} = (\tau_i)_{i=1}^{n+d+1}$ and $\mathbf{t} = (t_i)_{i=1}^{n+m+d+1}$ are two knot vectors for splines of degree d and $\boldsymbol{\tau}$ is a subsequence of $\mathbf{t}$, then $\mathcal{S}_{d,\tau} \subseteq \mathcal{S}_{d,\mathbf{t}}$. If we organize the two B-spline bases in the vectors $\mathbf{B}_{\boldsymbol{\tau}}$ and $\mathbf{B}_{\mathbf{t}}$, we therefore have the relation

$$\mathbf{B}_{\boldsymbol{\tau}}^T = \mathbf{B}_{\mathbf{t}}^T \mathbf{P} \quad (6.9)$$

for a suitable rectangular matrix $\mathbf{P}$ of dimension $(m+n) \times n$, the *knot insertion matrix* from $\boldsymbol{\tau}$ to $\mathbf{t}$. If $f = \mathbf{B}_{\boldsymbol{\tau}}^T \mathbf{c} = \mathbf{B}_{\mathbf{t}}^T \mathbf{b}$ is a spline lying in both spaces, it follows that $\mathbf{b} = \mathbf{Pc}$.

6.2.3 B-spline Gram matrices

In the following sections, inner products of B-splines play a crucial role both in the construction of spline wavelets and in the implementation of the corresponding algorithms. Therefore we have decided to provide the reader with some more detailed information about the computation of the integrals which form the entries of B-spline Gram matrices.

First we need a notation for matrices built from inner products of functions. If $\mathbf{f} = (f_1, \ldots, f_n)^T$ and $\mathbf{g} = (g_1, \ldots, g_m)^T$ are two vectors of functions on $[a,b]$, we denote their Gramian by

$$\langle \mathbf{g}, \mathbf{f}^T \rangle = \begin{pmatrix} g_1, \ldots, g_m \\ f_1, \ldots, f_n \end{pmatrix};$$

the $m \times n$ matrix having as (i,j)th entry the number $\int_a^b g_i f_j$.

Corresponding to nondecreasing sequences of knots $(\tau_i)_{i=n_1}^{n_2}$ and $(t_i)_{i=m_1}^{m_2}$, and nonnegative integers d, e with $n_2 - n_1 \geq d+2$ and $m_2 - m_1 \geq e+2$ we consider the integrals

$$I_{i,j}^{d,e} = \int_{-\infty}^{\infty} B_{i,d,\boldsymbol{\tau}}(x) B_{j,e,\mathbf{t}}(x) dx, \quad \text{for all } i, j.$$

Since the B-splines are piecewise polynomials, one way to compute this integral is to use numerical integration on each subinterval. Thus, if $(u_1, \ldots, u_{k+1})$ is an increasing arrangement of the distinct elements of

$$(\tau_i, \ldots, \tau_{i+d+1}, t_j, \ldots, t_{j+e+1}),$$

we have

$$I_{i,j}^{d,e} = \sum_{r=1}^{k} \int_{u_r}^{u_{r+1}} B_{i,d,\boldsymbol{\tau}}(x) B_{j,e,\mathbf{t}}(x) dx,$$

where each of the integrands is a polynomial of degree $d + e$ and can be computed exactly using N-point Gauss–Legendre quadrature provided $2N - 1 \geq d + e$. This method was implemented in de Boor et al. (1976) for $\boldsymbol{\tau} = \mathbf{t}$. It is numerically stable, but nodes and weights have to be precomputed and stored. In Phillips and Hanson (1974), specialized quadrature formulas using a B-spline as a weight function are considered.

An alternative to this method would be to convert the B-splines on $\boldsymbol{\tau}$ and $\mathbf{t}$ to their Bernstein–Bézier representations on each subinterval using knot insertion techniques. The integral on each subinterval can then be computed explicitly.

Yet another strategy is to use stable recurrence relations reminiscent of the recurrence relations for B-splines (Schumaker (1981)), and degree raising and products of B-splines (Mørken (1991)). Following de Boor et al. (1976), for each i, j, and positive integers k, l such that $\tau_{i+k} > \tau_i$ and $t_{j+l} > t_j$, we define quantities $T_{i,j}^{k,l}$ by

$$T_{i,j}^{k,l} = \frac{(k+l-1)!}{(k-1)!(l-1)!} \int_{-\infty}^{\infty} \frac{B_{i,k-1,\boldsymbol{\tau}}(x)}{\tau_{i+k} - \tau_i} \frac{B_{j,l-1,\mathbf{t}}(x)}{t_{j+l} - t_j} dx.$$

We set $T_{i,j}^{k,l} = 0$ if both $\tau_{i+k} - \tau_i = 0$ and $t_{j+l} - t_j = 0$, We set $T_{i,j}^{k,l} = 0$ if both $\tau_{i+k} - \tau_i = 0$ and $t_{j+l} - t_j = 0$, otherwise we define

$$T_{i,j}^{k,l} = \binom{k+l-1}{l} \frac{B_{i,k-1,\tau}^R(t_j)}{\tau_{i+k} - \tau_i} \quad \text{if} \quad t_{j+l} = t_j,$$

$$T_{i,j}^{k,l} = \binom{k+l-1}{l} \frac{B_{j,l-1,\mathbf{t}}^L(\tau_i)}{t_{j+l} - t_j} \quad \text{if} \quad \tau_{i+k} = \tau_i.$$

Here B_j^R and B_j^L are right and left continuous versions of the B-spline B_j, i.e., for degree $d = 0$

$$B_{i,0,\tau}^R(x) = \begin{cases} 1, & \text{if } \tau_j \leq x < \tau_{j+1}, \\ 0, & \text{otherwise} \end{cases}$$

and

$$B_{j,0,\mathbf{t}}^L(x) = \begin{cases} 1, & \text{if } t_j < x \leq t_{j+1}, \\ 0, & \text{otherwise.} \end{cases}$$

If $\tau_{i+k} > \tau_i$ and $t_{j+l} > t_j$ the following recurrence relations can be used for $k, l \geq 1$

$$T_{i,j}^{k,l} = \frac{(\tau_{i+k} - t_j)T_{i,j}^{k,l-1} + (t_{j+l} - \tau_{i+k})T_{i,j+1}^{k,l-1}}{t_{j+l} - t_j} + T_{i,j}^{k-1,l}, \quad (6.10)$$

$$T_{i,j}^{k,l} = \frac{(\tau_i - t_j)T_{i,j}^{k,l-1} + (t_{j+l} - \tau_i)T_{i,j+1}^{k,l-1}}{t_{j+l} - t_j} + T_{i+1,j}^{k-1,l}, \quad (6.11)$$

$$T_{i,j}^{k,l} = \frac{(t_j - \tau_i)T_{i,j}^{k-1,l} + (\tau_{i+k} - t_j)T_{i+1,j}^{k-1,l}}{\tau_{i+k} - \tau_i} + T_{i,j+1}^{k,l-1}, \quad (6.12)$$

$$T_{i,j}^{k,l} = \frac{(t_{j+l} - \tau_i)T_{i,j}^{k-1,l} + (\tau_{i+k} - t_{j+l})T_{i+1,j}^{k-1,l}}{\tau_{i+k} - \tau_i} + T_{i,j}^{k,l-1}. \quad (6.13)$$

For numerical accuracy it is advantageous to use a formula which computes the quantity on the left as a convex combination of the first two T-quantities on the right. The following list shows which formulas to use in all cases where $(\tau_i, \tau_{i+k}) \cap (t_j, t_{j+l}) \neq \emptyset$.

(i) $\tau_i \leq t_j \leq \tau_{i+k} \leq t_{j+l}$: use (6.10) or (6.12),
(ii) $\tau_i \leq t_j \leq t_{j+l} \leq \tau_{i+k}$: use (6.12) or (6.13),
(iii) $t_j \leq \tau_i \leq \tau_{i+k} \leq t_{j+l}$: use (6.10) or (6.11),
(iv) $t_j \leq \tau_i \leq t_{j+l} \leq \tau_{i+k}$: use (6.11) or (6.13).

We conclude that in general we need all four formulas, but there are always at least two stable choices. If $\boldsymbol{\tau} = \mathbf{t}$ and $d = e$ one needs only two of the formulas, see de Boor et al. (1976) for a numerically stable implementation.

6.3 Spline wavelets

An important and natural example of a multiresolution analysis on an interval is provided by spline spaces. We need a collection of knot vectors $(\mathbf{t}^j)_{j=0}^{\infty}$ such that $\mathbf{t}^{j-1}$ is a subsequence of $\mathbf{t}^j$ for $j \geq 1$, since this ensures that $\mathcal{S}_{d,\mathbf{t}^{j-1}}$ is a subspace of $\mathcal{S}_{d,\mathbf{t}^j}$. Each space has a Riesz-basis of B-splines, and knot insertion takes us from one space to the next. Consider two nested spline spaces $\mathcal{S}_{d,\boldsymbol{\tau}}$ and $\mathcal{S}_{d,\mathbf{t}}$ with general knot vectors such that $\boldsymbol{\tau}$ is a subsequence of $\mathbf{t}$. We assume that the dimension of $\mathcal{S}_{d,\boldsymbol{\tau}}$ is n with $n \geq d+1$, and that the dimension of $\mathcal{S}_{d,\mathbf{t}}$ is $n+m$ so $\mathbf{t}$ can be thought of as obtained from $\boldsymbol{\tau}$ by inserting m *new knots* $\mathbf{s} = (s_i)_{i=1}^m$. For simplicity we assume that

$$\tau_1 = \tau_{d+1}, \quad \tau_{n+1} = \tau_{n+d+1}, \quad \text{and} \quad \tau_{d+1} < s_1 \leq s_2 \leq \cdots \leq s_m < \tau_{n+1},$$

the general setting is discussed in Lyche and Mørken (1992). For notational convenience we rename the two sets of B-splines to

$$\phi_j = B_{j,d,\boldsymbol{\tau}}, \quad \text{for} \quad j = 1, \ldots, n,$$
$$\gamma_i = B_{i,d,\mathbf{t}}, \quad \text{for} \quad i = 1, \ldots, n+m.$$

We also set

$$V_0 = \mathcal{S}_{d,\boldsymbol{\tau}} \quad \text{and} \quad V_1 = \mathcal{S}_{d,\mathbf{t}},$$

and denote the orthogonal complement of V_0 in V_1 by W.

In Lyche and Mørken (1992) it was shown how to construct a minimally supported basis for W. Here we give an alternative presentation of this construction that clarifies some details. From this new approach it also follows quite easily that the constructed basis is the *only* minimally supported basis for W.

The basic idea underlying the construction is simple. Any function in W must be a linear combination of B-splines in V_1. Since we want the support of a typical basis function ψ to be minimal, we want a linear combination of as few consecutive B-splines in V_1 as possible, say q altogether. The constraint is that ψ must be orthogonal to all the s coarse B-splines whose supports intersect the support of ψ. Intuitively, the support of ψ should be minimal when the number of parameters is one greater than the number of conditions, i.e., when $q = s + 1$, and this turns out to be the case. An extra bonus is that in this case it is easy to write down an explicit formula for ψ. In this way the construction of minimally supported wavelets is reduced to the problem of finding minimal intervals that can support a wavelet, and we show that the total number of such intervals agrees exactly with the dimension of W.

6.3.1 Construction of minimally supported wavelets

A function ψ in V_1 with index support $[\ell : r]$, where $r \geq \ell$, is a linear combination of the $r - \ell + 1$ fine B-splines $\boldsymbol{\gamma} = (\gamma_\ell, \ldots, \gamma_r)^T$,

$$\psi = \sum_{j=\ell}^{r} w_j \gamma_j = \boldsymbol{\gamma}^T \mathbf{w},$$

and is zero off the interval $[t_\ell, t_{r+d+1}]$. Let us assume that there are $p = p(\ell, r) \geq 0$ old knots in the open interval (t_ℓ, t_{r+d+1}), namely $\tau_{k+1}, \ldots, \tau_{k+p}$, for some $k = k(\ell, r)$. There are then $p + d + 1$ coarse B-splines $\boldsymbol{\phi} = (\phi_{k-d}, \ldots, \phi_{k+p})^T$, whose supports intersect the interval (t_ℓ, t_{r+d+1}). Due to the local supports of B-splines, the spline ψ is orthogonal to V_0 if and only if

$$\langle \phi_i, \psi \rangle = 0, \quad \text{for } i = k - d, \ldots, k + p,$$

or in terms of the Gram matrix

$$\langle \boldsymbol{\phi}, \boldsymbol{\gamma}^T \rangle \mathbf{w} = 0.$$

As the orthogonality conditions are homogeneous, a non-trivial solution is guaranteed to exist if the number of parameters is strictly larger than the number of conditions imposed, namely $r - \ell + 1 > p(\ell, r) + d + 1$. In the hope of obtaining a wavelet ψ with smallest possible support we want this inequality to be satisfied with $r - \ell$ as small as possible. This motivates the following definition.

Definition 6.3.1 The index interval $[\ell : r]$ is said to be *minimal* if

$$r - \ell > p(\ell, r) + d \qquad (6.14)$$

and there is no true subinterval $[u : v]$ with this property, i.e., if $\ell \leq u < v \leq r$ and $v - u > p(u, v) + d$, then $u = \ell$ and $v = r$.

A minimal interval always exists for given $\boldsymbol{\tau}$ and $\mathbf{t}$, as the condition (6.14) is certainly satisfied for the index interval $[1 : n + m]$; a smallest subinterval for which (6.14) is true must therefore be minimal. Also, since a minimal interval cannot contain other minimal intervals than itself, we observe that if $[\ell_1 : r_1]$ and $[\ell_2 : r_2]$ are two distinct minimal intervals then

$$(\ell_2 - \ell_1)(r_2 - r_1) > 0. \tag{6.15}$$

Lemma 6.3.2 *For a minimal interval $[\ell : r]$ we have*

$$r - \ell = p(\ell, r) + d + 1,$$

so that for minimal intervals the number of parameters is exactly one larger than the number of orthogonality conditions imposed.

Proof Suppose the contrary is true, i.e., that $[\ell : r]$ is minimal, but $r - \ell > p(\ell, r) + d + 1$. We distinguish between two cases. First, if $t_{\ell+1} < \tau_{k+1}$, we have $p(\ell+1, r) = p(\ell, r)$. But then $r - (\ell+1) > p(\ell+1, r) + d$, contradicting the minimality of $[\ell : r]$.

If, however, $t_{\ell+1} = \tau_{k+1}$ and τ_{k+1} is a knot of multiplicity $s \geq 1$ in the knot sequence $\boldsymbol{\tau}$, then we have $p(\ell+1, r) = p(\ell, r) - s$, and therefore $r - (\ell+1) > p(\ell+1, r) + d + s$, again contradicting the minimality of $[\ell : r]$. $\square$

Our aim is to construct a wavelet with as small a support as possible, so let us clarify what exactly we mean by this.

Definition 6.3.3 A nonzero function ψ in V_1 with support $[u : v]$ is called a *B-wavelet* if it is orthogonal to $\mathcal{S}_{d,\boldsymbol{\tau}}$, and if it has *minimal support* in the sense that if g is another nonzero spline in W with support $[\tilde{u} : \tilde{v}]$, where $u \leq \tilde{u} \leq \tilde{v} \leq v$, then $u = \tilde{u}$ and $v = \tilde{v}$.

Our next result shows that any minimal interval $[\ell : r]$ is the support interval of a B-wavelet.

Theorem 6.3.4 *Let $[\ell : r]$ be a minimal interval. Then the spline function $\psi(x) = \sum_{i=\ell}^{r} w_i \gamma_i(x)$ given by*

$$\psi(x) = \det \begin{pmatrix} \langle \phi_{k-d}, \gamma_\ell \rangle & \langle \phi_{k-d}, \gamma_{\ell+1} \rangle & \cdots & \langle \phi_{k-d}, \gamma_r \rangle \\ \vdots & \vdots & & \vdots \\ \langle \phi_{k+p}, \gamma_\ell \rangle & \langle \phi_{k+p}, \gamma_{\ell+1} \rangle & \cdots & \langle \phi_{k+p}, \gamma_r \rangle \\ \gamma_\ell(x) & \gamma_{\ell+1}(x) & \cdots & \gamma_r(x) \end{pmatrix} \tag{6.16}$$

is a B-wavelet with index support $[\ell : r]$, its coefficients $(w_i)_{i=\ell}^{r}$ oscillate strictly in sign, and ψ itself has $r - \ell$ strong changes of sign in (t_ℓ, t_{r+d+1}).

Proof Expanding the determinant (6.16), we find that $\psi(x) = \sum_{i=\ell}^{r} w_i \gamma_i(x)$, where $w_i = (-1)^{r-i} v_i$, $v_i = \det G_i$, and

$$G_i = \begin{pmatrix} \phi_{k-d}, \ldots, \phi_{k+p} \\ \gamma_\ell, \ldots, \gamma_{i-1}, \gamma_{i+1}, \ldots, \gamma_r \end{pmatrix}.$$

Consequently, the spline ψ lies in V_1 and its support is included in $[\ell : r]$. Taking inner products of ψ in (6.16) with any of the coarse B-splines $\{\phi_{k+j}\}_{j=-d}^{p}$ yields a determinant with two identical rows and therefore establishes that ψ is indeed orthogonal to V_0.

We next show that $v_\ell > 0$. By a formula from Karlin (1968), p. 17, we have

$$v_\ell = \int_\Omega \det \begin{pmatrix} y_1, & \ldots, & y_{r-\ell} \\ \phi_{k-d}, & \ldots, & \phi_{k+p} \end{pmatrix} \det \begin{pmatrix} y_1, & \ldots, & y_{r-\ell} \\ \gamma_{\ell+1}, & \ldots, & \gamma_r \end{pmatrix},$$

where Ω denotes the set

$$\Omega = \{(y_1, \ldots, y_{r-\ell}) : a \leq y_1 \leq \cdots \leq y_{r-\ell} \leq b\}.$$

Theorem 6.2.1 immediately implies that $v_\ell \geq 0$. Since the B-splines are continuous functions, it is clear that v_ℓ will be positive if the integrand is positive at some point in the interior of Ω. But again, from Theorem 6.2.1, we know that this happens if

$$\text{supp}_0 \, \phi_{k-d-1+i} \cap \text{supp}_0 \, \gamma_{\ell+i} \neq \emptyset, \quad \text{for } i = 1, \ldots, r - \ell, \quad (6.17)$$

where $\text{supp}_0 f$ denotes the interior of the support of f. For $i = r - \ell$, we have that $\text{supp}_0 \phi_{k-d-1+i} = (\tau_{k-p}, \tau_{k+p+d+1})$ and $\text{supp}_0 \gamma_r = (t_r, t_{r+d+1})$. Since $\tau_{k+p} \in (t_\ell, t_{r+d+1})$ but $\tau_{k+p+1} \geq t_{r+d+1}$, we see that the condition in (6.17) is satisfied in this case.

Suppose now that (6.17) does not hold for some integer $i = j$ with $1 \leq j < r - \ell$, in other words

$$(\tau_{k-d-1+j}, \tau_{k+j}) \cap (t_{\ell+j}, t_{\ell+j+d+1}) = \emptyset.$$

As condition (6.17) holds for $i = r - \ell$, this is only possible if $\tau_{k+j} \leq t_{\ell+j}$. But since $\boldsymbol{\tau} \subset \mathbf{t}$, this implies that

$$\tau_{k+i} \leq t_{\ell+i}, \quad \text{for } i = j, j-1, \ldots, 1.$$

In particular, for $i = 1$ we then have

$$t_\ell < \tau_{k+1} = t_{\ell+1}.$$

From Lemma 6.3.2, we have $r - \ell = p(\ell, r) + d + 1$, but if τ_{k+1} has multiplicity $s \geq 1$ in $\boldsymbol{\tau}$, then $p(\ell+1, r) = p(\ell, r) - s$, and

$$r - (\ell + 1) = p(\ell + 1, r) + d + s > p(\ell + 1, r) + d,$$

contradicting the minimality of $[\ell : r]$. The conclusion is therefore that $v_\ell > 0$, and by an analogous proof we conclude that $v_i > 0$ for $i = \ell+1, \ldots, r$.

It remains to show that ψ is a B-wavelet. Note that $v_\ell > 0$ implies that the matrix G_ℓ is non-singular, so the only linear combination of $\gamma_{\ell+1}, \ldots, \gamma_r$ that satisfies all the orthogonality conditions is the trivial one. Similarly, since the last coefficient v_r is also positive, the submatrix G_r is also non-singular, which shows that there cannot be a non-trivial combination of $\gamma_\ell, \ldots, \gamma_{r-1}$ in W. This shows that ψ has minimal support.

That ψ has $r - \ell$ strong sign changes in (t_ℓ, t_{r+d+1}) follows from Lemma 2 in Lyche and Mørken (1992). □

Note at this point that as a consequence of the integral formulas of Section 6.2.3, the B-wavelet coefficients are in fact rational functions of the knots.

We cannot have more minimal intervals than new knots.

Lemma 6.3.5 *The B-wavelets in W generated by different minimal intervals are linearly independent and there are at most m of them.*

Proof Suppose that we have s B-wavelets with index supports $[\ell_j : r_j]$ for $j = 1, \ldots, s$. Equation (6.15) implies that these B-wavelets may be ordered so that $\ell_1 < \ell_2 < \cdots < \ell_s$, and from Lemma 6.2.2 we therefore know that they are linearly independent. But since $\dim W = m$ it follows that $s \leq m$. □

As we shall see shortly, there are exactly m minimal intervals, so the corresponding B-wavelets form a basis for W. The clue to this fact is an explicit procedure for constructing minimal intervals, one for each new knot.

The equation $r - \ell - p(\ell, r) = d + 1$ implied by Lemma 6.3.2 for a minimal interval can be rewritten as

$$r + d + 1 - \ell + 1 = p(\ell, r) + 2d + 3.$$

This relation can be interpreted as stating that, apart from the $p(\ell, r)$ old knots in (t_ℓ, t_{r+d+1}), there are $2d + 3$ knots in the index interval $[\ell : r]$. More precisely, there must be $2d + 3 - \rho_{\mathbf{t}}(\ell) - \lambda_{\mathbf{t}}(r + d + 1)$ new knots s_i in (t_ℓ, t_{r+d+1}). As the multiplicity of any given knot is at most $d + 1$, there has to be at least one such new knot.

Chapter 6   Nonuniform spline wavelets

If s_j is a new knot and $t_{\tilde{\ell}}$ and $t_{\tilde{r}}$ are two knots with $t_{\tilde{\ell}} < s_j < t_{\tilde{r}}$ we define the integer functions μ_j and ν_j by

$$\mu_j(\tilde{\ell}) = \#\{i < j : s_i \in (t_{\tilde{\ell}}, s_j]\},$$
$$\nu_j(\tilde{r}) = \#\{i > j : s_i \in [s_j, t_{\tilde{r}})\}.$$

We see that if $[\ell : r]$ is a minimal interval and s_j is a new knot in (t_ℓ, t_{r+d+1}) then the relation

$$\rho_{\mathbf{t}}(\ell) + \mu_j(\ell) + \nu_j(r+d+1) + \lambda_{\mathbf{t}}(r+d+1) = 2d+2 \quad (6.18)$$

must hold (the sum is $2d+2$ since s_j is not counted in (6.18)). However, equation (6.18) is not sufficient to ensure minimality, but the following proposition gives a "symmetric" construction of a minimal interval.

Proposition 6.3.6 *Let s_j be a new knot, let ℓ be the largest integer less than j such that*

$$\mu_j(\ell) + \rho_{\mathbf{t}}(\ell) = d+1, \quad (6.19)$$

and let r be the smallest integer greater than j such that

$$\nu_j(r+d+1) + \lambda_{\mathbf{t}}(r+d+1) = d+1. \quad (6.20)$$

Then $[\ell : r]$ is a minimal interval.

Proof We first show that the construction is well-defined in that integers ℓ and r that satisfy (6.19) and (6.20) exist. That (6.19) can be satisfied follows from three simple facts:

(i) $\mu_j(1) + \rho_{\mathbf{t}}(1) \geq d+1$;
(ii) if t_q is the largest q such that $t_q < s_j$, then $\mu_j(q) + \rho_{\mathbf{t}}(q) \leq d+1$;
(iii) if $t_p < s_j$ and $p > 1$ then $\mu_j(p-1) + \rho_{\mathbf{t}}(p-1) \leq \mu_j(p) + \rho_{\mathbf{t}}(p) + 1$, so the left-hand side in (6.19) increases by at most one when ℓ is reduced by one.

The first fact is immediate since $\rho_{\mathbf{t}}(1) = d+1$, while the second follows since no knots occur more than $d+1$ times in $\mathbf{t}$. The third fact is obvious if $t_{p-1} = t_p$, while if $t_{p-1} < t_p$ it follows from the three relations $\mu_j(p-1) \leq m_{\mathbf{t}}(t_p) + \mu_j(p)$, $\rho_{\mathbf{t}}(p-1) = 1$ and $\rho_{\mathbf{t}}(p) = m_{\mathbf{t}}(t_p)$. The existence of an integer r that satisfies (6.20) follows similarly.

The selection of ℓ and r ensures that $r - \ell = p(\ell, r) + 2d + 2$, but not yet that $r - \ell$ is minimal. So suppose that $[\ell : r]$ contains a minimal subinterval

$[\ell' : r']$. Without loss we assume that $\ell < \ell'$ and $r' \leq r$. If $s_j \in (t_{\ell'}, t_{r'+d+1})$ we must have

$$\mu_j(\ell') + \rho_{\mathbf{t}}(\ell') < d+1$$

since ℓ was chosen as the largest integer satisfying (6.19). Likewise we must have

$$\nu_j(r'+d+1) + \lambda_{\mathbf{t}}(r'+d+1) \leq d+1,$$

so

$$\rho_{\mathbf{t}}(\ell') + \mu_j(\ell') + \nu_j(r'+d+1) + \lambda_{\mathbf{t}}(r'+d+1) < 2d+2$$

contradicting (6.18) and thereby the minimality of $[\ell' : r']$.

Another possibility is that $s_j \leq t_{\ell'}$. As we noted above, there must be at least one new knot, say s_i, in $(t_{\ell'}, t_{r'+d+1})$. Since s_i also lies in (t_ℓ, t_{r+d+1}) we must have

$$\rho_{\mathbf{t}}(\ell) + \mu_i(\ell) = d + 1 - j + i,$$
$$\nu_i(r+d+1) + \lambda_{\mathbf{t}}(r+d+1) = d + 1 + j - i,$$

and ℓ and r must be the largest and smallest integers that satisfy these equations respectively. But if $\ell < \ell'$ and $r' \leq r$ we must have

$$\rho_{\mathbf{t}}(\ell') + \mu_i(\ell') < d + 1 - j + i$$
$$\nu_i(r'+d+1) + \lambda_{\mathbf{t}}(r'+d+1) \leq d + 1 + j - i,$$

so

$$\rho_{\mathbf{t}}(\ell') + \mu_i(\ell') + \nu_i(r'+d+1) + \lambda_{\mathbf{t}}(r'+d+1) < 2d+2$$

which again contradicts the minimality of $[\ell' : r']$. The final possibility is that $s_j \geq t_{r'+d+1}$ and this can be treated similarly. We therefore conclude that $[\ell : r]$ is minimal. □

By applying the construction in Proposition 6.3.6 to all m new knots we obtain m minimal intervals, and each of these gives rise to a B-wavelet as stated in Theorem 6.3.4.

Theorem 6.3.7 *Let $[\ell_j : r_j]$ be the m minimal intervals obtained by applying the construction in Proposition 6.3.6 to each of the m new knots $(s_j)_{j=1}^m$, with $s_1 \leq \cdots \leq s_m$, and let ψ_j denote the B-wavelet associated with $[\ell_j : r_j]$ for $j = 1, \ldots, m$. Then $\ell_1 < \cdots < \ell_m$, $r_1 < \cdots < r_m$, the m associated B-wavelets $\{\psi_j\}_{j=1}^m$ are the only B-wavelets in W (apart from scaling), and these B-wavelets form a basis for this space.*

Proof For fixed j with $2 \leq j \leq m$ we first show that $\ell_{j-1} < \ell_j$ and $r_{j-1} < r_j$. Since $s_{j-1} \leq s_j$ it follows from the definition of the ℓs and rs that $\ell_{j-1} \leq \ell_j$ and $r_{j-1} \leq r_j$. Suppose that $\ell_{j-1} = \ell_j = \ell$. By construction we then have

$$\mu_{j-1}(\ell) + \rho_{\mathbf{t}}(\ell) = d+1 = \mu_j(\ell) + \rho_{\mathbf{t}}(\ell).$$

But then $\mu_{j-1}(\ell) = \mu_j(\ell)$ which is impossible by the definition of the μs. Similarly it follows that $r_{j-1} < r_j$.

It remains to be shown that there are no other B-wavelets in W. Suppose that ψ is a B-wavelet. Since the B-wavelets $\{\psi_j\}_{j=1}^m$ form a basis for W we have

$$\psi = \sum_{j=j_1}^{j_2} c_j \psi_j$$

for certain numbers $(c_j)_{j=j_1}^{j_2}$ with c_{j_1} and c_{j_2} nonzero. But since the left and right ends of the index supports of $\{\psi_j\}_{j=1}^m$ are strictly increasing, we see that the index support of ψ is $[\ell_{j_1} : r_{j_2}]$. Now, ψ is a minimally supported wavelet in W. This implies $j_1 = j_2$ and $\psi = c\psi_{j_1}$ for some nonzero constant c. $\square$

Lemma 11 in Lyche and Mørken (1992) gives the following alternative representation of a B-wavelet.

Lemma 6.3.8 *Let $[\ell : r]$ be a minimal interval, let ψ be the associated B-wavelet, and let $\tau_{k+1}, \ldots, \tau_{k+p}$ be the old knots in (t_ℓ, t_{r+d+1}). Then $\psi = D^{d+1}\theta$, where θ is the spline of degree $2d+1$ given by*

$$\theta(x) = e \det \begin{pmatrix} x, \tau_{k+1}, \ldots, \tau_{k+p} \\ B_{\ell, 2d+1, \mathbf{t}}, \ldots, B_{\ell+p, 2d+1, \mathbf{t}} \end{pmatrix},$$

and e is a nonzero constant.

This lemma shows that B-wavelets can, if desired, be computed from smaller determinants than the one used in (6.16). Also since $\ell + p = r - d - 1$ the $p+1$ B-splines $(B_{i,2d-1,\mathbf{t}})_{i=\ell}^{\ell+p}$ of degree $2d+1$ only depend on the knots $t_\ell, \ldots t_{r+d+1}$.

Finally, we note that polynomial reproduction of degree d is obvious in our spline setting since the spline spaces contain polynomials of degree d and the approximation method from the fine space to the coarse space is orthogonal projection. In other words, polynomials of degree d can be projected from the fine space to the coarse space with zero error, i.e., the detail component in W will be zero. But this in turn means that all of W is orthogonal to polynomials of degree d, and in particular we have $\langle \psi, x^i \rangle = 0$ for $i = 0, \ldots, d$.

6.4 Algorithms for decomposition and reconstruction

To implement the complete wavelet transform, we must first decide how to choose the nested sequence of spaces $\{V_j\}_j$. If we are given a spline f in a fine space V_N, this means that we must determine which of the interior knots from V_N to keep in V_{N-1}, then which knots from V_{N-1} to keep in V_{N-2} and so on. We consider various approaches to selecting subspaces and then discuss reconstruction and decomposition using the B-wavelets that we constructed in Section 6.3, see also Chui and De Villiers (1998).

6.4.1 Choosing the subspaces

The simplest choice is to keep every other knot in analogy with classical wavelets. An obvious generalization is to choose a fraction μ/ν and keep μ out of every ν knots. The advantage of this approach is its simplicity.

An alternative is to use information about f when deciding which knots to pass on to V_{N-1}, as in the knot removal strategy in Lyche and Mørken (1988). In this strategy, given a spline function and a tolerance, the aim is to remove as many knots from the spline as possible, without letting the error exceed the tolerance. The knots are first ranked according to their significance in the representation of the spline. As many knots as possible are then removed, in the order of increasing significance, taking some uniformity constraints into account. The knots of the resulting approximation are then ranked again and removed according to the same criterion. This process is continued until no more knots can be removed.

Knot removal is clearly closely related to compression via wavelet decomposition, in that we compute in both cases successive approximations in coarser and coarser spaces. The difference is that in knot removal, the detail (error) functions are not stored, although they are computed (and represented in the fine space) in order to check the magnitude of the error. Using knot removal it is possible to ensure that the approximation in V_{N-1} will have certain desirable properties, for example that the approximation is so good that the error (the wavelet component of f) may be neglected. On the other hand, it may be argued that this mixes two stages of the wavelet analysis: the decomposition and the analysis. For efficiency, the choice of the spaces in the wavelet decomposition ought to be independent of f, but the following analysis must of course involve (the decomposed version of) f.

In the ranking part of knot removal in Lyche and Mørken (1988) each knot is given a weight indicating its significance. Since each new knot is associated with a spline wavelet one could instead use the wavelet coefficient as a weight for the corresponding knot. In particular, it is possible to use a wavelet decomposition in a knot removal strategy. We simply carry out a wavelet transform on some nested sequence of spaces, defined, say, by removing every second knot. The wavelets which remain after the thresholding using a certain tolerance then define the knots which are the candidates to keep.

6.4.2 Basis transformations

Our aim here is to sketch efficient algorithms for working with the spline wavelets that we constructed in Section 6.3. We use the same notation and let V_0 and V_1 be coarse and fine spline spaces of dimension n and $n+m$, respectively, spanned by B-splines of degree d that we organize into two vectors $\boldsymbol{\phi}$ and $\boldsymbol{\gamma}$. The complement space W of dimension m is spanned by the m B-wavelets $\boldsymbol{\psi} = (\psi_j)_{j=1}^m$.

Together the two bases $\boldsymbol{\phi}$ and $\boldsymbol{\psi}$ form a basis for V_1 so we have the standard wavelet relation

$$(\boldsymbol{\phi}^T, \boldsymbol{\psi}^T) = \boldsymbol{\gamma}^T(\mathbf{P}\ \mathbf{Q}) = \boldsymbol{\gamma}^T\mathbf{M} \qquad (6.21)$$

and its inverse

$$\boldsymbol{\gamma}^T = (\boldsymbol{\phi}^T, \boldsymbol{\psi}^T)\mathbf{M}^{-1} = (\boldsymbol{\phi}^T, \boldsymbol{\psi}^T)\begin{pmatrix}\mathbf{A}\\\mathbf{B}\end{pmatrix}.$$

The matrix $\mathbf{P}$ is the knot insertion matrix, while the matrix $\mathbf{Q}$ shows how the B-wavelets are related to the fine B-splines. It can be obtained from the explicit representation for B-wavelets in equation (6.16). When $\mathbf{P}$ and $\mathbf{Q}$ are known, the matrices $\mathbf{A}$ and $\mathbf{B}$ are also given, as a decomposition of the inverse of $\mathbf{M}$.

Due to the relation (6.21), any function $f_1 \in V_1$ has two different representations with uniquely determined coefficient vectors, namely

$$f_1 = \boldsymbol{\gamma}^T\mathbf{c}_1 = f_0 + g = \boldsymbol{\phi}^T\mathbf{c}_0 + \boldsymbol{\psi}^T\mathbf{w}.$$

The computation of the coarse spline coefficients $\mathbf{c}_0$ and the wavelet coefficients $\mathbf{w}$, when given the fine spline coefficients $\mathbf{c}_1$, is referred to as *decomposition*, while the converse procedure is referred to as *reconstruction*.

Reconstruction is accomplished via the reconstruction relation

$$\mathbf{c}_1 = \mathbf{M}\begin{pmatrix}\mathbf{c}_0\\\mathbf{w}\end{pmatrix} = \begin{pmatrix}\mathbf{P} & \mathbf{Q}\end{pmatrix}\begin{pmatrix}\mathbf{c}_0\\\mathbf{w}\end{pmatrix} = \mathbf{P}\mathbf{c}_0 + \mathbf{Q}\mathbf{w}, \qquad (6.22)$$

whereas decomposition is based on the inverse relation

$$\begin{pmatrix}\mathbf{c}_0\\\mathbf{w}\end{pmatrix} = \mathbf{M}^{-1}\mathbf{c}_1 = \begin{pmatrix}\mathbf{A}\\\mathbf{B}\end{pmatrix}\mathbf{c}_1 = \begin{pmatrix}\mathbf{A}\,\mathbf{c}_1\\\mathbf{B}\,\mathbf{c}_1\end{pmatrix}. \qquad (6.23)$$

Let us first see how we can perform reconstruction.

6.4.3 Reconstruction

Once the two matrices $\mathbf{P}$ and $\mathbf{Q}$ are known, reconstruction is straightforward with the relation (6.22), and since in many cases both $\mathbf{P}$ and $\mathbf{Q}$ are banded matrices, this is also efficient.

The matrix $\mathbf{P}$ can be computed efficiently by the Oslo algorithm, see Cohen et al. (1980), Lyche and Mørken (1986). Our primary concern here is therefore the efficient computation of $\mathbf{Q}$. From the relation $\boldsymbol{\psi}^T = \boldsymbol{\gamma}^T \mathbf{Q}$ we see that the jth column of $\mathbf{Q}$ gives the B-spline coefficients of ψ_j relative to the fine B-splines. More specifically, the jth B-wavelet ψ_j with support $[\ell_j : r_j]$ is given as in (6.16) with $\ell = \ell_j$, $r = r_j$ and $p = p_j = p(\ell_j, r_j)$. Note however that we are free to choose a different scaling of ψ_j. Due to the alternating coefficients of ψ_j according to Theorem 6.3.4, we have the following.

Lemma 6.4.1 *Let $[\ell_j : r_j]$ be the index support of the jth B-wavelet ψ_j^*, scaled such that the absolute values of its B-spline coefficients add to one. Let $p_j = p(\ell_j, r_j)$ be the number of old knots in $(t_{\ell_j}, t_{r_j+d+1})$. The nonzero B-spline coefficients $(q_{i,j})_{i=\ell_j}^{r_j}$ of ψ_j^* are given by the solution of the linear system*

$$\begin{pmatrix}\langle\phi_{k-d},\gamma_{\ell_j}\rangle & \langle\phi_{k-d},\gamma_{\ell_j+1}\rangle & \cdots & \langle\phi_{k-d},\gamma_{r_j}\rangle\\ \vdots & \vdots & & \vdots\\ \langle\phi_{k+p_j},\gamma_{\ell_j}\rangle & \langle\phi_{k+p_j},\gamma_{\ell_j+1}\rangle & \cdots & \langle\phi_{k+p_j},\gamma_{r_j}\rangle\\ 1 & -1 & \cdots & (-1)^{r_j-\ell_j}\end{pmatrix}\begin{pmatrix}q_{\ell_j,j}\\ \vdots\\ q_{r_j-1,j}\\ q_{r_j,j}\end{pmatrix} = \begin{pmatrix}0\\ \vdots\\ 0\\ 1\end{pmatrix}, \qquad (6.24)$$

and constitute the nonzero part of column j of $\mathbf{Q}$.

The linear system (6.24) can be solved by straightforward Gaussian elimination without pivoting since the matrix obtained by deleting the last row is totally positive. Recall that the computation of the inner product entries can be carried out as discussed in Section 6.2.3. Determination of $\mathbf{Q}$

is therefore straightforward once a scaling has been chosen. We will discuss scaling in some more detail in Section 6.4.5.

6.4.4 Decomposition

The algorithm for reconstruction is straightforward, but it is not so simple to devise a good decomposition algorithm. A direct computation of the inverse $\mathbf{M}^{-1}$, and thereby of the matrices $\mathbf{A}$ and $\mathbf{B}$ needed in (6.23), results in full matrices with corresponding loss of efficiency. The entries of these matrices typically decay away from the main diagonals, and so one might suggest setting all matrix elements below a certain threshold to zero in order to obtain banded matrices. However, practical experience shows that such a truncation leads to significant round-off errors.

An alternative is to treat the equation

$$\begin{pmatrix} \mathbf{P} & \mathbf{Q} \end{pmatrix} \begin{pmatrix} \mathbf{c}_0 \\ \mathbf{w} \end{pmatrix} = \mathbf{c}_1 \qquad (6.25)$$

as a linear system with $\begin{pmatrix} \mathbf{P} & \mathbf{Q} \end{pmatrix}$ as the coefficient matrix of size $n+m$, with the given coefficients $\mathbf{c}_1$ in V_1 as the right-hand side, and with $\mathbf{c}_0$ and $\mathbf{w}$ as the unknowns. If Gaussian elimination is used to solve this linear system we need to use some kind of reordering of equations and/or unknowns. This is because the matrix $\begin{pmatrix} \mathbf{P} & \mathbf{Q} \end{pmatrix}$ will typically have zero diagonal elements. One possibility is to reorder the unknowns and interlace the columns of $\mathbf{P}$ and $\mathbf{Q}$ to obtain a new (globally) banded coefficient matrix, allowing the use of a special banded system solver. Exactly how to do this interlacing is not clear, but it seems reasonable to place column i of $\mathbf{Q}$ in a position such that its nonzero entries are distributed as evenly as possible above and below the diagonal.

A third approach is based on the fact that, due to the orthogonality of the decomposition $f_1 = f_0 + g$, the functions f_0 and g are in fact the least squares best approximations to f_1 from V_0 and W, respectively, and can therefore be computed using normal equations. The normal equations can be derived quite simply. The wavelet function (the error) $g = f_1 - f_0$ should be orthogonal to V_0, i.e., to each coarse B-spline. Since $f_1 = \boldsymbol{\gamma}^T \mathbf{c}_1$ and $f_0 = \boldsymbol{\phi}^T \mathbf{c}_0$ these conditions can be expressed as a system of n linear equations in n unknowns,

$$\langle \boldsymbol{\phi}, \boldsymbol{\phi}^T \rangle \mathbf{c}_0 = \langle \boldsymbol{\phi}, \boldsymbol{\gamma}^T \rangle \mathbf{c}_1. \qquad (6.26)$$

The other set of normal equations are derived from the conditions that the approximation $f_0 = f_1 - g$ should be orthogonal to W, or equivalently to all the B-wavelets in W. With $g = \boldsymbol{\psi}^T \mathbf{w}$ this leads to m linear equations in the m unknowns $\mathbf{w}$,

$$\langle \boldsymbol{\psi}, \boldsymbol{\psi}^T \rangle \mathbf{w} = \langle \boldsymbol{\psi}, \boldsymbol{\gamma}^T \rangle \mathbf{c}_1. \tag{6.27}$$

To solve the systems (6.26) and (6.27) the four matrices involved must be computed. The two matrices $\langle \boldsymbol{\phi}, \boldsymbol{\phi}^T \rangle$ and $\langle \boldsymbol{\phi}, \boldsymbol{\gamma}^T \rangle$, consisting of inner products of B-splines, can be computed directly, using one of the techniques described in Section 6.2.3. The other two matrices are most conveniently computed by making use of the refinement equation (6.21) and relating the matrices to the matrix $\langle \boldsymbol{\gamma}, \boldsymbol{\gamma}^T \rangle$ of inner products of fine B-splines,

$$\langle \boldsymbol{\psi}, \boldsymbol{\psi}^T \rangle = \mathbf{Q}^T \langle \boldsymbol{\gamma}, \boldsymbol{\gamma}^T \rangle \mathbf{Q}, \qquad \langle \boldsymbol{\psi}, \boldsymbol{\gamma}^T \rangle = \mathbf{Q}^T \langle \boldsymbol{\gamma}, \boldsymbol{\gamma}^T \rangle.$$

For computational efficiency it is important to exploit the bandedness of all the matrices involved.

6.4.5 Stability considerations

It is common to require from a wavelet construction that the combined set of basis functions $(\boldsymbol{\phi}_0, \boldsymbol{\psi}_0, \boldsymbol{\psi}_1, \dots)$ form a Riesz basis as in (6.3), since this ensures that we have a basis for all of $L^2[a, b]$ and that computations with the combined basis can be performed without excessive loss of accuracy. For spline wavelets there are several interpretations of "Riesz basis". If the nested spaces $\{V_j\}_{j=0}^\infty$ are spline spaces of degree d, generated by knot vectors $(\mathbf{t}^j)_{j=0}^\infty$, the required inequalities are

$$K_1 \|(\mathbf{c}_0^T, \mathbf{w}_0^T, \mathbf{w}_1^T, \dots)\|_2 \leq \left\| \boldsymbol{\phi}_0^T \mathbf{c}_0 + \sum_{j=0}^\infty \boldsymbol{\psi}_j^T \mathbf{w}_j \right\|_2 \leq K_2 \|(\mathbf{c}_0^T, \mathbf{w}_0^T, \mathbf{w}_1^T, \dots)\|_2,$$

where K_1 and K_2 are two constants such that $K_1^{-1} K_2$ is not overly large. However, in the case of splines there is one added complication in that the constants may depend on the knots. It is therefore possible that we have a Riesz basis in the case where each new interior knot is inserted half way between two old knots, but that there are certain knot configurations where one or both of the constants become infinite. The ultimate form of stability would of course be that the constants are completely independent of the knots, as is the case for B-splines, see (6.4). To our knowledge, very little is known about the stability of spline wavelets at present; in fact it is not even known whether we have stability in the simplest nonuniform case where the knot intervals are halved each time.

In spite of the lack of results on stability of the wavelet construction, the stability of the B-spline basis also has consequences for wavelet computations. Recall that for B-splines $\mathbf{B}_{\boldsymbol{\tau}} = (B_{i,d})_{i=1}^{n}$ on a knot vector $\boldsymbol{\tau}$ we have the stability estimate

$$D_d^{-1}||\mathbf{c}||_2 \leq \left\|\tilde{\mathbf{B}}_{\boldsymbol{\tau}}^T\mathbf{c}\right\|_2 \leq ||\mathbf{c}||_2, \tag{6.28}$$

where the scaled B-spline $\tilde{B}_{i,d,\boldsymbol{\tau}}$ is defined by

$$\tilde{B}_{i,d,\boldsymbol{\tau}} = \left(\frac{d+1}{\tau_{i+d+1} - \tau_i}\right)^{1/2} B_{i,d,\boldsymbol{\tau}}. \tag{6.29}$$

Since

$$\left\|\tilde{\mathbf{B}}_{\boldsymbol{\tau}}^T\mathbf{c}\right\|_2^2 = \mathbf{c}^T \left\langle \tilde{\mathbf{B}}_{\boldsymbol{\tau}}, \tilde{\mathbf{B}}_{\boldsymbol{\tau}}^T \right\rangle \mathbf{c},$$

this implies that the largest eigenvalue of the Gram matrix $\langle \tilde{\mathbf{B}}_{\boldsymbol{\tau}}, \tilde{\mathbf{B}}_{\boldsymbol{\tau}}^T \rangle$ is 1, while the smallest is D_d^{-2}, so the condition number of the Gram matrix is D_d^2. In other words, if the B-splines are scaled as in (6.29), then the corresponding Gram matrix is always well-conditioned, while if the standard scaling is used, the condition number may be proportional to the ratio

$$\frac{\max_i(\tau_{i+d+1} - \tau_i)}{\min_i(\tau_{i+d+1} - \tau_i)}.$$

This means that if the wavelet decompositions are computed by solving normal equations, the coefficient matrix of the system (6.26) is well-conditioned if the B-splines are scaled as in (6.29).

When working with B-splines it is convenient, however, to work with the partition of unity normalization and rather scale the coefficient matrices in the linear systems so that they become well-conditioned. We introduce an $n \times n$ diagonal matrix $\mathbf{E}_{\boldsymbol{\tau}}$ with diagonal elements $(\tau_{i+d+1} - \tau_i)/(d+1)$ for $i = 1, \ldots, n$ and a similar $(n+m) \times (n+m)$ diagonal matrix $\mathbf{E}_{\mathbf{t}}$. Instead of (6.26) we can then solve the system

$$\mathbf{E}_{\boldsymbol{\tau}}^{-1/2}\langle \boldsymbol{\phi}, \boldsymbol{\phi}^T \rangle \mathbf{E}_{\boldsymbol{\tau}}^{-1/2}\left(\mathbf{E}_{\boldsymbol{\tau}}^{1/2}\mathbf{c}_0\right) = \mathbf{E}_{\boldsymbol{\tau}}^{-1/2}\langle \boldsymbol{\phi}, \boldsymbol{\gamma}^T \rangle \mathbf{c}_1,$$

which by the above analysis has a condition number bounded independently of the knot vectors.

The knot insertion matrix $\mathbf{P}$ can also be scaled so that its condition number is bounded. As before, we let $\mathbf{t} = (t_i)_{i=1}^{n+m}$ be a refined knot vector which contains $\boldsymbol{\tau}$ as a subsequence, and we let $\mathbf{P}$ be the matrix that relates the two bases as in (6.9). In Lyche and Mørken (1988) it was shown that two more inequalities can be fitted into (6.28),

$$D_d^{-1}||\mathbf{c}||_2 \leq \left\|\tilde{\mathbf{B}}_{\boldsymbol{\tau}}^T\mathbf{c}\right\|_2 \leq ||\tilde{\mathbf{P}}\mathbf{c}||_2 \leq ||\mathbf{c}||_2. \tag{6.30}$$

Here $\tilde{\mathbf{P}}$ is the scaled version of $\mathbf{P}$ given by
$$\tilde{\mathbf{P}} = \mathbf{E}_t^{1/2} \mathbf{P} \mathbf{E}_\tau^{-1/2}.$$
From the inequalities in (6.30) it follows that the condition number of $\tilde{\mathbf{P}}$ (defined as the ratio between its largest and smallest singular values) is bounded by D_d. This in turn means that some care should be observed when computing with the matrix $\langle \phi, \gamma^T \rangle = \mathbf{P} \langle \gamma, \gamma^T \rangle$. The lack of stability results for the matrix $\mathbf{Q}$ means that we have no guarantee that the system (6.27) can be solved accurately.

One suggestion for decomposition is to solve (6.25). Even if we use the scaled version $\tilde{\mathbf{P}}$ of $\mathbf{P}$, we cannot guarantee the stability of the system as long as the stability of $\mathbf{Q}$ is not under control.

So far we have only discussed stability with respect to the L^2-norm, but since the B-spline basis, scaled correctly, is stable in any L^p-space, see (6.4), one might try to establish stability in other norms as well. Stability with respect to the L^∞-norm would be of particular interest. Gram matrices are naturally related to the L^2-norm, so it is more difficult to prove that the systems (6.26) and (6.27) are stable in L^∞. For the system (6.25) it is known that $\mathbf{P}$ (without any further scaling) has bounded L^∞ condition number, and it appears likely that properly scaled versions of $\mathbf{Q}$ should be stable both in L^∞ and in L^2.

6.5 Tensor products and triangulations

In this section we briefly consider the bivariate setting, namely tensor product spline wavelets and (piecewise linear) spline wavelets on arbitrary bounded triangulations.

6.5.1 Tensor product spline wavelets

The generalization to the tensor product setting is fairly straightforward, but is included for the sake of completeness. Let $V_1 = V_0 \oplus W$ be one decomposition of a spline space as previously described, with corresponding matrices $\mathbf{P}, \mathbf{Q}, \mathbf{M}$, etc. and
$$\overline{V}_1 = \overline{V}_0 \oplus \overline{W}$$
another decomposition, as above, with dimensions $\overline{n}, \overline{m}$, and $\overline{n} + \overline{m}$, bases $\overline{\gamma}, \overline{\phi}$, and $\overline{\psi}$, and matrices $\overline{\mathbf{P}}, \overline{\mathbf{Q}}, \overline{\mathbf{M}}$, etc.

Then we have the decomposition of the tensor product space as
$$V_1 \times \overline{V}_1 = (V_0 \times \overline{V}_0) \oplus (V_0 \times \overline{W}) \oplus (W \times \overline{V}_0) \oplus (W \times \overline{W}),$$

i.e., three mutually orthogonal wavelet spaces representing functions that are (a) a coarse spline in the first and a wavelet in the second component, (b) a wavelet in the first and a coarse spline in the second component, and (c) a wavelet in both components.

A function f_1 in $V_1 \times \overline{V}_1$ can now be written using an $(n+m) \times (\overline{n}+\overline{m})$ coefficient matrix $\mathbf{C}^1$, namely

$$f_1(x,y) = \boldsymbol{\gamma}(x)^T \mathbf{C}^1 \overline{\boldsymbol{\gamma}}(y).$$

Similarly,

$$f_0(x,y) = \boldsymbol{\phi}(x)^T \mathbf{C}^0 \overline{\boldsymbol{\phi}}(y) \in V_0 \times \overline{V}_0$$
$$g^{(1)}(x,y) = \boldsymbol{\phi}(x)^T \mathbf{D}^{(1)} \overline{\boldsymbol{\psi}}(y) \in V_0 \times \overline{W}$$
$$g^{(2)}(x,y) = \boldsymbol{\psi}(x)^T \mathbf{D}^{(2)} \overline{\boldsymbol{\phi}}(y) \in W \times \overline{V}_0$$
$$g^{(3)}(x,y) = \boldsymbol{\psi}(x)^T \mathbf{D}^{(3)} \overline{\boldsymbol{\psi}}(y) \in W \times \overline{W}.$$

The matrix multiplication formula for reconstruction then becomes

$$\mathbf{C}^1 = \mathbf{P}(\mathbf{C}^0 \overline{\mathbf{P}}^T + \mathbf{D}^{(1)} \overline{\mathbf{Q}}^T) + \mathbf{Q}(\mathbf{D}^{(2)} \overline{\mathbf{P}}^T + \mathbf{D}^{(3)} \overline{\mathbf{Q}}^T).$$

Direct decomposition means first solving $\overline{n}+\overline{m}$ linear systems of size $n+m$ with $\mathbf{M} = \begin{pmatrix} \mathbf{P} & \mathbf{Q} \end{pmatrix}$ as coefficient matrix and the right-hand sides given by the columns of $\mathbf{C}^1$, as in

$$\begin{pmatrix} \mathbf{P} & \mathbf{Q} \end{pmatrix} \mathbf{R} = \mathbf{C}^1.$$

Splitting the resulting $(n+m) \times (\overline{n}+\overline{m})$ matrix $\mathbf{R}$ into two parts of sizes $n \times (\overline{n}+\overline{m})$ and $m \times (\overline{n}+\overline{m})$, i.e.,

$$\mathbf{R} = \begin{pmatrix} \mathbf{R}_s \\ \mathbf{R}_w \end{pmatrix}$$

shows that the second phase consists of first solving n systems of size $\overline{n}+\overline{m}$ with coefficient matrix $\overline{\mathbf{M}} = \begin{pmatrix} \overline{\mathbf{P}} & \overline{\mathbf{Q}} \end{pmatrix}$,

$$\begin{pmatrix} \overline{\mathbf{P}} & \overline{\mathbf{Q}} \end{pmatrix} \begin{pmatrix} (\mathbf{C}^0)^T \\ (\mathbf{D}^{(1)})^T \end{pmatrix} = \mathbf{R}_s^T,$$

in order to determine $\mathbf{C}^0$ and $\mathbf{D}^{(1)}$, and then solving m systems of size $\overline{n}+\overline{m}$, again with coefficient matrix $\overline{\mathbf{M}}$,

$$\begin{pmatrix} \overline{\mathbf{P}} & \overline{\mathbf{Q}} \end{pmatrix} \begin{pmatrix} (\mathbf{D}^{(2)})^T \\ (\mathbf{D}^{(3)})^T \end{pmatrix} = \mathbf{R}_w^T,$$

to compute $\mathbf{D}^{(2)}$ and $\mathbf{D}^{(3)}$.

On the other hand, the normal equation approach amounts to solving the matrix equations

$$\mathbf{FC}^0\overline{\mathbf{F}} = \mathbf{P}^T\mathbf{GC}^1\overline{\mathbf{GP}}$$
$$\mathbf{FD}^{(1)}\overline{\mathbf{H}} = \mathbf{P}^T\mathbf{GC}^1\overline{\mathbf{GQ}}$$
$$\mathbf{HD}^{(2)}\overline{\mathbf{F}} = \mathbf{Q}^T\mathbf{GC}^1\overline{\mathbf{GP}}$$
$$\mathbf{HD}^{(3)}\overline{\mathbf{H}} = \mathbf{Q}^T\mathbf{GC}^1\overline{\mathbf{GQ}}$$

for $\mathbf{C}^0$, $\mathbf{D}^{(1)}$, $\mathbf{D}^{(2)}$, and $\mathbf{D}^{(3)}$. Here $\mathbf{F} = \langle \boldsymbol{\phi}, \boldsymbol{\phi}^T \rangle$, $\overline{\mathbf{F}} = \langle \overline{\boldsymbol{\phi}}, \overline{\boldsymbol{\phi}}^T \rangle$, $\mathbf{H} = \langle \boldsymbol{\psi}, \boldsymbol{\psi}^T \rangle$, and $\overline{\mathbf{H}} = \langle \overline{\boldsymbol{\psi}}, \overline{\boldsymbol{\psi}}^T \rangle$ are univariate Gram matrices. As in the direct decomposition, this is achieved by solving linear systems, in this case in the first stage with the matrices $\mathbf{F}$ and $\mathbf{H}$ as coefficient matrices, and with the corresponding matrices $\overline{\mathbf{F}}$ and $\overline{\mathbf{H}}$ in the second.

6.5.2 Spline wavelets on triangulations

The bivariate non-tensor-product setting, i.e., the construction of compactly supported piecewise polynomial spline wavelets on triangulations, poses considerable challenges. This is especially true if the given triangulation is supposed to be bounded and arbitrary, and not an infinite uniform one, for which wavelets based on box splines have been constructed, de Boor et al. (1993).

Based on subdivision schemes it is possible to construct so-called surface wavelets, see Lounsbery et al. (1997), even for surface triangulations of arbitrary topology in 3D. These surface wavelets, however, possess in general no closed-form polynomial spline representations.

In the remainder of this section, we will consider only planar bounded triangulations in the following sense. A *triangle* is the convex hull of three non-collinear points $[x_1, x_2, x_3]$ in $\mathbb{R}^2$, and this triangle has three edges $[x_1, x_2]$, $[x_2, x_3]$ and $[x_3, x_1]$. Let $\mathcal{T}_0 = \{T_1, \ldots, T_M\}$ be a set of triangles and let $\Omega = \bigcup_{i=1}^M T_i$ be their union. Then we say that $\mathcal{T}_0$ is a *triangulation* if

(i) the intersection $T_i \cap T_j$, $i \neq j$, is either empty or corresponds to a common vertex or a common edge,
(ii) the number of boundary edges incident on a boundary vertex is two,
(iii) the region Ω is simply connected.

Associated with $\mathcal{T}_0$ are the set $\mathcal{E}_0$ which consists of all edges of triangles in $\mathcal{T}_0$, and the set $\mathcal{V}_0$ which consists of all the vertices of $\mathcal{T}_0$. By a *boundary vertex* or *boundary edge* we mean a vertex or edge contained in the boundary of Ω. All other vertices and edges are *interior*.

While it is possible to use very general refinement procedures in the univariate setting we described earlier, it is necessary to restrict our attention to uniform refinement of a triangulation in order to obtain the explicit results outlined later on.

Given a triangulation $\mathcal{T}_0$ we consider its *uniform refinement* $\mathcal{T}_1$, the triangulation formed by dividing each triangle $[x_1, x_2, x_3]$ in $\mathcal{T}_0$ into four congruent subtriangles. Specifically, if y_1, y_2, y_3 are the midpoints of the edges $[x_2, x_3]$, $[x_3, x_1]$, $[x_1, x_2]$ respectively, then the subtriangles are

$$[x_1, y_2, y_3], \quad [y_1, x_2, y_3], \quad [y_1, y_2, x_3], \quad [y_1, y_2, y_3].$$

We clearly have $\mathcal{V}_0 \subset \mathcal{V}_1$, and if we set $\mathcal{V}_* = \mathcal{V}_1 \setminus \mathcal{V}_0$, we see that a vertex in $\mathcal{V}_1$ is either a *coarse vertex*, namely an element of $\mathcal{V}_0$, or a *fine vertex*, an element of $\mathcal{V}_*$.

Contrary to the univariate setting, where the polynomial degree of the splines was arbitrary, we are only aware of explicit results on triangular spline wavelets of polynomial degree 1, i.e., piecewise linear functions.

Let V_j be the linear space of piecewise linear functions over $\mathcal{T}_j$, in other words, the set of functions which are linear over each triangle in $\mathcal{T}_j$ and continuous over Ω. It is then clear that

$$V_0 \subset V_1.$$

As basis functions for the spaces V_0 and V_1, respectively, we choose the *nodal* (or *hat*) functions, namely for each $v \in \mathcal{V}_0$, $\phi_v \in V_0$ such that $\phi_v(w) = \delta_{vw}$ for $w \in \mathcal{V}_0$, and analogously, for each $v \in \mathcal{V}_1$, $\gamma_v \in V_1$ such that $\gamma_v(w) = \delta_{vw}$ for $w \in \mathcal{V}_1$. The basis sets of nodal functions are then $\boldsymbol{\phi} = \{\phi_v\}_{v \in \mathcal{V}_0}$ and $\boldsymbol{\gamma} = \{\gamma_v\}_{v \in \mathcal{V}_1}$. Their cardinalities, and thus the dimensions of the spaces V_0 and V_1, are $|\mathcal{V}_0|$ and $|\mathcal{V}_1|$, respectively.

Note that the support of ϕ_v is the union of all triangles which contain the vertex $v \in \mathcal{V}_0$, called the *cell of v*, which we denote by

$$\mathcal{C}_v := \bigcup_{\substack{T \in \mathcal{T}_0 \\ v \in T}} T.$$

We can define an inner product $\langle \cdot, \cdot \rangle$ on $L^2(\Omega)$ by

$$\langle f, g \rangle = \sum_{T \in \mathcal{T}_0} \frac{1}{\mathrm{a}(T)} \int_T f(x) g(x) \, dx, \qquad f, g \in L^2(\Omega), \tag{6.31}$$

where $\mathrm{a}(T)$ is the area of triangle T. With respect to this (weighted) inner product, the spaces V_j become Hilbert spaces, with corresponding (weighted)

2-norm $\|f\|_2 = \langle f, f \rangle^{1/2}$. This norm is equivalent to the unweighted L^2-norm. In principle, it is also possible to use the standard, nonweighted inner product, and obtain similar results, but the use of the inner product in (6.31) results in a considerable reduction of the computations necessary to determine the wavelet coefficients, which is especially important for applications in computer graphics (see Lounsbery et al. (1997)).

Let W denote the relative orthogonal complement of the coarse space V_0 in the fine space V_1 with respect to the inner product (6.31), so that

$$V_1 = V_0 \oplus W.$$

The dimension of W is $|\mathcal{E}_0| = |\mathcal{V}_1| - |\mathcal{V}_0|$. Therefore, a basis of W is typically described by associating one basis element with each edge in the triangulation $\mathcal{T}_0$, or equivalently, with each midpoint of such an edge.

We let the coarse vertices $v, v^* \in \mathcal{V}_0$ be the endpoints of the edge $[v, v^*] \in \mathcal{E}_0$, which has the midpoint $u \in \mathcal{V}_*$. Finding a wavelet basis $\{\psi_u\}_{u \in \mathcal{V}_*}$ of W thus amounts to first describing the construction of an element $\psi_u \in V_1$ that actually lies in W, and then showing that the whole set of these elements in W is linearly independent. Additionally, the total collection should be stable, i.e., satisfy inequality (6.3).

The surface wavelet construction in Lounsbery et al. (1997), when specialized to this planar piecewise linear setting, starts with a fine hat function γ_u. Letting $L_0(\gamma_u)$ denote the orthogonal projection of γ_u onto V_0 using the inner product (6.31), the difference $\gamma_u - L_0(\gamma_u)$ is a wavelet in W. This function, however, has global support on all of Ω, while for computational reasons it would be desirable to have a locally supported basis for W.

We are aware of only two (different and independently developed) approaches for obtaining a basis for W with local support. The one in Stevenson (1998) originates from the numerical treatment of PDEs and can be generalized to higher dimensions, but produces piecewise linear spline wavelets, whose supports are somewhat larger than for the wavelets studied in Floater and Quak (2000) and Floater et al. (2000). The latter construction is based on an approach for computing directly the orthogonality conditions that are imposed for minimally supported elements of the wavelet space.

More specifically, the two different strategies share the common approach that for each coarse vertex $v \in \mathcal{V}_0$, some kind of auxiliary function in V_1 is defined whose support lies in the cell $\mathcal{C}_v$, and which satisfies most, but not all orthogonality conditions needed for an element in W. A linear combination of auxiliary functions corresponding to several coarse vertices is then used to produce an element in V_1 satisfying all orthogonality conditions, i.e., a true wavelet.

The approach in Stevenson (1998) uses an auxiliary function $\tilde{\phi}_v \in V_1$ for a given vertex $v \in \mathcal{V}_0$, with support in $\mathcal{C}_v$, that is orthogonal to all but one coarse hat function, namely the one for v itself, i.e.,

$$\tilde{\phi}_v \perp \phi_w, \quad w \in \mathcal{V}_0, \quad w \neq v.$$

In addition to the endpoints of the edge $[v, v^*]$ containing u, let $a, b \in \mathcal{V}_0$ be the other two coarse vertices of the quadrilateral in $\mathcal{T}_0$ containing $[v, v^*]$, if the edge is interior, and just let $a \in \mathcal{V}_0$ be the remaining vertex of the triangle in $\mathcal{T}_0$ containing $[v, v^*]$, when the edge is a boundary one. A wavelet ψ_u^* is then constructed in Stevenson (1998) by a linear combination of the fine hat function γ_u and the auxiliary functions $\tilde{\phi}_v, \tilde{\phi}_{v^*}, \tilde{\phi}_a$, and $\tilde{\phi}_b$ in the interior case, and just $\tilde{\phi}_v, \tilde{\phi}_{v^*}$, and $\tilde{\phi}_a$ in the boundary case. Consequently,

$$\operatorname{supp}(\psi_u^*) \subset (\mathcal{C}_v \cup \mathcal{C}_{v^*} \cup \mathcal{C}_a \cup \mathcal{C}_b),$$

or in the boundary case,

$$\operatorname{supp}(\psi_u^*) \subset (\mathcal{C}_v \cup \mathcal{C}_{v^*} \cup \mathcal{C}_a).$$

Different auxiliary functions $\sigma_{v,u}$ for a coarse vertex $v \in \mathcal{V}_0$ and a fine vertex $u \in [v, v^*]$ are called semi-wavelets in Floater and Quak (2000). Here, the support of $\sigma_{v,u} \in V_1$ is also included in $\mathcal{C}_v$, but the function satisfies all but two orthogonality conditions, namely, the ones imposed by v and v^*,

$$\sigma_{v,u} \perp \phi_w, \quad w \in \mathcal{V}_0, \quad w \neq v, v^*.$$

Then it is possible to just add $\sigma_{v,u}$ and $\sigma_{v^*,u}$ to obtain a true wavelet ψ_u, resulting in a smaller support than for Stevenson's approach, as

$$\operatorname{supp}(\psi_u) \subset (\mathcal{C}_v \cup \mathcal{C}_{v^*}).$$

Note that wavelets whose support the dimension of the space of all wavelets whose support is contained in $\mathcal{C}_v \cup \mathcal{C}_{v^*}$, is not only larger than one, but the dimension even depends on the specific local topology of the triangulation, see Floater and Quak (1999). This, maybe, leaves a larger choice of basis elements, but it also becomes possible to make a wrong choice, i.e., one can easily find a choice of a nonzero element in W for each $u \in [v, v^*]$, with support in $\mathcal{C}_v \cup \mathcal{C}_{v^*}$, so that the total collection does not form a basis. In Floater and Quak (2000) it is shown that it is in fact possible to choose specific elements ψ_u of the given support, so that the total collection $\{\psi_u\}_{u \in \mathcal{V}_*}$ is a stable basis. The algorithmic aspects of this particular approach, including examples from terrain modeling, are presented in Floater et al. (2000).

6.6 Discussion of some examples

Perhaps the simplest situation is the one where **t** is obtained from $\boldsymbol{\tau}$ by adding one knot. The corresponding wavelet space is one-dimensional and the only B-wavelet is globally supported. The wavelet coefficient of a given function f then gives an indication of the importance of the new knot for the representation of f.

In recent years there has been considerable interest in so-called multiwavelets. Spline multiwavelets result from wavelet constructions on uniform knot vectors where each knot has a fixed multiplicity r. If the knots are at the integers, the complete B-spline basis is generated by r distinct B-splines and their integer translates. This is clearly a special case of our construction. The classical wavelet theory generalizes nicely to multiwavelets, see for example Strela (1996) and the references therein.

In some cases, for example for closed parametric curves, it is useful to deal with splines which are periodic on some interval $[a,b]$. For $n, d \in \mathbb{N}$ with $n \geq d+2$ let $\mathbf{t} = (t_j)_{j=1}^{n+2d+1}$ be a knot sequence with $t_{d+1} = a$, $t_{n+d+1} = b$ and $t_{i+n} = t_i + b - a$ for $i = 1, 2, \ldots, 2d+1$. On this $(b-a)$-periodic knot sequence we can define periodic B-splines which form a basis for a space of periodic splines. The dimension of this space is n. We refer to Schumaker (1981) for further details. Spline wavelets can be constructed for nested periodic spline spaces, see Lyche and Schumaker (2000) for the quadratic, trigonometric spline case. In general the matrices **P** and **Q** will be as in the non-periodic case except that we get a few corner elements due to wraparound of some columns.

The standard adaptation of any kind of wavelet approach to an interval, as mentioned already in the introduction, is to keep most of the interior scaling and wavelet functions, i.e., functions whose supports lie completely within the interval. Special boundary scaling and wavelet functions are then introduced using approaches depending on the specific type of multiresolution analysis. For spline wavelets of degree d, this means that uniform dyadic knots are used in the interior of the interval, with knots of multiplicity $d+1$ at the interval endpoints, i.e., for nonnegative integers d, n the knot sequence is

$$\mathbf{u}_{d,n} = \{\overbrace{0, \ldots, 0}^{d+1}\} \cup \left\{\frac{k}{2^n} : 1 \leq k \leq 2^n - 1\right\} \cup \{\overbrace{1, \ldots, 1}^{d+1}\}.$$

This setting was first considered in Chui and Quak (1992). The corresponding decomposition and reconstruction algorithms were investigated in Quak and Weyrich (1994a), their tensor-product versions in Quak and

Weyrich (1994b), and a spline wavelet packet approach in Quak and Weyrich (1997).

As a typical and very frequently used example, let us consider the cubic case $d = 3$, for which we want to provide all numerical values needed to implement the algorithms of Section 6.4. Consider specifically the knot sequences $\tau = \mathbf{u}_{3,3}$ and its uniform refinement $\mathbf{t} = \mathbf{u}_{3,4}$. The corresponding spline spaces V_0 and V_1 are of course nested, have dimensions 11 and 19, respectively, and contain all polynomials of degree 3. The corresponding matrices $\mathbf{P}$ and $\mathbf{Q}$ thus have dimensions 19×11 and 19×8 and are given below. These matrices contain the information that is necessary also for any $n \geq 4$, as further refinement only leads to more interior columns, while the number of boundary columns as well as their entries remain unchanged.

$$16\mathbf{P} = \begin{pmatrix} 16 & 0 & 0 & 0 & 0 & 0 & 0 & 0 & 0 & 0 & 0 \\ 8 & 8 & 0 & 0 & 0 & 0 & 0 & 0 & 0 & 0 & 0 \\ 0 & 12 & 4 & 0 & 0 & 0 & 0 & 0 & 0 & 0 & 0 \\ 0 & 3 & 11 & 2 & 0 & 0 & 0 & 0 & 0 & 0 & 0 \\ 0 & 0 & 8 & 8 & 0 & 0 & 0 & 0 & 0 & 0 & 0 \\ 0 & 0 & 2 & 12 & 2 & 0 & 0 & 0 & 0 & 0 & 0 \\ 0 & 0 & 0 & 8 & 8 & 0 & 0 & 0 & 0 & 0 & 0 \\ 0 & 0 & 0 & 2 & 12 & 2 & 0 & 0 & 0 & 0 & 0 \\ 0 & 0 & 0 & 0 & 8 & 8 & 0 & 0 & 0 & 0 & 0 \\ 0 & 0 & 0 & 0 & 2 & 12 & 2 & 0 & 0 & 0 & 0 \\ 0 & 0 & 0 & 0 & 0 & 8 & 8 & 0 & 0 & 0 & 0 \\ 0 & 0 & 0 & 0 & 0 & 2 & 12 & 2 & 0 & 0 & 0 \\ 0 & 0 & 0 & 0 & 0 & 0 & 8 & 8 & 0 & 0 & 0 \\ 0 & 0 & 0 & 0 & 0 & 0 & 2 & 12 & 2 & 0 & 0 \\ 0 & 0 & 0 & 0 & 0 & 0 & 0 & 8 & 8 & 0 & 0 \\ 0 & 0 & 0 & 0 & 0 & 0 & 0 & 2 & 11 & 3 & 0 \\ 0 & 0 & 0 & 0 & 0 & 0 & 0 & 0 & 4 & 12 & 0 \\ 0 & 0 & 0 & 0 & 0 & 0 & 0 & 0 & 0 & 8 & 8 \\ 0 & 0 & 0 & 0 & 0 & 0 & 0 & 0 & 0 & 0 & 16 \end{pmatrix}$$

Note that the entries of $\mathbf{Q}$ near the boundary differ somewhat from those derived in Quak and Weyrich (1994a). There, the boundary spline wavelets were constructed in order to preserve as many coefficients from the interior elements as possible. Consequently, the boundary spline wavelets in Quak and Weyrich (1994a) are not B-wavelets, in the sense that their index supports are not minimal. Their actual support intervals, however, are the same as for the B-wavelets.

The matrix **Q** as given here offers also a nice illustration of the as yet unsolved problem of how the various B-wavelets, i.e., the columns of **Q**, should be scaled. Here, more or less arbitrarily and just to save some space, each column has 1 and -124 as its first (or last) two nonzero entries. To achieve the scaling as in Lemma 6.4.1, for example, the two interior columns need to be multiplied by $1/80640$, the first and last one by $27877/5025860410$, the second and next to last by $82900002/876051996025$, and the third and third to last by $1584/130442935$.

$$\mathbf{Q} = \begin{pmatrix}
-\frac{1136914560}{27877} & 0 & 0 & 0 & 0 & 0 & 0 & 0 \\
\frac{1655323200}{27877} & \frac{9450650880}{1381667} & 0 & 0 & 0 & 0 & 0 & 0 \\
-\frac{1321223960}{27877} & -\frac{77583612430}{4145001} & -\frac{153545}{396} & 0 & 0 & 0 & 0 & 0 \\
\frac{633094403}{27877} & \frac{409599117799}{16580004} & \frac{6643465}{3168} & 1 & 0 & 0 & 0 & 0 \\
-\frac{229000092}{27877} & -\frac{36869700393}{1381667} & -\frac{738445}{88} & -124 & 0 & 0 & 0 & 0 \\
\frac{46819570}{27877} & \frac{157389496903}{8290002} & \frac{29839177}{1584} & 1677 & 1 & 0 & 0 & 0 \\
-124 & -\frac{32916268667}{4145001} & -\frac{19335989}{792} & -7904 & -124 & 0 & 0 & 0 \\
1 & \frac{27809640281}{16580004} & \frac{58651607}{3168} & 18482 & 1677 & 1 & 0 & 0 \\
0 & -124 & -\frac{521819}{66} & -24264 & -7904 & -124 & 0 & 0 \\
0 & 1 & \frac{442733}{264} & 18482 & 18482 & \frac{442733}{264} & 1 & 0 \\
0 & 0 & -124 & -7904 & -24264 & -\frac{521819}{66} & -124 & 0 \\
0 & 0 & 1 & 1677 & 18482 & \frac{58651607}{3168} & \frac{27809640281}{16580004} & 1 \\
0 & 0 & 0 & -124 & -7904 & -\frac{19335989}{792} & -\frac{32916268667}{4145001} & -124 \\
0 & 0 & 0 & 1 & 1677 & \frac{29839177}{1584} & \frac{157389496903}{8290002} & \frac{46819570}{27877} \\
0 & 0 & 0 & 0 & -124 & -\frac{738445}{88} & -\frac{36869700393}{1381667} & -\frac{229000092}{27877} \\
0 & 0 & 0 & 0 & 1 & \frac{6643465}{3168} & \frac{409599117799}{16580004} & \frac{633094403}{27877} \\
0 & 0 & 0 & 0 & 0 & -\frac{153545}{396} & -\frac{77583612430}{4145001} & -\frac{1321223960}{27877} \\
0 & 0 & 0 & 0 & 0 & 0 & \frac{9450650880}{1381667} & \frac{1655323200}{27877} \\
0 & 0 & 0 & 0 & 0 & 0 & 0 & -\frac{1136914560}{27877}
\end{pmatrix}$$

In this special case, it is possible to determine a strategy for interlacing the columns of **P** and **Q** to obtain a banded square matrix, allowing to implement decomposition by solving the (permuted) system by means of a special banded system solver. For each column of **P** and **Q**, respectively, the index of the row which contains the element of largest absolute value is determined. This index then provides the column number for the overall square matrix where the original column is to be placed. For the given example matrices, the columns of **P** thus become columns $1, 3, 4, 6, 8, 10, 12, 14, 16, 17, 19$

of the permuted square matrix, while the columns of **Q** become the columns $2, 5, 7, 9, 11, 13, 15, 18$.

In order to implement decomposition via the linear system (6.26), we need the Gram matrices of the B-splines. Note again that in this special case, also these Gram matrices differ only in the number of interior columns, while the blocks corresponding to the boundary elements need to be computed only once and can be used for all refinement levels.

The 19×19 inner product matrix is given by $40320 \langle \boldsymbol{\gamma}, \boldsymbol{\gamma}^T \rangle =$

$$\begin{pmatrix}
720 & 441 & 93 & 6 & 0 & 0 & 0 & 0 & 0 & 0 & 0 & 0 & 0 & 0 & 0 & 0 & 0 & 0 & 0 \\
441 & 1116 & \frac{1575}{2} & 174 & \frac{3}{2} & 0 & 0 & 0 & 0 & 0 & 0 & 0 & 0 & 0 & 0 & 0 & 0 & 0 & 0 \\
93 & \frac{1575}{2} & 1647 & 1132 & \frac{239}{2} & 1 & 0 & 0 & 0 & 0 & 0 & 0 & 0 & 0 & 0 & 0 & 0 & 0 & 0 \\
6 & 174 & 1132 & 2416 & 1191 & 120 & 1 & 0 & 0 & 0 & 0 & 0 & 0 & 0 & 0 & 0 & 0 & 0 & 0 \\
0 & \frac{3}{2} & \frac{239}{2} & 1191 & 2416 & 1191 & 120 & 1 & 0 & 0 & 0 & 0 & 0 & 0 & 0 & 0 & 0 & 0 & 0 \\
0 & 0 & 1 & 120 & 1191 & 2416 & 1191 & 120 & 1 & 0 & 0 & 0 & 0 & 0 & 0 & 0 & 0 & 0 & 0 \\
0 & 0 & 0 & 1 & 120 & 1191 & 2416 & 1191 & 120 & 1 & 0 & 0 & 0 & 0 & 0 & 0 & 0 & 0 & 0 \\
0 & 0 & 0 & 0 & 1 & 120 & 1191 & 2416 & 1191 & 120 & 1 & 0 & 0 & 0 & 0 & 0 & 0 & 0 & 0 \\
0 & 0 & 0 & 0 & 0 & 1 & 120 & 1191 & 2416 & 1191 & 120 & 1 & 0 & 0 & 0 & 0 & 0 & 0 & 0 \\
0 & 0 & 0 & 0 & 0 & 0 & 1 & 120 & 1191 & 2416 & 1191 & 120 & 1 & 0 & 0 & 0 & 0 & 0 & 0 \\
0 & 0 & 0 & 0 & 0 & 0 & 0 & 1 & 120 & 1191 & 2416 & 1191 & 120 & 1 & 0 & 0 & 0 & 0 & 0 \\
0 & 0 & 0 & 0 & 0 & 0 & 0 & 0 & 1 & 120 & 1191 & 2416 & 1191 & 120 & 1 & 0 & 0 & 0 & 0 \\
0 & 0 & 0 & 0 & 0 & 0 & 0 & 0 & 0 & 1 & 120 & 1191 & 2416 & 1191 & 120 & 1 & 0 & 0 & 0 \\
0 & 0 & 0 & 0 & 0 & 0 & 0 & 0 & 0 & 0 & 1 & 120 & 1191 & 2416 & 1191 & \frac{239}{2} & \frac{3}{2} & 0 & 0 \\
0 & 0 & 0 & 0 & 0 & 0 & 0 & 0 & 0 & 0 & 0 & 1 & 120 & 1191 & 2416 & 1132 & 174 & 6 & 0 \\
0 & 0 & 0 & 0 & 0 & 0 & 0 & 0 & 0 & 0 & 0 & 0 & 1 & \frac{239}{2} & 1132 & 1647 & \frac{1575}{2} & 93 & 0 \\
0 & 0 & 0 & 0 & 0 & 0 & 0 & 0 & 0 & 0 & 0 & 0 & 0 & \frac{3}{2} & 174 & \frac{1575}{2} & 1116 & 441 & 0 \\
0 & 0 & 0 & 0 & 0 & 0 & 0 & 0 & 0 & 0 & 0 & 0 & 0 & 0 & 6 & 93 & 441 & 720 & 0
\end{pmatrix}$$

References

Ahlberg, J.H. and Nilson, E.N. (1965). Orthogonality properties of spline functions. *J. Math. Anal. Appl.*, **11**, 321–337.

Boor, C. de (1976). Splines as linear combinations of B-splines. A survey. In *Approximation Theory, II*, ed. G.G. Lorentz, C.K. Chui and L.L. Schumaker, pp. 1–47. Academic Press, New York.

Boor, C. de, Höllig, K. and Riemenschneider, S. (1993). *Box Splines*. Springer-Verlag, New York.

Boor, C. de, Lyche, T. and Schumaker, L.L. (1976). On calculating with B-splines II. Integration. In *Numerische Methoden der Approximationstheorie Vol. 3, ISNM 30*, ed. L. Collatz, G. Meinardus and H. Werner, pp. 123–146. Birkhäuser Verlag, Basel.

Buhmann, M. and Micchelli, C.A. (1992). Spline prewavelets for non-uniform knots. *Numer. Math.*, **61**, 455–474.

Carnicer, J.M., Dahmen, W. and Peña, J.M. (1996). Local decomposition of refinable spaces and wavelets. *Appl. Comput. Harmonic Anal.*, **3**, 127–153.

Chui, C. (1992). *An Introduction to Wavelets*. Academic Press, Boston.

Chui, C.K. and Quak, E.G. (1992). Wavelets on a bounded interval. In *Numerical Methods in Approximation Theory, ISNM 105*, ed. D. Braess, L.L. Schumaker, pp. 53–75. Birkhäuser, Basel.

Chui, C.K. and De Villiers, J. (1998). Spline-wavelets with arbitrary knots on a bounded interval: Orthogonal decomposition and computational algorithms. *Comm. in Appl. Anal.*, **4**, 457–486.

Cohen, A., Daubechies, I. and Vial, P. (1993). Wavelets on the interval and fast wavelet transforms. *Appl. Comput. Harmonic Anal.*, **1**, 54–81.

Cohen, E., Lyche, T.and Riesenfeld, R. (1980). Discrete B-splines and subdivision techniques in computer-aided geometric design and computer graphics. *Comp. Graphics and Image Proc.*, **14**, 87–111.

Dahmen, W., Kunoth, A. and Urban, K. (1999). Biorthogonal spline wavelets on the interval—stability and moment conditions. *Appl. Comput. Harmonic Anal.*, **6**, 132–196.

Daubechies, I. (1992). *Ten Lectures on Wavelets*. CBMS Conf. Series in Appl. Math., Vol. 61, SIAM, Philadelphia.

Daubechies, I., Guskov, I., Schröder, P. and Sweldens, W. (1999). Wavelets on irregular point sets. *Phil. Trans. R. Soc. Lon. A.*, **357**, 2397–2413.

Floater, M.S. and Quak, E.G. (1999). Piecewise linear prewavelets on arbitrary triangulations. *Numer. Math*, **82**, 221–252.

Floater, M.S. and Quak, E.G. (2000). Linear independence and stability of piecewise linear prewavelets on arbitrary triangulations. *SIAM J. Numer. Anal.*, **38**, 58–79.

Floater, M.S., Quak, E.G. and Reimers, M. (2000). Filter bank algorithms for piecewise linear prewavelets on arbitrary triangulations. *J. Comp. and Appl. Math.*, to appear.

Karlin, S. (1968). *Total Positivity*. Stanford University Press, Stanford.

Lounsbery, M., DeRose, T.D. and Warren, J. (1997). Multiresolution analysis for surfaces of arbitrary topological type. *ACM Trans. Graphics*, **16**, 34–73.

Lyche, T. and Mørken, K. (1986). Making the Oslo algorithm more efficient. *SIAM J. Numer. Anal.*, **23**, 663–675.

Lyche, T. and Mørken, K. (1988). A data reduction strategy for splines. *IMA J. Numer. Anal.*, **8**, 185–208.

Lyche, T. and Mørken, K. (1992). Spline-wavelets of minimal support. In *Numerical Methods in Approximation Theory, ISNM 105*, ed. D. Braess, L.L. Schumaker, pp. 177–194. Birkhäuser, Basel.

Lyche, T. and Schumaker, L.L. (1994). L-spline wavelets. In *Wavelets: Theory, Algorithms, and Applications*, ed. C. Chui, L. Montefusco, and L. Puccio, pp. 197–212. Academic Press, New York.

Lyche, T. and Schumaker, L.L. (2000). A multiresolution tensor spline method for fitting functions on the sphere. *SIAM J. Scient. Comp.*, to appear.

Meyer, Y. (1991). Ondelettes sur l'interval. *Rev. Mat. Iberoamericana*, **7**, 115–133.
Mørken, K. (1991). Some identities for products and degree raising of splines. *Constr. Approx.*, **7**, 195–208.
Phillips, J.L. and Hanson, R.J. (1974). Gauss quadrature rules with B-spline weight functions. *Math. Comp.*, **28**, 666.
Quak, E.G. and Weyrich, N. (1994a). Decomposition and reconstruction algorithms for spline wavelets on a bounded interval. *Appl. Comput. Harmonic Anal.*, **1**, 217–231.
Quak, E.G. and Weyrich, N. (1994b). Decomposition and reconstruction algorithms for bivariate spline wavelets on the unit square. In *Wavelets, Images, and Surface Fitting*, ed. P.J. Laurent, A. Le Méhauté, and L.L. Schumaker, pp. 419–428. A.K. Peters, Boston.
Quak, E.G. and Weyrich, N. (1997). Algorithms for spline wavelet packets on an interval. *BIT*, **37**, 76–95.
Schumaker, L.L. (1981). *Spline Functions: Basic Theory*. Wiley–Interscience, New York.
Stevenson, R. (1998). Piecewise linear (pre-)wavelets on non-uniform meshes. In *Multigrid Methods V*, ed. W. Hackbusch and G. Wittum, pp. 306–319. Springer, Berlin.
Strela, V. (1996). *Multiwavelets: Theory and Applications*. Ph.D. thesis, MIT.
Sweldens, W. and Schröder, P. (1996). Building your own wavelets at home. In *Wavelets in Computer Graphics*, ACM SIGGRAPH Course Notes.

7
Applied and computational aspects of nonlinear wavelet approximation

A. COHEN

Abstract

Nonlinear approximation has recently found computational applications such as data compression, statistical estimation and adaptive schemes for partial differential and integral equations, especially through the development of wavelet-based methods. The goal of this chapter is to provide a short survey of nonlinear wavelet approximation from the perspective of these applications, as well as to highlight some remaining open questions.

7.1 Introduction

Numerous problems in approximation theory have in common the following general setting: we are given a family of subsets $(S_N)_{N \geq 0}$ of a normed space X, and for $f \in X$, we consider the *best approximation error*

$$\sigma_N(f) := \inf_{g \in S_N} \|f - g\|_X. \tag{7.1}$$

Typically, N represents the number of parameters needed to describe an element in S_N, and in most cases of interest, $\sigma_N(f)$ goes to zero as this number tends to infinity.

For a given f we can then study the *rate of approximation*, i.e., the range of $r \geq 0$ for which there exists a $C > 0$ such that

$$\sigma_N(f) \leq C N^{-r}. \tag{7.2}$$

Note that in order to study such an asymptotic behavior, we can use a sequence of *near best approximation*, i.e., $f_N \in S_N$ such that

$$\|f - f_N\|_X \leq C \sigma_N(f),$$

with $C > 1$ independent of N. Such a sequence always exists even when the infimum is not attained in (7.1), and clearly (7.2) is equivalent to the same estimate with $\|f - f_N\|_X$ in place of $\sigma_N(f)$.

Linear approximation deals with the situation when the S_N are linear subspaces. Classical instances of linear approximation families are the following:

(1) polynomial approximation: $S_N := \Pi_N$, the space of algebraic polynomials of degree N;
(2) spline approximation with uniform knots: with some integer $0 \leq k < m$ fixed, S_N is the spline space on $[0, 1]$ consisting of C^k piecewise polynomial functions of degree m on the intervals $[j/N, (j+1)/N]$, $j = 0, \ldots, N-1$;
(3) linear approximation in a basis: given a basis $(e_k)_{k \geq 0}$ in a Banach space, $S_N := \mathrm{span}(e_0, \ldots, e_N)$.

In all these situations N is typically the dimension of S_N, possibly up to some multiplicative constant. In contrast, *nonlinear approximation* addresses the situation where the S_N are not linear spaces, but are still typically characterized by $\mathcal{O}(N)$ parameters. Instances of nonlinear approximation families are the following:

(1) rational approximation: $S_N := \{\frac{p}{q} \,:\, p, q \in \Pi_N\}$, the set of rational functions of degree N;
(2) free-knot spline approximation: with some integer $0 \leq k < m$ fixed, S_N is the spline space on $[0, 1]$ with N free knots consisting of C^k piecewise polynomial functions of degree m on intervals $[x_j, x_{j+1}]$, for all partitions $0 = x_0 < x_1 < \cdots < x_{N-1} < x_N = 1$;
(3) N-term approximation in a basis: given a basis $(e_k)_{k \geq 0}$ in a Banach space, S_N is the set of all possible combinations $\sum_{k \in E} x_k e_k$ with $\#(E) \leq N$, where $\#(E)$ is the cardinality of E.

Note that these examples are in some sense nonlinear generalizations of the previous linear examples, since they include each of them as particular subsets. Also note that in all of these examples (except for the splines with uniform knots), we have the natural property $S_N \subset S_{N+1}$, which expresses the fact that the approximation is "refined" as N grows.

On a theoretical level a basic problem both for linear and nonlinear approximation can be stated as follows:

Problem 1 *Given a nonlinear family* $(S_N)_{N \geq 0}$, *what are the analytic properties of a function* f *which ensure a prescribed rate* $\sigma_N(f) \leq CN^{-r}$?

By "analytic properties", we typically have in mind smoothness, since we know that in many contexts a prescribed rate r can be achieved provided that f belongs to some smoothness class $X_r \subset X$. Ideally, one might hope to identify the *maximal class* X_r such that the rate r is ensured, i.e., have a sharp result of the type

$$f \in X_r \Leftrightarrow \sigma_N(f) \leq CN^{-r}.$$

Another basic problem, perhaps on a slightly more applied level, is the effective construction of near best approximants.

Problem 2 *Given a nonlinear family $(S_N)_{N \geq 0}$, find a simple implementable procedure $f \mapsto f_N \in S_N$ such that $\|f - f_N\|_X \leq C\sigma_N(f)$ for all $N \geq 0$.*

In the case of linear approximation, this question is usually solved if we can find a sequence of projectors $P_N : X \mapsto S_N$ such that $\|P_N\|_{X \to X} \leq K$ with K independent of N (in this case, simply take $f_N = P_N f$ and note that $\|f - f_N\|_X \leq (1 + K)\sigma_N(f)$). It is in general a more difficult problem in the case of nonlinear methods. Since the 1960s, research in approximation theory has evolved significantly toward nonlinear methods, in particular solving the above two problems for various spaces S_N.

More recently, nonlinear approximation has become attractive on a more applied level, as a tool for understanding and analyzing the performance of *adaptive methods* in signal and image processing, statistical estimation and numerical simulation. This is in part due to the emergence of *wavelet bases* for which simple N-term approximations (derived by thresholding the coefficients) yield in some sense optimal adaptive approximations. In such applications the problems that arise are typically the following.

Problem 3 (data compression) *How can we exploit the reduction of parameters in the approximation of f by $f_N \in S_N$ from the perspective of optimally encoding f by a small number of bits? This raises the question of a proper quantization of these parameters, namely, a proper encoding of these parameters by a finite number of bits.*

Problem 4 (statistical estimation) *Can we use nonlinear approximation as a denoising scheme? Here we need to understand the interplay between the approximation process and the presence of noise.*

Problem 5 (numerical simulation) *How can we compute a proper nonlinear approximation of a function u which is not given to us as data but as the solution of some problem $F(u) = 0$? This is in particular the goal of adaptive refinement strategies in the numerical treatment of PDEs.*

The goal of this chapter is to briefly survey the subject of nonlinear approximation, with particular focus on these last more applied questions, and some emphasis on wavelet-based methods. We would like to point out that these questions are also addressed in the survey paper DeVore (1997) which contains a more substantial development of the theoretical aspects of nonlinear approximation. We hope that this chapter will be useful to the non-expert who wants to obtain a first general and intuitive view of the subject, motivated by its various applications, before going into a more detailed study.

The chapter is organized as follows. To start, we discuss in Section 7.2 a simple example, based on piecewise constant functions, which illustrates the differences between linear and nonlinear approximation. In Section 7.3, we address the important case of N-term wavelet approximation and thresholding algorithms. Applications to signal compression and estimation are discussed in Section 7.4 and Section 7.5. Applications to adaptive numerical simulation are briefly described in Section 7.6. Finally, we conclude in Section 7.7 with some remarks and open problems which naturally arise in the multivariate setting.

7.2 A simple example

Let us consider the approximation of functions defined on the unit interval $I = [0,1]$ by piecewise constant functions. More precisely, given a disjoint partition of I into N subintervals $I_0, \ldots, I_{N-1}$, and a function f in $L^1(I)$, we shall approximate f on each I_k by its average $a_{I_k}(f) = |I_k|^{-1} \int_{I_k} f(t) dt$. The resulting approximant can thus be written as

$$f_N := \sum_{k=1}^{N} a_{I_k}(f) \chi_{I_k},$$

where χ_A is the characteristic function of a set A, i.e., $\chi_A(x) = 1$ if $x \in A$, 0 elsewhere.

If the I_k are fixed independently of f, then f_N is simply the orthogonal projection of f onto the space of piecewise constant functions on the partition I_k, i.e., a *linear approximation* of f. A natural choice is the uniform partition $I_k := [k/N, (k+1)/N]$. With such a choice, let us now consider the error between f and f_N, for example in the L^∞ (uniform) metric. For this, we shall assume that f is in $C(I)$, the space of continuous functions on I. It is then clear that on each I_k we have

$$|f(t) - f_N(t)| = |f(t) - a_{I_k}(f)| \leq \sup_{t,u \in I_k} |f(t) - f(u)|.$$

We thus have the error estimate

$$\|f - f_N\|_\infty \leq \sup_k \sup_{t,u \in I_k} |f(t) - f(u)|. \tag{7.3}$$

This can be converted into an estimate in terms of N, under some additional smoothness assumptions on f. In particular, if f has a bounded first derivative, we have $\sup_{t,u \in I_k} |f(t) - f(u)| \leq |I_k| \|f'\|_\infty = N^{-1} \|f'\|_\infty$, and thus

$$\|f - f_N\|_\infty \leq N^{-1} \|f'\|_\infty. \tag{7.4}$$

Similarly, if f is in the Hölder space C^α for some $\alpha \in (0,1)$, we obtain an estimate in $\mathcal{O}(N^{-\alpha})$. By considering simple examples such as $f(x) = x^\alpha$ for $0 < \alpha \leq 1$, one can easily check that these rates are actually sharp.

If we now consider an *adaptive partition* where the I_k depend on the function f itself, we enter the topic of *nonlinear approximation*. In order to understand the potential gain in switching from uniform to adaptive partitions, let us consider a function f such that f' is integrable. Since we have $\sup_{t,u \in I_k} |f(t) - f(u)| \leq \int_{I_k} |f'(t)| dt$, we see that a natural choice of the I_k can be made by equalizing the quantities $\int_{I_k} |f'(t)| dt = N^{-1} \int_0^1 |f'(t)| dt$, so that, in view of the basic estimate (7.3), we obtain the error estimate

$$\|f - f_N\|_\infty \leq N^{-1} \|f'\|_1.$$

In comparison with the uniform/linear situation, we thus have obtained the same rate as in (7.4) for a larger class of functions, since f' is not assumed to be bounded but only integrable. From a slightly different perspective, the nonlinear approximation rate might be significantly better than the linear rate for a fixed function f. For example, the function $f(x) = x^\alpha$, $0 < \alpha \leq 1$, has the linear rate $N^{-\alpha}$ and the nonlinear rate N^{-1} since $f'(x) = \alpha x^{\alpha-1}$ is in $L^1(I)$.

The above construction of an adaptive partition based on balancing the L^1 norm of f' is somewhat theoretical. In particular a suitable adaptive approximation algorithm should also operate on functions which are not in $W^{1,1}$. Let us describe two natural algorithms for building an adaptive partition.

The first one is sometimes known as *adaptive splitting* (see, e.g., DeVore and Yu (1990)). Given a partition of $[0,1]$, and any interval I_k of this partition, we split I_k into two subintervals of equal size if $\|f - a_{I_k}(f)\|_{\infty, I_k} \geq \varepsilon$, otherwise we leave it as it was. Starting this procedure from the single interval $I = [0,1]$ and using a fixed tolerance $\varepsilon > 0$ at each step, we end up with an adaptive partition into dyadic intervals and a corresponding piecewise constant approximation f_N with $N = N(\varepsilon)$ pieces such that $\|f - f_N\|_\infty \leq \varepsilon$.

Another simple algorithm uses *wavelet thresholding*. Consider the decomposition of f in the Haar basis

$$f = a_I(f) + \sum_{j \geq 0} \sum_{k=0}^{2^j - 1} \langle f, \psi_{j,k} \rangle \psi_{j,k}. \tag{7.5}$$

We recall that $\psi_{j,k} = 2^{j/2} \psi(2^j \cdot - k)$ with $\psi = \chi_{[0,1/2]} - \chi_{[1/2,1]}$. For each $j \geq 0$, the sum $\sum_{k=0}^{2^j-1} \langle f, \psi_{j,k} \rangle \psi_{j,k}$ describes the fluctuation $P_{j+1}f - P_j f$ where P_j is the projector onto the space V_j of piecewise constant functions on the intervals $I_{j,k} = [2^{-j}k, 2^{-j}(k+1)]$, $k = 0, \ldots, 2^j - 1$. A natural adaptive approximation can be obtained by retaining in (7.5) only the N largest contributions in the L^∞ metric, i.e., the indices (j, k) corresponding to the N largest $\|d_{j,k} \psi_{j,k}\|_\infty = 2^{j/2} |d_{j,k}|$, where $d_{j,k} := \langle f, \psi_{j,k} \rangle$. Indeed, since $\psi_{j,k}$ has zero mean, we have $|\langle f, \psi_{j,k}\rangle| \leq 2^{-j/2} \inf_{c \in \mathbb{R}} \|f - c\|_{\infty, I_{j,k}}$ so that fine scale contributions will be retained only in the regions of important fluctuation, i.e., where they are needed to refine the approximation.

Both methods give rise to specific partitions and adaptive approximations f_N. Ideally, one would hope that these partitions are optimal in the sense that $\|f - f_N\|_\infty$ is up to a constant the best approximation error of f from piecewise constants on N free intervals. This is not true for both algorithms, but they are somehow close to this property in many senses. For example, if $f' \in L^p$ for some $p > 1$, one can prove the rates $\mathcal{O}(N^{-q})$ for any $q < 1$. Also note that both algorithms have natural generalizations to the case of approximation in the L^p metric, by using $\|f - a_{I_k}(f)\|_{p, I_k} \geq \varepsilon$ as a splitting criterion or keeping the N largest $\|d_{j,k} \psi_{j,k}\|_p = 2^{j(1/2 - 1/p)} |d_{j,k}|$.

In this chapter we focus on the wavelet approach for the following reason. In this approach the nonlinearity is reduced to a very simple operation (thresholding according to the size of the coefficients), resulting in simple and efficient algorithms for dealing with many applications, as well as a relatively simple analysis of these applications.

7.3 Wavelets and thresholding

Wavelet bases offer a simple and powerful setting for building nonlinear approximations of functions. *Simple* because a near-optimal approximation is often achieved by a straightforward thresholding procedure; *powerful* because it is in some sense optimal for a large variety of smoothness classes modeling the function to be approximated.

We first recall some general notations and features for these bases (see Daubechies (1992) and Cohen (1999) for more details). They are usually associated with multiresolution approximation spaces $(V_j)_{j \geq 0}$ such that $V_j \subset V_{j+1}$ and V_j is generated by a local basis $(\varphi_\lambda)_{\lambda \in \Gamma_j}$, where Γ_j is a set of indices typically associated to a spatial discretization of mesh size $h_j \sim 2^{-j}$. By *local* we mean that the supports are controlled by

$$\mathrm{diam}\big(\mathrm{supp}(\varphi_\lambda)\big) \leq C 2^{-j}$$

if $\lambda \in \Gamma_j$ and are "almost disjoint" in the sense that

$$\#\Big(\big\{\mu \in \Gamma_j \text{ s.t. } \mathrm{supp}(\varphi_\lambda) \cap \mathrm{supp}(\varphi_\mu) \neq \emptyset\big\}\Big) \leq C$$

with C independent of λ and j. A complement space W_j of V_j into V_{j+1} is generated by a similar local basis $(\psi_\lambda)_{\lambda \in \nabla_j}$, $\nabla_j = \Gamma_{j+1} \setminus \Gamma_j$. The full multiscale wavelet basis $(\psi_\lambda)_{\lambda \in \nabla}$, $\nabla = \cup_{j \geq 0} \nabla_j$ allows one to expand an arbitrary function f into $f = \sum_{\lambda \in \nabla} d_\lambda \psi_\lambda$, where $\nabla := \cup_{j \geq 0} \nabla_j$ and with the convention that we incorporate the functions $(\varphi_\lambda)_{\lambda \in \Gamma_0}$ into the first "layer" $(\psi_\lambda)_{\lambda \in \nabla_0}$.

In the standard constructions of wavelets on the Euclidean space $\mathbb{R}^d$, the scaling functions have the form $\varphi_\lambda = \varphi_{j,k} = 2^{jd/2}\varphi(2^j \cdot -k)$, $k \in \mathbb{Z}^d$, and similarly for the wavelets, so that Γ_j is naturally viewed as the uniform mesh $2^{-j}\mathbb{Z}^d$. In the case of a general domain $\Omega \in \mathbb{R}^d$, special adaptations of the basis functions are required near the boundary $\partial\Omega$, which are accounted for in the general notations Γ_j and ∇_j.

Wavelets need not be orthonormal, but one often requires that they constitute a Riesz basis of $L^2(\Omega)$, i.e., their finite linear combinations are dense in L^2 and for all sequences $(d_\lambda)_{\lambda \in \nabla}$ we have the norm equivalence

$$\bigg\| \sum_{\lambda \in \nabla} d_\lambda \psi_\lambda \bigg\|_2^2 \sim \sum_{\lambda \in \nabla} |d_\lambda|^2.$$

In such a case, the coefficients d_λ in the expansion of f are obtained by an inner product $d_\lambda = \langle f, \tilde\psi_\lambda \rangle$, where the dual wavelet $\tilde\psi_\lambda$ is an L^2-function. In the standard biorthogonal constructions, the dual wavelet system $(\tilde\psi_\lambda)_{\lambda \in \nabla}$ is also built from nested spaces $\tilde V_j$ and has similar local support properties as the primal wavelets ψ_λ. The practical advantage of such a setting is the possibility of "switching" between the "standard" (or "nodal") discretization of $f \in V_j$ in the basis $(\varphi_\lambda)_{\lambda \in \Gamma_j}$ and its "multiscale" representation in the basis $(\psi_\lambda)_{|\lambda| < j}$ by means of fast $\mathcal{O}(N)$ decomposition and reconstruction algorithms, where $N \sim 2^{dj}$ denotes the dimension of V_j in the case where Ω is bounded.

An important feature is the possibility of characterizing the smoothness of a function f through the numerical properties of its wavelet coefficients or of the error of (linear) approximation $\|f - P_j f\|$, where $P_j f := \sum_{|\lambda| \leq j} \langle f, \tilde{\psi}_\lambda \rangle \psi_\lambda$ is a projector onto V_j. In particular, for functions of d variables, Sobolev spaces are characterized by

$$\|f\|_{H^s}^2 \sim \|P_0 f\|_2^2 + \sum_{j \geq 0} 2^{2sj} \|f - P_j f\|_2^2 \sim \sum_{\lambda \in \nabla} 2^{2s|\lambda|} |d_\lambda|^2, \quad (7.6)$$

and fractional Hölder classes by

$$\|f\|_{C^s} \sim \|P_0 f\|_\infty + \sup_{j \geq 0} 2^{sj} \|f - P_j f\|_\infty \sim \sup_{\lambda \in \nabla} 2^{(s+d/2)|\lambda|} |d_\lambda|. \quad (7.7)$$

Such norm equivalences are essentially valid under the restriction that the wavelet ψ_λ itself has slightly more than H^s or C^s smoothness. They reflect the intuitive idea that the linear approximation error $\|f - P_j f\|_p$ decays like $\mathcal{O}(2^{-sj})$ or $\mathcal{O}(N^{-s/d})$ provided that f has "s derivatives in L^p". In order to give a precise general statement for any p, one needs to introduce the Besov spaces $B^s_{p,q}$ which measure smoothness of order $s > 0$ in L^p according to

$$\|f\|_{B^s_{p,q}} := \|f\|_p + \left\| \left(2^{sj} \omega_m(f, 2^{-j})_p \right)_{j \geq 0} \right\|_{\ell^q},$$

where $\omega_m(f, t)_p := \sup_{|h| \leq t} \|\Delta_h^m f\|_p = \sup_{|h| \leq t} \left\| \sum_{k=0}^{m} \binom{m}{k} (-1)^k f(\cdot - kh) \right\|_p$, is the mth order L^p modulus of smoothness and m is any integer strictly larger than s. Recall that we have $H^s \sim B^s_{2,2}$ for all $s > 0$, $C^s \sim B^s_{\infty,\infty}$ and $W^{s,p} \sim B^s_{p,p}$ for all non-integer $s > 0$ and $p \neq 2$. For such classes, the characterization by approximation error or wavelet coefficients takes the following form which generalizes (7.6) and (7.7):

$$\begin{aligned}\|f\|_{B^s_{p,q}} &\sim \|P_0 f\|_p + \left\| \left(2^{sj} \|f - P_j f\|_p \right)_{j \geq 0} \right\|_{\ell^q} \\ &\sim \left\| \left(2^{(s+d/2-d/p)j} \|(d_\lambda)_{|\lambda|=j}\|_{\ell^p} \right)_{j \geq 0} \right\|_{\ell^q}.\end{aligned} \quad (7.8)$$

Let us now turn to nonlinear wavelet approximation. Denoting by

$$S_N := \left\{ \sum_{\lambda \in E} c_\lambda \psi_\lambda \; ; \; \#(E) \leq N \right\},$$

the set of all possible N-term combinations of wavelets, we are interested in the behavior of $\sigma_N(f)$ as defined in (7.1) for some given error norm X. We first consider the case $X = L^2$ and assume for simplicity that the ψ_λ constitute an orthonormal basis. In this case, it is a straightforward computation that the best N-term approximation of a function f is achieved by its truncated expansion

$$f_N := \sum_{\lambda \in E_N(f)} d_\lambda \psi_\lambda,$$

where $E_N(f)$ contains the indices corresponding to the N largest $|d_\lambda|$. The approximation error is thus given by

$$\sigma_N(f) = \|f - f_N\|_2 = \left(\sum_{\lambda \notin E_N(f)} |d_\lambda|^2 \right)^{1/2} = \left(\sum_{n \geq N} d_n^2 \right)^{1/2},$$

where $(d_n)_{n \geq 0}$ is defined as the *decreasing rearrangement* of $\{|d_\lambda|, \lambda \in \nabla\}$ (i.e., d_{n-1} is the nth largest $|d_\lambda|$).

Consider now the Besov spaces $B^s_{\tau,\tau}$ where $s > 0$ and τ are linked by $1/\tau = 1/2 + s/d$. According to the norm equivalence (7.8) we note that these spaces are simply characterized by

$$\|f\|_{B^s_{\tau,\tau}} \sim \|(d_\lambda)_{\lambda \in \nabla}\|_{\ell^\tau}.$$

Thus if $f \in B^s_{\tau,\tau}$, we find that the decreasing rearrangement $(d_n)_{n \geq 0}$ satisfies

$$n|d_n|^\tau \leq \sum_{k=0}^{n-1} |d_k|^\tau \leq \sum_{k \geq 0} |d_k|^\tau = \sum_{\lambda \in \nabla} |d_\lambda|^\tau \leq C\|f\|^\tau_{B^s_{\tau,\tau}} < \infty,$$

and therefore

$$|d_n| \leq C n^{-1/\tau} \|f\|_{B^s_{\tau,\tau}}. \tag{7.9}$$

It follows that the approximation error is bounded by

$$\sigma_N(f) \leq C\|f\|_{B^s_{\tau,\tau}} \left(\sum_{n \geq N} n^{-\frac{2}{\tau}} \right)^{1/2} \leq C N^{\frac{1}{2} - \frac{1}{\tau}} \|f\|_{B^s_{\tau,\tau}} = C N^{-s/d} \|f\|_{B^s_{\tau,\tau}}. \tag{7.10}$$

At this stage let us make some remarks.

- As was previously noted, the rate $N^{-s/d}$ can be achieved by linear approximation for functions having s derivative in L^2, i.e., functions in H^s. Just as in the simple example of Section 7.2, the gain in switching to nonlinear approximation is that the class $B^s_{\tau,\tau}$ is larger than H^s. In particular $B^s_{\tau,\tau}$ contains discontinuous functions for arbitrarily large values of s, while functions in H^s are necessarily continuous if $s > d/2$.
- Note that $0 < \tau < 2$ and is monotonically decreasing in s.

- The rate (7.10) is implied by $f \in B^s_{\tau,\tau}$. On the other hand it is easy to check that (7.10) is equivalent to the property (7.9), which in itself is equivalent to the property that the sequence $(d_\lambda)_{\lambda \in \nabla}$ is in the weak space ℓ^τ_w, i.e.,

$$\ell^\tau_w = \Big\{ (d_\lambda)_{\lambda \in \nabla} : \#(\{\lambda : |d_\lambda| \leq \varepsilon\}) \leq C\varepsilon^{-\tau} \Big\}.$$

This shows that the property $f \in B^s_{\tau,\tau}$ is almost equivalent to the rate (7.10). One can easily check that the exact characterization of $B^s_{\tau,\tau}$ is given by the stronger property $\sum_{N\geq 1} (N^{s/d} \sigma_N(f))^\tau N^{-1} < \infty$.

- The space $B^s_{\tau,\tau}$ is *critically embedded in* L^2 in the sense that the injection is not compact. This can be viewed as an instance of the Sobolev embedding theorem, or directly checked in terms of the non-compact embedding of ℓ^τ into ℓ^2 when $\tau \leq 2$. In particular, $B^s_{\tau,\tau}$ is not contained in any Sobolev space H^s for $s > 0$. Therefore no convergence rate can be expected for the linear approximation of functions in $B^s_{\tau,\tau}$.

The general theory of nonlinear wavelet approximation developed by DeVore and his collaborators extends these results to various error norms, for which the analysis is far more difficult than for the L^2 norm. This theory is fully detailed in DeVore (1997), and we would like to summarize it by stressing three main types of results, the first two answering, respectively, problems 1 and 2 as described in the introduction.

Approximation and smoothness spaces Given an error norm $\|\cdot\|_X$ corresponding to some smoothness space in d dimensions, the space Y of those functions such that $\sigma_N(f) = \text{dist}_X(f, S_N) \leq CN^{-s/d}$ has a typical description in terms of another smoothness space. Typically, if X represents r orders of smoothness in L^p, then Y will represent $r+s$ orders of smoothness in L^τ with $1/\tau = 1/p + s/d$ and its injection in X is not compact. This generic result has a graphical interpretation displayed in Figure 7.1. In this figure, a point $(r, 1/p)$ represents function spaces with smoothness r in L^p, and the point Y sits s levels of smoothness above X on the critical embedding line of slope d emanating from X. Of course in order to obtain rigorous results, one needs to specify in each case the exact meaning of "s derivatives in L^p" and/or slightly modify the property $\sigma_N(f) \leq CN^{-s/d}$. For example, if $X = L^p$ for some $p \in (1, \infty)$, then $f \in B^s_{\tau,\tau} = Y$ with $1/\tau = 1/p + s/d$ if and only if $\sum_{N\geq 1} [N^{s/d} \sigma_N(f)]^\tau N^{-1} < \infty$. One also needs to assume that the wavelet basis has sufficient smoothness, since it should be at least contained in Y.

Realization of a near-best approximation For various error metrics X, a near best approximation of f in S_N is achieved by $f_N := \sum_{\lambda \in \Lambda_N(f)} d_\lambda \psi_\lambda$

where $d_\lambda := \langle f, \tilde{\psi}_\lambda \rangle$ are the wavelet coefficients of f and $\Lambda_N(f)$ is the set of indices corresponding to the N largest contributions $\|d_\lambda \psi_\lambda\|_X$. This fact is rather easy to prove when X is a Besov space, by using (7.8). A much more elaborate proof is needed to show that it is also true for spaces such as L^p and $W^{m,p}$ for $1 < p < \infty$, and for the Hardy spaces H^p when $p \leq 1$ (see Temlyakov (1998)).

Connections with other types of nonlinear approximation In the univariate setting, the smoothness spaces Y characterized by a certain rate of nonlinear approximation in X are essentially the same if we replace N-term combinations of wavelets by splines with N free knots or by rational functions of degree N. The similarity between wavelets and free-knot splines is intuitive since both methods allow the same kind of adaptive refinement, either by inserting knots or by adding wavelet components at finer scales. The similarities between free-knot spline approximation and rational approximation were elucidated on by Petrushev (1988). However, the equivalence between wavelets and these other types of approximation is not valid in the multivariate context (see Section 7.7). Also closely related to N-term approximations are *adaptive splitting procedures*, which are generalizations of the splitting procedure proposed in Section 7.2 to higher order piecewise polynomial approximation (see e.g. DeVore and Yu (1990) and DeVore (1997)). In the case of the example of Section 7.2, we remark that the piecewise constant approximation resulting from the adaptive splitting procedure can always be viewed as an N-term approximation in the Haar system, in which the active coefficients have a certain *tree structure*: if $\lambda = (j,k)$ is used in the approximation, then $(j-1, [k/2])$ is also used at the previous coarser level. Therefore the performances of adaptive splitting approximations are essentially equivalent to those of N-term approximations with the additional tree structure restriction. These performances have been studied in Cohen et al. (1999b) where it is shown that the tree structure restriction does not affect the order $N^{-s/d}$ of N-term approximation in $X \sim (1/p, r)$ if the space $Y \sim (1/\tau, r+s)$ is replaced by $\tilde{Y} \sim (1/\tilde{\tau}, r+s)$ with $1/\tilde{\tau} < 1/\tau = 1/p + s/d$.

7.4 Data compression

There exist many interesting applications of wavelets to signal processing and we refer to Mallat (1998) for a detailed overview. In this section and in the following one, we would like to discuss two applications which exploit the fact that certain signals – in particular images – have a sparse representation in wavelet bases. Nonlinear approximation theory allows us to

Chapter 7 Nonlinear wavelet approximation   199

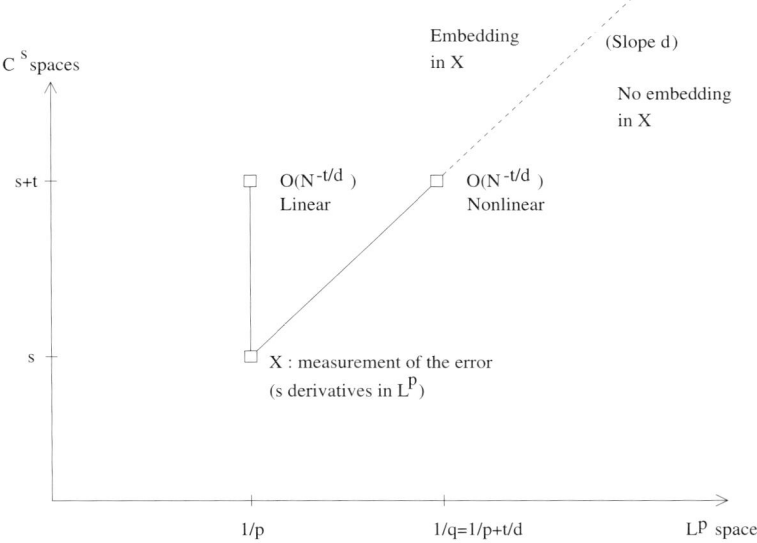

Fig. 7.1. Graphical interpretation of linear and nonlinear approximation

"quantify" the level of sparsity in terms of the decay of the error of N-term approximation.

From a mathematical point of view, the N-term approximation of a signal f can already be viewed as a "compression" algorithm since we are reducing the number of degrees of freedom which represent f. However, practical compression means that the approximation of f is represented by a *finite number of bits*. Wavelet-based compression algorithms are a particular case of transform coding algorithms which have the following general structure:

- Transformation: the original signal f is transformed into its representation **d** (in our case of interest, the wavelet coefficients $\mathbf{d} = (d_\lambda)$) by an invertible transform $\mathcal{R}$.
- Quantization: the representation **d** is replaced by an approximation $\tilde{\mathbf{d}}$ which can only take a finite number of values. This approximation can be encoded with a finite number of bits.
- Reconstruction: from the encoded signal, one can reconstruct $\tilde{\mathbf{d}}$ and therefore an approximation $\tilde{f} = \mathcal{R}^{-1}\tilde{\mathbf{d}}$ of the original signal f.

Therefore, a key issue is the development of appropriate quantization strategies for the wavelet representation and the analysis of the error produced by quantizing the wavelet coefficients. Such strategies should in some sense minimize the distortion $\|f - \tilde{f}\|_X$ for a prescribed number of bits N and error metric X.

Of course this program only makes sense if we refer to a certain model of the signal. In a deterministic context, one considers the error $\sup_{f \in Y} \|f - \tilde{f}\|_X$ for a given class Y, while in a stochastic context, one considers the error $E(\|f - \tilde{f}\|_X)$ where the expectation is taken over the realizations f of a stochastic process. In the following we shall indicate some results in the deterministic context.

We shall discuss here the simple case of *scalar quantization* which amounts to quantizing independently the coefficients d_λ into approximations $\tilde{d}_\lambda$ in order to produce $\tilde{\mathbf{d}}$. Similarly to the distinction between linear and nonlinear approximation, we can distinguish between two types of quantization strategies.

- Non-adaptive quantization: the number of bits which are used to represent $\tilde{d}_\lambda$ depends only on the index λ. In practice they typically depend on the scale level $|\lambda|$. Fewer bits are allocated to the fine scale coefficients which have smaller values than the coarse scale coefficients *in an averaged sense*.
- Adaptive quantization: the map $d_\lambda \mapsto \tilde{d}_\lambda$ and the number of bits which are used to represent $\tilde{d}_\lambda$ depends both on λ and on the amplitude value $|d_\lambda|$. In practice they typically only depend on $|d_\lambda|$. More bits are allocated to the large coefficients which correspond to different indices from one signal to another.

The second strategy is clearly more appropriate for exploiting the sparsity of the wavelet representation, since a large number of bits will be used only for a small number of numerically significant coefficients.

In order to analyze this idea more precisely, let us consider the following specific strategy. For a fixed $\varepsilon > 0$, we allocate no bits to the details for which $|d_\lambda| \leq \varepsilon$ by setting $\tilde{d}_\lambda = 0$. This amounts to thresholding them. We allocate j bits to a detail for which $2^{j-1}\varepsilon < |d_\lambda| \leq 2^j \varepsilon$. If the 2^j possible values of $\tilde{d}_\lambda$ are taken equally spaced within the range $(-2^j \varepsilon, -2^{j-1}\varepsilon) \cup (2^{j-1}\varepsilon, 2^j \varepsilon)$, we ensure that for all λ

$$|d_\lambda - \tilde{d}_\lambda| \leq \varepsilon.$$

If we measure the error in $X = L^2$, assuming the ψ_λ are a Riesz basis, we find that

$$\|f - \tilde{f}\|_2^2 \sim \sum_{\lambda \in \nabla} |d_\lambda - \tilde{d}_\lambda|^2 \leq \varepsilon^2 \#(\{\lambda : |d_\lambda| \geq \varepsilon\}) + \sum_{\{\lambda : |d_\lambda| \leq \varepsilon\}} |d_\lambda|^2. \quad (7.11)$$

Note that the second term on the right-hand side is simply the error of nonlinear approximation obtained by thresholding at the level ε, while the first term corresponds to the effect of quantizing the significant coefficients.

Let us now assume that the class of signals Y has a sparse wavelet representation in the sense that there exists $\tau \leq 2$ and $C > 0$ such that for all $f \in Y$ we have $\mathbf{d} = (d_\lambda)_{\lambda \in \nabla} \in \ell_w^\tau(\nabla)$, with $\|\mathbf{d}\|_{\ell_w^\tau} \leq C$, i.e., for all $f \in Y$

$$\#(\{\lambda : |d_\lambda(f)| > \eta\}) \leq C\eta^{-\tau}. \quad (7.12)$$

We have seen in the previous section that this property is satisfied when $\|f\|_{B_{\tau,\tau}^s} \leq C$ for all $f \in Y$ with $1/\tau = 1/2 + s/d$ and that it is equivalent to the nonlinear approximation property $\sigma_N \leq CN^{-s/d}$. Using (7.12), we can estimate both terms in (7.11) as follows: for the quantization term, we simply obtain

$$\varepsilon^2 \#(\{\lambda : |d_\lambda| \geq \varepsilon\}) \leq C\varepsilon^{2-\tau},$$

while for the thresholding term we have

$$\sum_{\{\lambda : |d_\lambda| \leq \varepsilon\}} |d_\lambda|^2 \leq \sum_{j \geq 0} 2^{-2j} \varepsilon^2 \#(\{\lambda : |d_\lambda| \geq \varepsilon 2^{-j-1}\}) \leq C\varepsilon^{2-\tau}. \quad (7.13)$$

Recall that $0 < \tau < 2$, so that the bounds tend to zero with ε. Therefore we find that the compression error is estimated by $C\varepsilon^{1-\tau/2}$. We can also estimate the number of bits N_q which are used to quantize the d_λ according to

$$N_q = \sum_{j>0} j \#(\{\lambda : 2^{j-1}\varepsilon < |d_\lambda| \leq 2^j \varepsilon\}) \leq C\varepsilon^{-\tau} \sum_{j>0} j 2^{-\tau j} \leq C\varepsilon^{-\tau}.$$

Comparing N_q and the compression error estimate, we find the striking result that

$$\|f - \tilde{f}\|_2 \leq CN_q^{(\tau/2-1)/\tau} = CN_q^{-s/d},$$

with C a constant independent of $f \in Y$. At first sight, it seems that we obtain with only N bits the same rate as for nonlinear approximation which requires N real coefficients.

However, a specific additional difficulty of adaptive quantization is that we also need to encode the *addresses* λ such that $2^{j-1}\varepsilon < |d_\lambda| \leq 2^j \varepsilon$. The bit cost N_a of this addressing can be significantly close to N_q or even higher. If the class of signals is modeled by (7.12), we actually find that N_a is infinite since the large coefficients could be located anywhere. In order to have $N_a \leq C\varepsilon^{-\tau}$ as well, and thus obtain the desired estimate $\|f - \tilde{f}\|_2 \leq CN^{-s/d}$ with $N = N_q + N_a$, it is necessary to make some minor additional assumptions on Y which restrict the location of the large coefficients, and to develop a suitable addressing strategy.

The most efficient wavelet-compression algorithms, such as the one developed in Shapiro (1993), typically apply addressing strategies based on *tree structures* within the indices λ. We also refer to Cohen et al. (1999b) where it is proved that such a strategy allows us to recover optimal rate/distortion bounds, i.e., optimal behaviors of the compression error with respect to the number of bits N, for various deterministic classes Y modeling the signals.

In practice such results can only be observed for a certain range of N, since the original signal itself is most often given by a finite number of bits N_o, e.g., a digital image. Therefore modeling the signal by a function class and deriving rate/distortion bounds from this model is usually relevant only for a low bit rate $N << N_0$, i.e., high compression ratio. One should then of course address the questions of "what are the natural deterministic classes which model real signals" and "what can one say about the sparsity of wavelet representations for these classes". An interesting example is given by real images which are often modeled by the space BV of functions with bounded variation. This function space represents functions which have one order of smoothness in L^1 in the sense that both components of their gradient are finite measures. This includes in particular functions of the type χ_Ω for domains Ω with boundaries of finite length. In Cohen et al. (1999c) it is proved that the wavelet coefficients of a function $f \in BV$ are sparse in the sense that they are in ℓ_w^1. This allows us to expect a nonlinear approximation error of the order of $N^{-1/2}$ for images, and a similar rate for compression provided that we can handle the addressing with a reasonable number of bits. This last task turns out to be feasible, thanks to some additional properties such as the L^∞-boundedness of images.

7.5 Statistical estimation

In recent years, wavelet-based thresholding methods have been widely applied to a large range of problems in statistics such as density estimation, white noise removal, nonparametric regression, diffusion estimation, (see the pioneering work of Donoho et al. (1995)). In some sense the growing interest in thresholding strategies represents a significant "switch" from linear to nonlinear/adaptive methods.

Here we shall consider the simple white noise model. That is, given a function $f(t)$ we observe on $[0,1]$

$$dg(t) = f(t)dt + \varepsilon dw(t),$$

where $w(t)$ is a Brownian motion. In other words, we observe the function f with an additive white Gaussian noise of variance ε^2. This model can of course be generalized to higher dimensions. We are now interested in constructing an estimator $\tilde{f}$ from the data g. The most common measure of the estimation error is in the mean square sense. Assuming that $f \in L^2$ we are interested in the quantity $E(\|\tilde{f} - f\|_{L^2}^2)$. As with data compression, the design of an optimal estimation procedure in order to minimize the mean square error is relative to a specific model of the signal f either by a deterministic class Y or by a stochastic process.

Linear estimation methods define $\hat{f}$ by applying a linear operator to g. In many practical situations this operator is translation invariant and amounts to a filtering procedure, i.e., $\tilde{f} = h * g$. For example, in the case of a second order stationary process, the *Wiener filter* gives an optimal solution in terms of $\hat{h}(\omega) := \hat{r}(\omega)/(\hat{r}(\omega) + \varepsilon^2)$ where $\hat{r}(\omega)$ is the power spectrum of f, i.e., the Fourier transform of $r(u) := E(f(t)f(t+u))$. Another frequently used linear method is by projection on some finite-dimensional subspace V, i.e., $\tilde{f} = Pg = \sum_{n=1}^{N} \langle g, \tilde{e}_n \rangle e_n$, where $(e_n)_{n=1}^N$, $(\tilde{e}_n)_{n=1}^N$ are biorthogonal bases of V and $N := \dim(V)$. In this case, using the fact that $E(\tilde{f}) = Pf$ (since the expectation of the noise and thus of its projection are zero), we can estimate the error as follows:

$$\begin{aligned}E(\|\tilde{f} - f\|_2^2) &= E(\|Pf - f\|_2^2) + E(\|P(g - f)\|_2^2) \\ &\leq E(\|Pf - f\|_2^2) + CN\varepsilon^2.\end{aligned}$$

If P is an orthonormal projection, we can assume that $e_n = \tilde{e}_n$ is an orthonormal basis so that $E(\|P(g-f)\|_2^2) = \sum_n E(|\langle f-g, e_n \rangle|^2) = \sum_n \varepsilon^2$, and therefore the above constant C is equal to 1. Otherwise this constant depends on the "angle" of the projection P. In the above estimation, the first term $E(\|Pf - f\|_2^2)$ is the *bias* of the estimator. It reflects the approximation property of the space V for the model, and typically decreases with the dimension of V. Note that in the case of a deterministic class Y, it is simply given by $\|Pf - f\|_2^2$. The second term $CN\varepsilon^2$ represents the *variance* of the estimator which increases with the dimension of V. A good estimator should find an optimal balance between these two terms.

Consider for instance the projection on the multiresolution space V_j, i.e., $\tilde{f} := \sum_{|\lambda| \leq j} \langle g, \tilde{\psi}_\lambda \rangle \psi_\lambda$, together with a deterministic model: the functions f

satisfy

$$\|f\|_{H^s} \leq C, \tag{7.14}$$

where H^s is the Sobolev space of smoothness s. Then we can estimate the bias by the linear approximation estimate $C2^{-2sj}$ and the variance by $C2^j\varepsilon^2$ since the dimension of V_j adapted to $[0,1]$ is of order 2^j. Assuming an *a priori* knowledge of the level ε of the noise, we find that the scale level balancing the bias and variance terms is $j(\varepsilon)$, where $2^{j(\varepsilon)(1+2s)} \sim \varepsilon^{-2}$. We thus select as our estimator

$$\tilde{f} := P_{j(\varepsilon)}g.$$

With such a choice,

$$E\big(\|\tilde{f} - f\|_2^2\big) \leq C\varepsilon^{4s/(1+2s)}.$$

Let us make a few comments on this simple result.

- The convergence rate $4s/(1+2s)$ of the estimator, as the noise level tends to zero, improves with the smoothness of the model. It can be shown that this is actually the optimal or *minimax* rate, in the sense that for any estimation procedure, there always exist an f in the class (7.14) for which we have $E(\|\tilde{f} - f\|_2^2) \geq c\varepsilon^{4s/(1+2s)}$.
- One of the main limitations of the above estimator is that it depends not only on the noise level (which in practice can often be evaluated), but also on the modeling class itself since $j(\varepsilon)$ depends on s. A better estimator should give an optimal rate for a large variety of function classes.
- The projection $P_{j(\varepsilon)}$ is essentially equivalent to low pass filtering which eliminates those frequencies larger than $2^{j(\varepsilon)}$. The practical drawbacks of such denoising strategies are well known. While they remove the noise, low-pass filters tend to blur the singularities of the signals, such as the edge in an image. This problem is implicitly reflected in the fact that signals with edges correspond to a value of s which cannot exceed $1/2$ and therefore the convergence rate is at most $\mathcal{O}(\varepsilon)$.

Let us now turn to nonlinear estimation methods based on wavelet thresholding. The simplest thresholding estimator is defined by

$$\tilde{f} := \sum_{\lambda \in \Gamma(g;\eta)} \langle g, \tilde{\psi}_\lambda \rangle \psi_\lambda, \tag{7.15}$$

where $\Gamma(g;\eta) = \{\lambda : |\langle g, \tilde{\psi}_\lambda \rangle| \geq \eta\}$, i.e., discarding coefficients of the data of size less than some $\eta > 0$. Let us remark that the wavelet coefficients of the observed data can be expressed as

$$\langle g, \tilde{\psi}_\lambda \rangle = \langle f, \tilde{\psi}_\lambda \rangle + \varepsilon b_\lambda$$

where the b_λ are Gaussian variables. These variables have variance 1 if we assume (which is always possible up to a renormalization) that $\|\tilde{\psi}_\lambda\|_2^2 = 1$. In the case of an orthonormal basis the b_λ are independent. Therefore the observed coefficients appear as those of the real signal perturbed by an additive noise of level ε. It thus seems at first sight that a natural choice for a threshold is to simply fix $\eta := \varepsilon$, i.e., we can hope to remove most of the noise, while preserving the most significant coefficients of the signal, which is particularly appropriate if the wavelet decomposition of f is sparse.

In order to understand the rate that we could expect from such a procedure, we shall again consider the class of signals described by (7.12). For the moment, let us assume that we have an *oracle* which gives us the knowledge of those λ such that the wavelet coefficients of the real signal are larger than ε, so that we can build the modified estimator

$$\overline{f} := \sum_{\lambda \in \Gamma(f;\varepsilon)} \langle g, \tilde{\psi}_\lambda \rangle \psi_\lambda. \tag{7.16}$$

In this case, $\overline{f}$ can be viewed as the projection Pg of g onto the space $V(f,\varepsilon)$ spanned by $\{\psi_\lambda : \lambda \in \Gamma(f;\varepsilon)\}$, so that we can estimate the error by a sum of bias and variance terms according to

$$\begin{aligned} E(\|\tilde{f} - f\|_2^2) &= \|f - Pf\|_2^2 + E(\|P(f-g)\|_2^2) \\ &\leq C\left[\sum_{\lambda \in \Gamma(f;\varepsilon)} |\langle f, \tilde{\psi}_\lambda \rangle|^2 + \varepsilon^2 \#(\Gamma(f;\varepsilon))\right]. \end{aligned}$$

For the bias term, we recognize the nonlinear approximation error which is bounded by $C\varepsilon^{2-\tau}$ according to (7.13). From the definition of the class (7.12) we find that the variance term is also bounded by $C\varepsilon^{2-\tau}$. Thus, we obtain for the oracle estimator the convergence rate $\varepsilon^{2-\tau}$. In particular, if we consider the model

$$\|f\|_{B^s_{\tau,\tau}} \leq C,$$

with $1/\tau = 1/2 + s$, we obtain

$$E(\|\tilde{f} - f\|_2^2) \leq C\varepsilon^{2-\tau} = C\varepsilon^{4s/(1+2s)}.$$

Let us again make a few comments.

- In a similar way to approximation rates, nonlinear methods achieve the same estimation rate as linear methods but for much weaker models. The exponent $4s/(1+2s)$ was achieved by the linear estimator for the class (7.14) which is more restrictive than (7.12).
- In contrast to the linear estimator, we see that the nonlinear estimator does not need to be tuned according to the value of τ or s. In this sense, it is very robust.
- Unfortunately, (7.16) is unrealistic since it is based on the "oracle assumption". In practice, we are thresholding according to the values of the observed coefficients $\langle g, \tilde{\psi}_\lambda \rangle = \langle f, \tilde{\psi}_\lambda \rangle + \varepsilon^2 b_\lambda$, and we need to consider the possible effect of the noise $\varepsilon^2 b_\lambda$. Another unrealistic aspect, also in (7.15), is that one cannot evaluate the full infinite set of coefficients $(\langle g, \tilde{\psi}_\lambda \rangle)_{\lambda \in \nabla}$.

The strategy proposed in Donoho et al. (1995) solves the above difficulties as follows. A realistic estimator is built by (i) a systematic truncation of the estimator (7.15) above a scale $j(\varepsilon)$ such that $2^{-2\alpha j(\varepsilon)} \sim \varepsilon^2$ for some fixed $\alpha > 0$, and (ii) a choice of threshold slightly above the noise level according to

$$\eta(\varepsilon) := C(\alpha)\varepsilon |\log(\varepsilon)|^{1/2}.$$

It is then possible to prove that the resulting, more realistic, estimator

$$\tilde{f} := \sum \langle g, \tilde{\psi}_\lambda \rangle \psi_\lambda,$$

where the summation is taken over all $\lambda \in \Gamma(g; \eta(\varepsilon)) \cap \{\lambda : |\lambda| \leq j(\varepsilon)\}$, has the rate $[\varepsilon |\log(\varepsilon)|^{1/2}]^{4s/(1+2s)}$ (i.e., almost the same asymptotic performance as the oracle estimator) both for the functions which are in the class (7.12) and in the Sobolev class H^α. The "minimal" Sobolev smoothness α, which is needed to allow the truncation of the estimator, can be taken arbitrarily close to zero up to a change of the constants in the threshold and in the convergence estimate.

7.6 Adaptive numerical simulation

Numerical simulation is nowadays an essential tool for the understanding of physical processes modeled by partial differential or integral equations. In many instances, the solutions of these equations exhibit singularities, resulting in a slower convergence of the numerical schemes as the discretization tends to zero. Moreover, such singularities might be physically significant such as shocks in fluid dynamics or local accumulation of stress in elasticity, and therefore they should be well approximated by the numerical method. In order to maintain memory size and computational cost at a reasonable level, it is necessary to use adaptive discretizations which should typically be more refined near the singularities.

In the finite element context, such discretizations are produced by *mesh refinement*. Starting from an initial coarse triangulation, we allow further subdivision of certain elements into finer triangles, and we define the discretization space according to this locally refined triangulation. This is of course subject to certain rules, in particular that of preserving the conformity of the discretization, e.g., when continuity of the finite elements is required. The use of wavelet bases as an alternative to finite elements is still in its infancy (some first surveys are Cohen (1999) and Dahmen (1997)), and is strongly motivated by the possibility of producing simple adaptive approximations. In the wavelet context, the more appropriate terminology is *space refinement*. We directly produce an approximation space

$$V_\Lambda := \text{span}\{\psi_\lambda \ : \ \lambda \in \Lambda\},$$

by selecting a set Λ such that V_Λ is well-adapted to describe the solution of our problem. If N denotes the cardinality of the adapted finite element or wavelet space, i.e., the number of degrees of freedom which are used in the computations, we see that in both cases the numerical solution u_N can be viewed as an adaptive approximation of the solution u in a space Σ_N, which depends on u.

A specific difficulty of adaptive numerical simulation is that the solution u is unknown at the beginning, except for some rough *a priori* information such as global smoothness. In particular the location and structure of the singularities are often unknown, and therefore the design of an optimal discretization for a prescribed number of degrees of freedom is a much more difficult task than simple compression of fully available data.

This difficulty has motivated the development of *adaptive strategies* based on *a posteriori analysis*, i.e., using the currently computed numerical solution to update the discretization and derive a better adapted numerical solution. In the finite element setting, such an analysis has been developed since the 1970s (see Babuska and Reinboldt (1978) or Verfürth (1994)) in terms of *local error indicators* which aim to measure the contribution of each element to the error. The rule of thumb is then to refine the triangles which exhibit the largest error indicators. More recently, similar error indicators and refinement strategies have been also proposed in the wavelet context (see Bertoluzza (1995) and Dahlke et al. (1997)).

Nonlinear approximation can be viewed as a benchmark for adaptive strategies. If the solution u can be adaptively approximated in Σ_N with an error $\sigma_N(u)$ in a certain norm X, we would ideally like that the adaptive strategy produces an approximation $u_N \in \Sigma_N$ such that the error $\|u-u_N\|_X$ is of the same order as $\sigma_N(u)$. In the case of wavelets this means that the error produced by the adaptive scheme should be of the same order as the error produced by keeping the N largest coefficients of the exact solution. In most instances, unfortunately, such a program cannot be achieved by an adaptive strategy and a more reasonable goal is to obtain an optimal asymptotic rate. That is, if $\sigma_N(u) \leq CN^{-s}$ for some $s > 0$, then an *optimal adaptive strategy* should produce an error $\|u - u_N\|_X \leq \tilde{C}N^{-s}$. An additional important aspect is the computational cost of deriving u_N. A *computationally optimal strategy* should produce u_N in a number of operations which is proportional to N. A typical instance of a computationally optimal algorithm, for a fixed discretization, is the multigrid method for linear elliptic PDEs.

It should be noted that very often the norm X in which one can hope for an optimal error estimate is dictated by the problem at hand. For example, in the case of an elliptic problem, this will typically be a Sobolev norm equivalent to the energy norm (e.g., the H^1 norm when solving the Laplace equation).

Most existing wavelet adaptive schemes have in common the following general structure. At some step n of the computation, a set Λ_n is used to represent the numerical solution $u_{\Lambda_n} = \sum_{\lambda \in \Lambda_n} d_\lambda^n \psi_\lambda$. In the context of an *initial value problem* of the type

$$\partial_t u = E(u), \quad u(x,0) = u_0(x),$$

the numerical solution at step n is typically an approximation to u at time $n\Delta t$ where Δt is the time step of the resolution scheme. In the context of a *stationary problem* of the type

$$F(u) = 0,$$

the numerical solution at step n is typically an approximation to u which should converge to the exact solution as n tends to ∞. Here E and F are typically partial differential or integral operators, not necessarily linear. In both cases, the derivation of $(\Lambda_{n+1}, u_{\Lambda_{n+1}})$ from $(\Lambda_n, u_{\Lambda_n})$ usually contains three basic steps.

- *Refinement:* a larger set $\tilde{\Lambda}_{n+1}$ with $\Lambda_n \subset \tilde{\Lambda}_{n+1}$ is derived from an *a posteriori* analysis of the computed coefficients d_λ^n, $\lambda \in \Lambda_n$.
- *Computation:* an intermediate numerical solution $\tilde{u}_{n+1} = \sum_{\lambda \in \tilde{\Lambda}_{n+1}} d_\lambda^{n+1} \psi_\lambda$ is computed from u_n and the data of the problem.
- *Coarsening:* the smallest coefficients of $\tilde{u}_{n+1}$ are thresholded, resulting in

the new approximation $u_{n+1} = \sum_{\lambda \in \Lambda_{n+1}} d_\lambda^{n+1} \psi_\lambda$ supported on the smaller set $\Lambda_{n+1} \subset \tilde{\Lambda}_{n+1}$.

Of course the precise description and tuning of these operations strongly depends on the type of equation at hand, as well as on the type of wavelets which are being used.

In the case of linear elliptic problems, it was recently proved in Cohen et al. (1999a) that an appropriate tuning of these three steps results in an optimal adaptive wavelet strategy both in terms of approximation properties and computational time. It should be remarked that such optimality results are still open questions in the more classical context of adaptive mesh refinement in finite element spaces. Many open problems also remain in the wavelet context concerning the possibility of deriving optimal adaptive schemes for more complicated problems, e.g., mixed formulations, singular perturbations, nonlinear equations.

7.7 The curse of dimensionality

The three applications which were discussed in the previous sections exploit the sparsity of wavelet decompositions for certain classes of functions, or equivalently the convergence properties of nonlinear wavelet approximations of these functions. Nonlinear adaptive methods in such applications are typically relevant if these functions have isolated singularities in which case there might be a substantial gain of convergence rate when switching from linear to nonlinear wavelet approximation.

However, a closer look at some simple examples shows that this gain tends to decrease for multivariate functions. Consider the L^2-approximation of the characteristic function $f = \chi_\Omega$ of a smooth domain $\Omega \subset [0,1]^d$. Due to the fact that f is discontinuous across the boundary $\partial\Omega$, one can easily check that linear approximation cannot behave better than

$$\sigma_N(f) = \|f - P_j f\|_2 \sim \mathcal{O}(2^{-j/2}) \sim \mathcal{O}(N^{-1/2d}),$$

where $N = \dim(V_j) \sim 2^{dj}$ and P_j is the orthogonal projection into V_j. Turning to nonlinear approximation, we notice that since $\int \tilde{\psi}_\lambda = 0$, all the coefficients d_λ are zero except for those for which the support of $\tilde{\psi}_\lambda$ overlaps the boundary. At scale level j there is thus at most $K2^{d-1}j$ nonzero coefficients, where K depends on the support of the $\{\psi_\lambda\}$ and on the $(d-1)$-dimensional measure of $\partial\Omega$. For such coefficients, we have the estimate

$$|d_\lambda| \leq \|\tilde{\psi}_\lambda\|_1 \leq C 2^{-dj/2}.$$

In the univariate case, i.e., when Ω is a simple interval, the number of nonzero coefficients up to scale j is bounded by jK. Therefore, using N nonzero coefficients at the coarsest levels gives an error estimate with exponential decay

$$\sigma_N(f) \leq \left[\sum_{j \geq N/K} K|C2^{-j/2}|^2\right]^{1/2} \leq \tilde{C}2^{-N/2K},$$

which is a spectacular improvement over the linear rate. In the multivariate case, the number of nonzero coefficients up to scale j is bounded by $\sum_{l=0}^{j} K2^{(d-1)l}$ and thus by $\tilde{K}2^{(d-1)j}$. Therefore, using N nonzero coefficients at the coarsest levels gives an error estimate

$$\sigma_N(f) \leq \left[\sum_{\tilde{K}2^{(d-1)j} \geq N} K2^{(d-1)j}|C2^{-dj/2}|^2\right]^{1/2} \leq \tilde{C}N^{-1/(2d-2)},$$

which is much less of an improvement. For example, in the 2D case, we only go from $N^{-1/4}$ to $N^{-1/2}$ by switching to nonlinear wavelet approximation.

This simple example illustrates the *curse of dimensionality* in the context of nonlinear wavelet approximation. The main reason for the degradation of the approximation rate is the large number, $K2^{(d-1)j}$, of wavelets which are needed to refine the boundary from level j to level $j+1$. On the other hand, if we view the boundary itself as the graph of a smooth function, it is clear that approximating this graph with accuracy 2^{-j} should require significantly fewer parameters than $K2^{(d-1)j}$. This reveals the fundamental limitation of wavelet bases. They fail to exploit the smoothness of the boundary and therefore cannot capture the simplicity of f in a small number of parameters. Another way of describing this limitation is by noting that nonlinear wavelet approximation allows local refinement of the approximation, but imposes some *isotropy* on this refinement process. In order to capture the boundary with a small number of parameters, one would typically need to refine more in the normal direction than in the tangential directions, i.e., apply *anisotropic local refinement*.

In this context, other approximation tools outperform wavelet bases. It is easy to check that the use of piecewise constant functions on an adaptive partition of N triangles in 2D will produce the rate $\sigma_N(f) \sim \mathcal{O}(N^{-1})$, precisely because one is allowed to use arbitrarily anisotropic triangles to match the boundary. In the case of rational functions we have an even more spectacular result. If $\partial\Omega$ is C^∞, then $\sigma_N(f) \leq C_r N^{-r}$ for any $r > 0$. These remarks reveal that, in contrast to the 1D case, free triangulations or rational approximation outperform N-term approximation, and could be considered as a better tool in view of applications such as those which were discussed throughout this chapter. This is however not really true in practice. In numerical simulation, rational functions are difficult to use and free triangulations are often limited by shape constraints which restrict their anisotropy. Neither method is being used in statistical estimation or data compression, principally due to the absence of fast and robust algorithms which would produce an optimal adaptive approximation in a similar manner to wavelet thresholding.

The development of new approximation and representation tools, which could both capture anisotropic features, such as edges with a very small number of parameters, and be implemented by fast and robust procedures, is currently the object of active research. A significant breakthrough was recently achieved by Donoho and Candés who developed representations into *ridgelet* bases which possess the scale-space localization of wavelets together with some directional selection (see e.g., Candés and Donoho (1999)). Such bases allow us, for example, to recover with a simple thresholding procedure the rate $\mathcal{O}(N^{-1})$ for a bivariate function which is smooth except along a smooth curve of discontinuity.

References

Babuska, I. and Reinboldt, W.C. (1978). Error estimates for adaptive finite element computations. *SIAM J. Numer. Anal.*, **15**, 736–754.

Bertoluzza, S. (1995). A posteriori error estimates for the wavelet Galerkin method. *Appl. Math. Lett.*, **8**, 1–6.

Candés, E.J. and Donoho, D.L. (1999). Ridgelets: a key to higher-dimensional intermittency? *Phil. Trans. Roy. Soc. A*, **357**, 2459–2509.

Cohen, A. (1999). Wavelets in numerical analysis. In *Handbook of Numerical Analysis*, vol. VII, ed. P.G. Ciarlet and J.L. Lions. Elsevier. Amsterdam.

Cohen, A., Dahmen, W. and DeVore, R.A. (1999a). Adaptive wavelet methods for elliptic operator equations–convergence rate. *Math. Comp.*, to appear.

Cohen, A., Dahmen, W., Daubechies, I. and DeVore, R.A. (1999b). Tree approximation and optimal encoding. Preprint.

Cohen, A., DeVore, R.A., Petrushev, P. and Xu, H. (1999c). Nonlinear approximation and the space $BV(\mathbb{R}^2)$. *Amer. Jour. of Math.*, **121**, 587–628.

Dahlke, S., Dahmen, W., Hochmuth, R. and Schneider, R. (1997). Stable multiscale bases and local error estimation for elliptic problems. *Appl. Numer. Math.*, **23**, 21–47.

Dahmen, W. (1997). Wavelet and multiscale methods for operator equations. *Acta Numerica*, **6**, 55–228.

Daubechies, I. (1992). *Ten Lectures on Wavelets*. SIAM, Philadelphia.

DeVore, R.A. (1997). Nonlinear approximation. *Acta Numerica*, **7**, 51–150.

DeVore R.A. and Yu, X.M. (1990). Degree of adaptive approximation. *Math. Comp.*, **55**, 625–635.

Donoho, D., Johnstone, I., Kerkyacharian, G, and Picard, D. (1995). Wavelet shrinkage: asymptotia? (with discussion). *Jour. Roy. Stat. Soc.*, **57**, Series B, 301–369.

Mallat, S. (1998). *A Wavelet Tour of Signal Processing*. Academic Press, New York.

Petrushev, P. (1988). Direct and converse theorems for spline and rational approximation and Besov spaces. In *Function Spaces and Applications*, ed. M. Cwikel, J. Peetre, Y. Sagher and H. Wallin, pp. 363–377. Lecture Notes in Math. **1302**, Springer, Berlin.

Shapiro, J. (1993). Embedded image coding using zero trees of wavelet coefficients. *IEEE Signal Processing*, **41**, 3445–3462.

Temlyakov, V. (1998). The best m-term approximation and greedy algorithms. *Adv. Comput. Math.*, **8**, 249–265.

Verfürth, R. (1994). A posteriori error estimation and adaptive mesh-refinement techniques. *J. Comp. Appl. Math.*, **50**, 67–83.

8
Subdivision, multiresolution and the construction of scalable algorithms in computer graphics

P. SCHRÖDER

Abstract

Multiresolution representations are a critical tool in addressing complexity issues (time and memory) for the large scenes typically found in computer graphics applications. Many of these techniques are based on classical subdivision techniques and their generalizations. In this chapter we review two exemplary applications from this area, multiresolution surface editing and semi-regular remeshing. The former is directed towards building algorithms which are fast enough for interactive manipulation of complex surfaces of arbitrary topology. The latter is concerned with constructing smooth parameterizations for arbitrary topology surfaces as they typically arise from 3D scanning techniques. Remeshing such surfaces then allows the use of classical subdivision ideas. We focus in particular on the practical aspects of making the well-understood mathematical machinery applicable and accessible to the very general settings encountered in practice.

8.1 Introduction

Many applications in computer graphics are dominated by the need to deal efficiently with large datasets. This need arises from the goal of providing the user with a responsive system (ideally) allowing interactive work, i.e., update rates of several frames per second. However, compelling or realistic datasets such as topographical maps or geometric models appearing in entertainment and engineering tend to have large amounts of geometric detail. One measure of the latter is the number of triangles or patches necessary to provide a good approximation of a given complex surface. Not only must these datasets be displayed and transmitted rapidly, but in many cases expensive numerical computations must be performed on this data. Examples include surface denoising, fairing, compression, editing, optimization, and finite element (FEM) computations. The target machines for such computations are typically small workstations or home PCs with hardware accelerated graphics adapters.

These needs provide the motivation to search for efficient representations and algorithms which perform a variety of discrete and continuous computations. Some of the properties these representations and algorithms should satisfy are:

- **Scalability** their time and space complexity should ideally be linear in the input size. Conversely algorithms which require quadratic space or time are generally impractical as input sizes reach into the tens of thousands or more primitives. Note that scalability can also mean the ability to take advantage of large parallel computing resources. We will not discuss this issue further as our attention is directed towards small, widely available computing platforms.
- **Speed/Fidelity tradeoffs** to provide the user with an interactive experience, algorithms that are capable of providing approximate answers quickly, as well as more precise answers with additional time (and storage) are preferred over algorithms that always provide highly accurate answers after long computing times. For example, this requirement favors iterative solvers over direct solvers, or wavelet methods which provide "control knobs" for approximation quality versus computational time and storage.
- **Intuitive control** most applications in computer graphics involve a possibly naive user, thus the parameters exposed by the system must be intuitive and should not require a deep understanding of the underlying representations and algorithms. For example, it is very difficult to manipulate Fourier coefficients directly to achieve a desired effect, while locally-supported functions such as B-splines provide an easy-to-grasp relationship between cause and effect.
- **Robustness** the larger the input dataset the higher the probability that all manner of "screw cases" will appear in a given dataset. Algorithms must handle these in a robust fashion and ideally be insensitive to them.
- **Uniformity** a system which attempts to deal with many different representations concurrently tends to be more complicated and fragile than one which is based on a single, and very general representation.

One general concept which is now widely accepted as fundamental to the construction of scalable algorithms is the idea of multiresolution. In the context of computer graphics multiresolution comes in two fundamental flavors: (a) constructions which are based on classical notions of multiresolution as

Chapter 8   Scalable algorithms in computer graphics   215

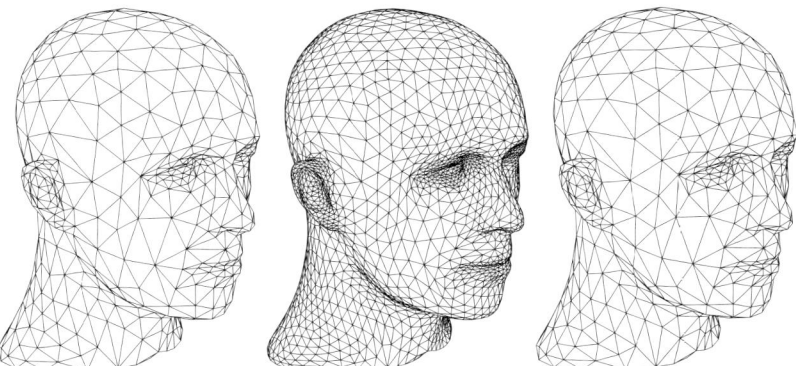

Fig. 8.1. Multiresolution as it is defined through subdivision. Starting with a coarsest level mesh of arbitrary connectivity successive levels of a multiresolution representation are produced through subdivision. (Original dataset courtesy Hugues Hoppe.)

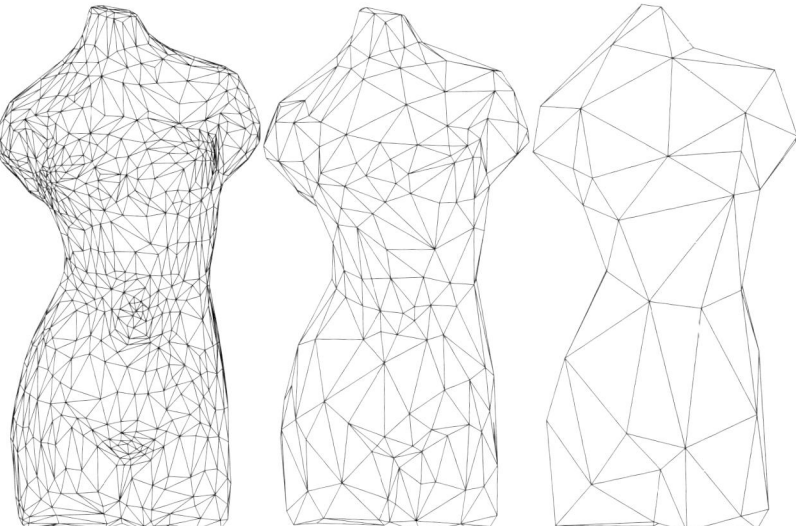

Fig. 8.2. Multiresolution as it is defined through mesh simplification. Starting with a finest level mesh of arbitrary connectivity, successive levels of a multiresolution representation are produced through simplification. (Original dataset courtesy Avalon.)

they appear in wavelets and subdivision; and (b) constructions based on mesh simplification (see Figures 8.1 and 8.2, respectively).

In the following we study two representative examples of these ideas. The first concerns the construction of a multiresolution surface editor which allows the manipulation of intricate, arbitrary topology surfaces at interactive rates on low end platforms. It uses classical subdivision surfaces (Loop) and extends them with the help of a Burt–Adelson type pyramid to a true multiresolution setting, i.e., a setting with non-trivial details at finer resolutions. The second application concerns the construction of smooth parameterizations for arbitrary connectivity meshes. Today, meshes that describe detailed geometric objects are typically built through 3D scanning. Before these meshes become amenable to a number of numerical algorithms they must be provided with a smooth parameterization. In particular this is the first step in transforming ("remeshing") them into a setting suitable for subdivision-based multiresolution approaches.

The work discussed in the following was performed jointly with my (former) student Denis Zorin and long term collaborator Wim Sweldens. David Dobkin and his student Aaron Lee participated in this work as well and I am grateful to all of them.

8.2 Interactive multiresolution surface editing

Applications such as industrial and engineering design as well as special effects and animation require creation and manipulation of intricate geometric models of arbitrary global topology (genus, number of boundaries, and connected components). Like real world geometry, these models can carry detail at many scales (see Figure 8.3). Such models are often created with 3D range sensing devices such as laser scanners (see for example the work of Curless and Levoy (1996)), and the resulting meshes can be composed of as many as a million or more triangles. Manipulating such fine meshes is difficult, especially when they are to be edited or animated. Even without accounting for any computation on the mesh itself, available rendering resources alone may not be able to cope with the sheer size of the data.

A popular approach for addressing the complexity challenge posed by such geometry is based on polygonal simplification (Heckbert and Garland (1997)). This aims to produce a lower resolution polygonal model which closely approximates the original geometry based on various error measures, such as Hausdorff distance or Sobolev norms. While these methods can produce approximations using fewer polygons, ultimately a *smooth* rather

Fig. 8.3. Before the Armadillo started working out he was flabby, complete with a double chin. Now he exercises regularly. The original is on the right (courtesy Venkat Krishnamurthy, Stanford University). The edited version on the left illustrates large scale edits, such as his belly, and smaller scale edits such as his double chin; all edits were performed at about 5 frames per second on an Indigo R10000 Solid Impact.

than piecewise linear representations is desired. For example, a set of spline patches which approximates the original geometry. Finding such approximations is made difficult by the fact that we are dealing with arbitrary topology initial meshes. In this setting parametric methods are not straightforward to apply, since a priori no simple parameterization is given and certainly none over a convenient domain such as the unit square.

Hoppe et al. (1994) have described an algorithm which can produce a set of (piecewise) smooth patches in the arbitrary topology setting using subdivision. However, such an approximation is typically associated with a loss of high frequency details or suffers from a very large number of patches to maintain high accuracy in the presence of fine scale detail. Lost detail can be reintroduced by combining patches with displacement maps (Krishnamurthy and Levoy (1996)). Unfortunately such hybrid representations are difficult to manage in the arbitrary topology setting.

Certainly, having a compact representation is useful when considering storage and transmission costs, but we seek more. It is desirable to work with representations which also support editing. Effective editing strategies must account for the fact that a designer will typically desire to make coarse smooth changes with the same ease as detailed changes to the geometry. This requires some form of multiresolution. In the traditional spline patch-based editing setting such editing semantics were first supported by hierarchical splines (H-splines) proposed by Forsey and Bartels (1988). H-splines are constructed from regular splines (typically bi-cubic) by adding finer resolution B-splines onto an existing coarser resolution spline patch. By repeating this process, one can build very complicated shapes which are entirely parameterized over the unit square. A critical observation first made by Forsey and Bartels concerns the parameterization of the finer level basis functions. The most natural choice, using a global coordinate frame, leads to non-intuitive behavior of the surface during editing. Instead Forsey and Bartels suggested parameterizing the details with respect to coordinate frames induced by the coarser level geometry, i.e., through its partial derivatives and normal. Even though the resulting representations now become nonlinear they are preferable since they exhibit the right editing behavior. A feature attached to the local surface normal will "travel" in a natural way with the surface as the coarse shape of the surface is adjusted (see Figure 8.4).

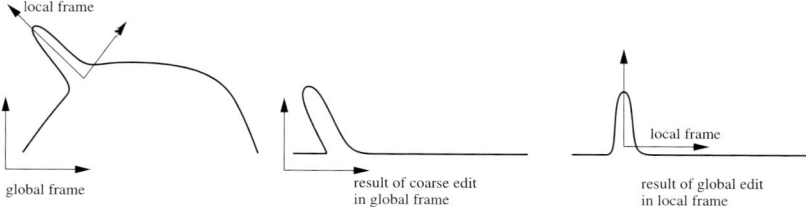

Fig. 8.4. Finer level details should be parameterized with respect to a local frame induced by the coarser level surface to achieve intuitive editing behavior.

Forsey and Bartels' original work focused on the ab initio design setting, i.e., building surfaces from "scratch." In this setting the user's help is enlisted in defining what is meant by different levels of resolution. The user decides where to add detail and manipulates the corresponding controls. To edit an a priori given model it is important to have a general procedure for defining coarser levels and computing details between levels. We refer to this as the *analysis* algorithm in analogy to the wavelet setting. An H-spline analysis algorithm based on weighted least squares was introduced (Forsey and Wong (1995)), but is too expensive to run interactively. Note that even in an ab initio design setting, analysis is needed since a long sequence of editing steps often leads to overly-refined H-splines which need to be consolidated for the sake of performance.

One particular avenue for making rigorous these notions of hierarchy, multiresolution approximation, and projection of fine scale geometry into coarse scales, is through the use of wavelets. Finkelstein and Salesin (1994), for example, used B-spline wavelets to describe multiresolution editing of curves.

As in H-splines, parameterization of details with respect to a coordinate frame induced by the coarser level approximation is required to get correct editing semantics. Gortler and Cohen (1995) pointed out that wavelet representations of details tend to behave in undesirable ways during editing and returned to a pure B-spline representation as used in H-splines. In general we find that critically sampled representations tend to break down under editing operations. One way to understand this is to consider the fact that in a critically sampled wavelet setting there exists a very carefully tuned balance between the low pass (scaling function) and high pass (wavelet) filters. One cancels the aliases of the other. When a curve is edited, however, this balance is disturbed in a nonlinear fashion because of the deformation of the underlying frames which are attached to the curve. As a consequence one observes in practice that the alias cancellation property is lost and undesirable "wiggles" appear in the geometry. These problems can be remedied through the use of overrepresentations such as Burt–Adelson type pyramids (Burt and Adelson (1983)) (see Section 8.4).

All the approaches mentioned above rely on a setting which admits parameterization over a trivial domain such as the unit interval (curve editing) or the unit square. Carrying these ideas over into the arbitrary topology surface framework is not straightforward. The first to do so were Lounsbery et al. (1997) who exploited the connection between wavelets and subdivision to define different levels of resolution for arbitrary topology geometry. While subdivision provides the scaling functions, corresponding wavelets can be constructed with the help of the lifting scheme (Sweldens (1996); Sweldens and Schröder (1996); Schröder and Sweldens (1995)). The original constructions were limited to polyhedral wavelets, i.e., those constructed with the help of interpolating scaling functions such as linear splines or the limit functions of the (modified) butterfly subdivision scheme (Dyn et al. (1990); Zorin et al. (1996)). The latter is not as suitable for editing purposes because of negative weights in the subdivision masks. During editing operations, when one adjusts ("pulls on") a control point, undesired ripples will appear on the surface.

8.2.1 Goals

We are seeking a geometric representation which unifies the advantages of the approaches discussed above and remedies their shortcomings. It should address the arbitrary topology challenge, support hierarchical editing semantics and span the spectrum from patches to finely detailed triangle meshes. Subdivision surfaces based on generalizations of spline patches, combined with detail based extensions as in the H-spline approach provide such a representation. They provide a path from patches to fine meshes using a single primitive (hierarchical triangle mesh), and support the patch-type semantics of manipulation *and* finest-level-detail polyhedral edits equally well. Because they are the limit of successive refinement, subdivision surfaces support multiresolution algorithms, such as level-of-detail rendering, compression, wavelets, and numerical multigrid in a natural way. The basic computational kernels are simple, resulting in easy-to-implement and efficient codes. Boundaries and features such as creases can also be included through modified rules (Hoppe et al. (1994); Schweitzer (1996); DeRose et al. (1998); Levin (1999)), reducing the need for trimming curves. Even though the surfaces are of arbitrary topology they possess simple parameterizations and efficient algorithms exist for exact evaluation of values and derivatives of the limit surface at arbitrary parameter locations (Stam (1998)). The main challenge is to make the basic algorithms fast enough to escape the exponential time and space growth of naive subdivision.

8.2.2 Notation

Before beginning with the description of the details of our algorithms we fix some notation. We need enough machinery to talk about meshes both in terms of their topology and geometry, i.e., their embedding in $\mathbf{R}^3$. For this purpose it is useful to carefully distinguish between the topological and geometric information. We denote a triangle mesh as a pair $(\mathcal{P}, \mathcal{K})$, where $\mathcal{P}$ is a set of N point positions $p_a = (x_a, y_a, z_a) \in \mathbf{R}^3$ with $1 \leq a \leq N$, and $\mathcal{K}$ is an *abstract simplicial complex* which contains all the topological, i.e., adjacency information. The complex $\mathcal{K}$ is a set of subsets of $\{1, \ldots, N\}$. These subsets are called simplices and come in 3 types: vertices $v = \{a\} \in V \subset \mathcal{K}$; edges $e = \{a, b\} \in E \subset \mathcal{K}$; and faces $f = \{a, b, c\} \in T \subset \mathcal{K}$, so that any non-empty subset of a simplex of $\mathcal{K}$ is again a simplex of $\mathcal{K}$, e.g., if a face is present then so are its edges and vertices.

Let δ_i denote the standard ith basis vector in $\mathbf{R}^N$. For each simplex s, its *topological realization* $|s|$ is the strictly convex hull of $\{\delta_i \mid i \in s\}$. Thus $|\{a\}| = \delta_a$, $|\{a,b\}|$ is the open line segment between δ_a and δ_b, and $|\{a, b, c\}|$ is an open equilateral triangle. The topological realization $|\mathcal{K}|$ is defined as $\bigcup_{s \in \mathcal{K}} |s|$. The *geometric realization* $\varphi(|\mathcal{K}|)$ relies on a linear map $\varphi : \mathbf{R}^N \to \mathbf{R}^3$ defined by $\varphi(\delta_a) = p_a$. Abusing notation a bit we will abbreviate the geometric realization by writing $p_a = p(a)$ below. The resulting polyhedron consists of points, segments, and triangles in $\mathbf{R}^3$.

Two vertices $\{a\}$ and $\{b\}$ are *neighbors* if $\{a, b\} \in E$. A set of vertices is *independent* if no two vertices are neighbors. A set of vertices is *maximally independent* if no larger independent set contains it (see Figure 8.18, left side). The 1-ring neighborhood of a vertex $\{a\}$ is the set

$$\mathcal{N}(a) = \{b \mid \{a,b\} \in E\}.$$

The *outdegree* K_a of a vertex is its number of neighbors. The *star* of a vertex $\{a\}$ is the set of simplices

$$\text{star}(a) = \bigcup_{a \in s,\, s \in \mathcal{K}} s.$$

We say that $|\mathcal{K}|$ is a two-dimensional manifold (or 2-manifold) with boundaries if for each a, $|\text{star}(a)|$ is homeomorphic to a disk (interior vertex) or half-disk (boundary vertex) in $\mathbf{R}^2$. An edge $e = \{a,b\}$ is called a *boundary edge* if there is only one face f with $e \subset f$.

8.3 Subdivision

For our purposes subdivision has two distinct components, one given by topological operations on the simplicial complex and the other by operations on the geometric realization of the mesh, i.e., the association of a *point* in 3D with every *vertex* in the complex. Starting with an initial mesh $(\mathcal{P}^0, \mathcal{K}^0 = (V^0, E^0, T^0))$ subdivision builds a new one $(\mathcal{P}^1, \mathcal{K}^1)$ by refining each face $\{a,b,c\} \in T^0$ into four faces $\{a, c_e, b_e\}$, $\{b, a_e, c_e\}$, $\{c, b_e, a_e\}$, $\{a_e, b_e, c_e\} \in T^1$. The new vertices $\{a_e\}, \{b_e\}, \{c_e\} \in V^1$ can be thought of as children of the edges across from the vertices $\{a\}, \{b\}, \{c\} \in V^0$, respectively. Continuing this construction recursively results in meshes $(\mathcal{P}^i, \mathcal{K}^i)$. The superscript i indicates the *level* of triangles and vertices, respectively. The vertex sets are nested as $V^j \subset V^i$ if $j < i$. We define *odd* vertices on level i as $M^i = V^{i+1} \setminus V^i$. The set V^{i+1} consists of two disjoint sets: *even* vertices (V^i) and *odd* vertices (M^i). We define the *level* of a vertex a as the smallest i for which $a \in V^i$. The level of a is $i+1$ if and only if $a \in M^i$.

With each set V^i we associate a geometric realization $p^i(a) \in \mathbf{R}^3$, for all $a \in V^i$. The set p^i contains all points on level i, $p^i = \{p^i(a) \mid a \in V^i\}$. Finally, a *subdivision scheme* is a linear operator S which takes the points from level i to points on the *finer* level $i+1$: $p^{i+1} = S p^i$, defining the geometric realization at level $i+1$.

Assuming that the subdivision converges, we can define a limit surface σ as

$$\sigma = \lim_{k \to \infty} S^k p^0.$$

$\sigma(a) \in \mathbf{R}^3$ denotes the point on the limit surface associated with vertex a.

In order to later define detail offsets with respect to a local frame we also need tangent vectors and a normal. For the subdivision schemes that we use, such vectors can be defined through the application of linear operators D^1 and D^2 acting on p^i so that $\partial_1^i(a) = (D^1 p^i)(a)$ and $\partial_2^i(a) = (D^2 p^i)(a)$ are linearly independent tangent vectors at $\sigma(a)$. Together with an orientation they define a local orthonormal frame $F^i(a) = (n^i(a), \partial_1^i(a), \partial_2^i(a))$. For reasons of efficiency, it is important to note that in general it is not necessary to use precise normals and tangents during editing; as long as the frame vectors are affinely related to the positions of vertices of the mesh, we can expect intuitive editing behavior.

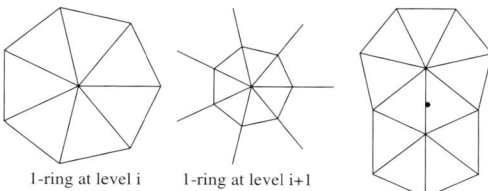

Fig. 8.5. An even vertex has a 1-ring of neighbors at each level of refinement (left/middle). Odd vertices—in the middle of edges—have 1-rings around each of the vertices at either end of their edge (right).

We are particularly interested in subdivision schemes which belong to the class of *1-ring schemes*. In these schemes points at level $i + 1$ depend only on 1-ring neighborhoods of points at level i. Let $a \in V^i$ (a even) then the point $p^{i+1}(a)$ is a function of $p^i(a_n)$ for $a_n \in \mathcal{N}^i(a)$, i.e., only the immediate neighbors of a (see Figure 8.5 left/middle). If $m \in M^i$ (m odd), it is the vertex inserted when splitting an edge of the mesh. In this case the point $p^{i+1}(m)$ is a function of the 1-rings around the vertices at the ends of the edge (see Figure 8.5 right).

Two examples of such schemes are Loop's (Loop (1987)), which is an approximating scheme generalizing quartic box splines, and the (modified) butterfly scheme (Dyn et al. (1990); Zorin et al. (1996); Prautzsch and Umlauf (1998)), which is interpolating. The stencil weights for these schemes are shown in Figure 8.6.

Discussion There are many possible choices of subdivision schemes to use for a surface editor, each of which has certain properties which make it more suitable for one task than another. Broadly, we can distinguish interpolating and approximating schemes, those based on triangles or quadrilaterals,

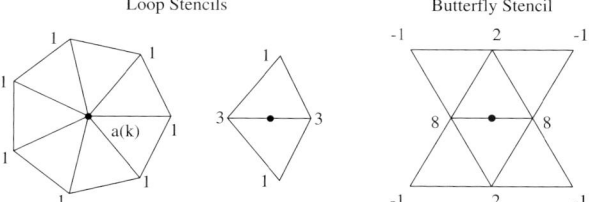

Fig. 8.6. Stencils for Loop subdivision with unnormalized weights for even and odd vertices (left) and for the regular butterfly (right). Note that the butterfly scheme, since it is interpolating, only has a stencil for odd vertices.

as well as primal versus dual schemes. Interpolating schemes such as the butterfly scheme (Dyn et al. (1990); Zorin et al. (1996)) (triangles) or Kobbelt's quad scheme (Kobbelt (1996a)) have the advantage that any vertex has only one associated point, which is a sample of the limit surface. This simplifies the implementation and makes adaptive subdivision particularly easy. For example, it suffices to place a simple restriction criterion on the quadtrees that hold the point positions. A disadvantage of interpolating schemes arises from the fact that the fundamental solutions of the subdivision process by necessity have negative lobes. As a consequence the surface develops oscillations when a control point is moved. In general, interpolating schemes tend to produce surfaces which lack fairness and can respond in non-intuitive ways when a user manipulates the control points. An exception are variational subdivision schemes (Kobbelt (1996b)) which are defined to produce fair surfaces that are also interpolating. Unfortunately they are computationally intensive, as each subdivision step requires the inversion of a (sparse) system of linear equations. Additionally it is not clear how to support adaptive subdivision in this setting.

In contrast approximating schemes, which are largely based on generalizations of splines, such as Loop (1987), Catmull–Clark (Catmull and Clark (1978)), and Doo–Sabin (Doo and Sabin (1978)) tend to result in much nicer editing behavior. Since the basis functions do not exhibit any undulations, pulling on a control point results in a smooth, bump-like surface change. Implementation of these schemes is slightly more involved as each vertex now has to hold a number of point positions, one for every level of the subdivision process on which it exists. For an efficient implementation this requires dynamically adjustable storage at each vertex. Adaptive subdivision is still possible through the use of a restriction criterion, but care must be taken to reference the point position at the proper level. Alternatively each vertex can carry one extra point position which is a sample of the limit surface (easily computed through the application of an associated mask corresponding to the left eigenvector of the eigenvalue 1).

The choice between triangles and quadrilaterals as basic shape is mostly dictated by the application domain. Triangles are somewhat more general as primitives. For example, when drawing the surface any graphics rendering hardware will convert arbitrary polygons to triangles. On the other hand, many objects exhibit two major directions ("left-right/top-bottom," "north-south/west-east"), which favors quadrilaterals as basic primitives. If the shapes are to be used in finite element analysis one may prefer triangles or quadrilaterals, depending on considerations such as approximation properties.

Finally we point out the difference between primal and dual schemes. Consider the generalization of bi-quadratic splines in the form of Doo–Sabin's subdivision scheme (a similar argument applies to the schemes studied in Peters and Reif (1997)). When applying subdivision to the control mesh, each vertex at the coarser level becomes the parent of a face at the next finer level. Similarly each edge generates a face at the next finer level and so does each face. They are corner cutting schemes. We call these dual since they can be thought of as subdivision on the dual complex. Consider the complex of the coarser mesh. Take its dual and refine it by quadrisection, i.e., each quadrilateral face receives a new vertex, as do all edges. Finally, dualize the resulting complex again to arrive at the mesh of the finer level of the original scheme. Unfortunately this observation only applies to the topology, not the geometry. The resulting data structures are awkward and it is not clear how to support adaptive subdivision in this framework. This is the main reason for our preference of primal schemes. The latter include Catmull–Clark and Loop.

Another consideration in implementations concerns the support size of the stencils and basis functions, respectively. For example, the interpolating constructions mentioned above have larger support leading to larger support stencils when computing tangent vectors. Aside from performance considerations, larger stencils also imply many more special cases near the boundary and near extraordinary vertices, i.e., those with valence other than 6 (triangles) or 4 (quadrilaterals). All these considerations lead us to prefer Loop's scheme. It is based on triangles, the support for limit point evaluation or tangent evaluation is a 1-ring, the basis functions are non-negative, and adaptive subdivision is easily supported through a restriction criterion imposed on the quadtrees which hold the point positions of the control meshes at different levels of resolution. The latter is particularly critical in escaping the exponential time and memory growth of naive subdivision.

8.4 Multiresolution extensions of subdivision

Using subdivision alone one could easily build a very efficient geometric modeling system. The user manipulates the coarsest level control points and thus shapes the limit surface. Suitably high performance can be delivered through the use of adaptive subdivision, for example, allocating larger triangles in areas which are relatively flat and smaller triangles in areas of high curvature. Figure 8.7 shows an instance of such adaptation which is naturally supported by the recursive nature of subdivision. Simply stop a particular branch of the recursion when a local approximation criterion has been satisfied. These criteria can also be varied across the surface so as to allocate resources in regions of particular interest without compromising overall performance (see Figure 8.8).

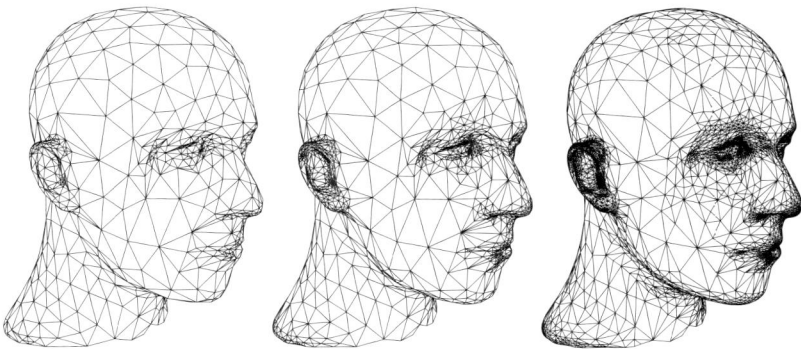

Fig. 8.7. Subdivision describes a smooth surface as the limit of a sequence of refined polyhedra. The meshes show several levels of an adaptive Loop surface.

The capabilities of such an editing system as presented to the user are very similar to what one would find in more traditional patch-based modelers, the only difference being that the system can deal with arbitrary topology surfaces and not just those describable over a tensor product domain. What is missing is the ability to define detail at many different levels of a hierarchy. Figure 8.9 shows an example of the head of Armadillo man. Here the top level control points were subdivided four times, yielding the surface on the left side. It is very smooth as one would expect. What the designer would

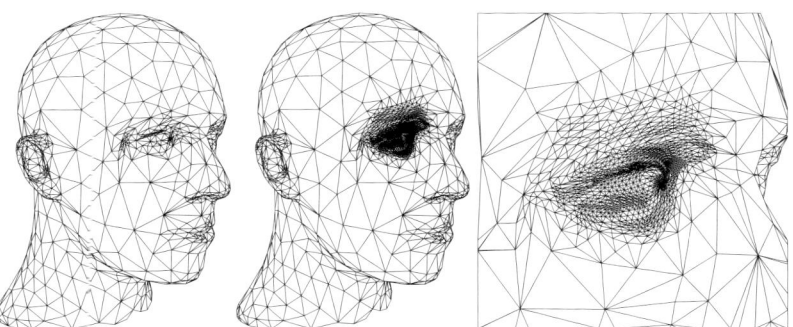

Fig. 8.8. It is easy to change the surface approximation quality criterion locally. Here a "lens" was applied to the right eye of the Mannequin head through a locally decreased approximation quality ϵ to force very fine resolution of the mesh around the eye. This example shows a pure subdivision surface with no multiresolution details.

like is the ability to include details at finer levels. Conceptually this is particularly easy to accomplish in the subdivision setting. Simply allow the control mesh points to be perturbed by the addition of "detail" vectors after each subdivision step. The results of allowing such a perturbation are demonstrated in the middle and right picture in Figure 8.9. The middle picture shows the effect of allowing additional displacements to be added after the first subdivision step, while the right image shows the effect of allowing displacements to be added after the first subdivision step and once more after the second subdivision step. In this way the introduction of detail in an "octave-like" fashion of finer and finer scales becomes straightforward.

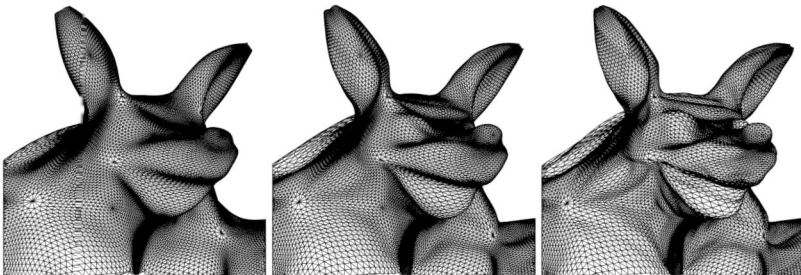

Fig. 8.9 Wireframe renderings of virtual surfaces representing the first three levels of control points.

To put these ideas on a solid basis we must discuss the analysis operator. Subdivision, or synthesis, goes from coarse to fine, while analysis goes from fine to coarse and computes the detail vectors. We first need *smoothing* and downsampling, i.e., a linear operation H to build a smooth coarse mesh at level $i-1$ from a fine mesh at level i:

$$p^{i-1} = H \, p^i.$$

Several options are available here.

- **Least squares** One could define analysis to be optimal in the least squares sense,

$$\min_{p^{i-1}} \|p^i - S \, p^{i-1}\|^2.$$

  The solution may have unwanted undulations and is too expensive to compute interactively (Forsey and Wong (1995)).
- **Fairing** A coarse surface could be obtained as the solution to a global variational problem. This is too expensive as well. An alternative is presented by Taubin (1995), who uses a *local* smoothing approach.

Because of its computational simplicity we decided to use a version of Taubin smoothing. As before let $a \in V^i$ have K neighbors $a_k \in V^i$. Use the average, $\bar{p}^i(v) = K^{-1} \sum_{k=1}^{K} p^i(a_k)$, to define the discrete Laplacian $\mathcal{L}(a) = \bar{p}^i(v) - p^i(a)$. On this basis Taubin gives a Gaussian-like smoother with reduced shrinkage problems

$$H := (I + \mu \mathcal{L})(I + \lambda \mathcal{L}),$$

with μ and λ tuned to ameliorate the shrinkage problem inherent in straight Gaussian smoothing (Taubin (1995)). With subdivision and smoothing in place, we can describe the transform needed to support multiresolution editing.

Recall that for multiresolution editing we want the difference between successive levels expressed with respect to a frame induced by the coarser level, i.e., the offsets are relative to the smoother level. With each vertex a and each level $i > 0$ we associate a *detail vector*, $d^i(a) \in \mathbf{R}^3$. The set d^i contains all detail vectors on level i, $d^i = \{d^i(a) \mid a \in V^i\}$. As indicated in Figure 8.10 the detail vectors are defined as

$$d^i = (F^i)^T (p^i - S \, p^{i-1}) = (F^i)^T (I - S \, H) \, p^i,$$

i.e., the detail vectors at level i record how much the points at level i differ from the result of subdividing the points at level $i-1$. This difference is

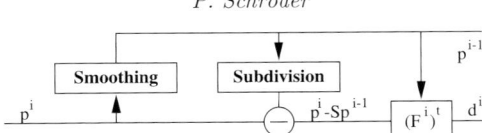

Fig. 8.10. Wiring diagram of the multiresolution transform.

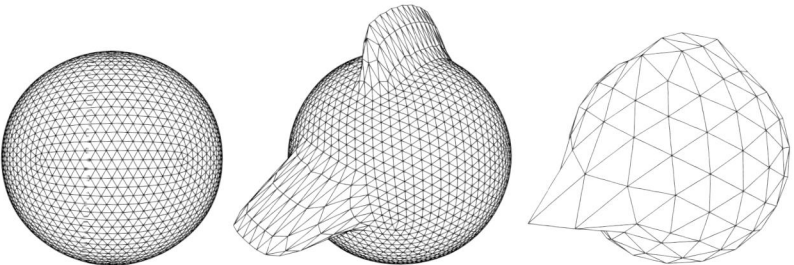

Fig. 8.11. Analysis propagates the changes on finer levels to coarser levels, keeping the magnitude of details under control. Left: The initial mesh. Center: A simple edit on level 3. Right: The effect of the edit on level 2. A significant part of the change was absorbed by higher level details.

then represented with respect to the local frame F^i to obtain coordinate independence.

Since detail vectors are sampled on the fine level mesh V^i, this transformation yields an overrepresentation in the spirit of the Burt–Adelson pyramid (Burt and Adelson (1983)). The only difference is that the smoothing filter (Taubin) is not the dual of the subdivision filter (Loop). Theoretically it would be possible to subsample the detail vectors and only record a detail per odd vertex (elements of M^{i-1}). This is what happens in the wavelet transform. However, subsampling the details severely restricts the family of smoothing operators that can be used and is not desirable in an editing environment in either case (see our comments in Section 8.2).

Figure 8.11 shows an example of analysis applied to a shape with a strong feature at a fine level. This feature yields details at coarser levels which are smoothed versions of the finer level feature.

Figure 8.12 shows two triangle mesh approximations of the Armadillo head with multiresolution details. Approximately the same number of triangles are used for the adaptive and uniform mesh. The adaptive rendering took

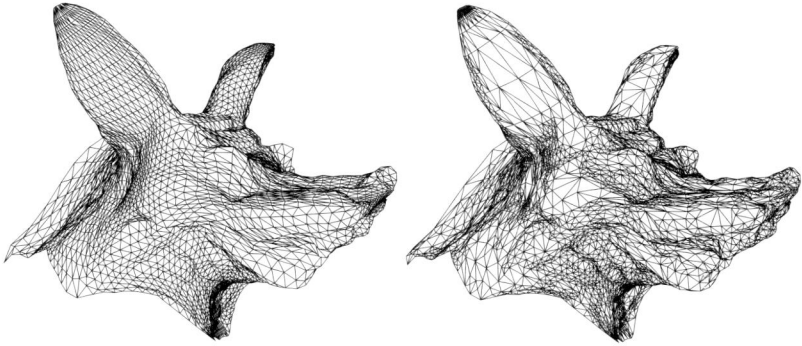

Fig. 8.12. On the left a uniformly subdivided mesh with approximately 11k triangles. On the right a mesh with the same number of triangles but subdivided non-uniformly for a better approximation within the same rendering budget.

local surface approximation quality and magnitudes of details at finer levels into account.

The performance of the system is good enough to allow the editing of intricate shapes such as Armadillo man on a PC. The Armadillo man has approximately 172,000 triangles on 6 levels of subdivision. Display list creation took 3 seconds on the PC for the full model. We adjusted the approximation criteria so that the model would render at 5 frames per second. On the PC this amounted to a model approximation with 35,000 triangles. Note that even though only an approximation is rendered on the screen, the entire model is maintained in memory. Thanks to the surface-relative local frames for the detail vectors, any edits performed at coarse levels, even if fine levels are not displayed, are correctly reflected at fine levels. Figure 8.13 shows a coarse level edit in which control points associated with the tip of the ears are pulled, while Figure 8.14 shows a fine level edit in which the user moves control points at a fine subdivision level in the vicinity of the eye.

8.5 Irregular meshes

The approach presented so far relies on the special structure of the underlying meshes for its efficiency. We assumed that there exists a coarsest level mesh which is recursively subdivided. The resulting type of meshes have a *semi-regular* structure, i.e, all but the coarsest level vertices have valence 6.

Fig. 8.13. An example of a coarse level edit in which a group of control points at the tip of the ears is pulled up and then subsequently only the control points at the tip of one ear are pushed down. Note that fine level detail on the shape moves along with the surface as one would intuitively expect.

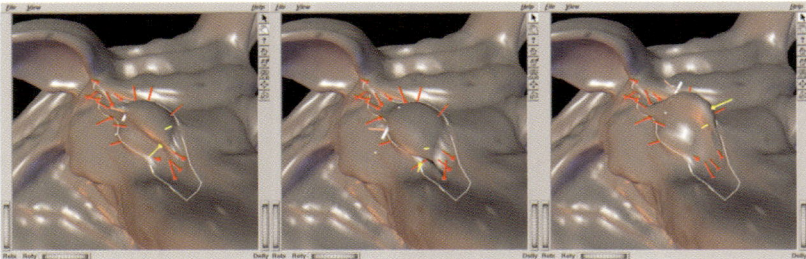

Fig. 8.14. An example of a fine level edit. Here control points at the third subdivision level are exposed and the user selects a group of them around the eye lid, affecting only a very small region around the eye.

When geometry is built "from scratch" it is an easy matter to enforce this requirement. The situation changes when one wants to work with an arbitrary input mesh coming from some other application. For example, meshes coming from 3D acquisition devices typically are completely unstructured, or irregular. If we wish to apply multiresolution analysis and associated algorithms to these we must first convert them to semi-regular meshes, or *remesh* them. A first step in such a procedure, and for that matter in many other numerical algorithms which aim to deal with such meshes, is the establishment of a smooth parameterization. In a very fundamental sense we are seeking methods of building a proper manifold structure, i.e., charts and an atlas, for such settings. While an arbitrary triangle mesh can always be thought of as parameterized over itself this is not a useful observation

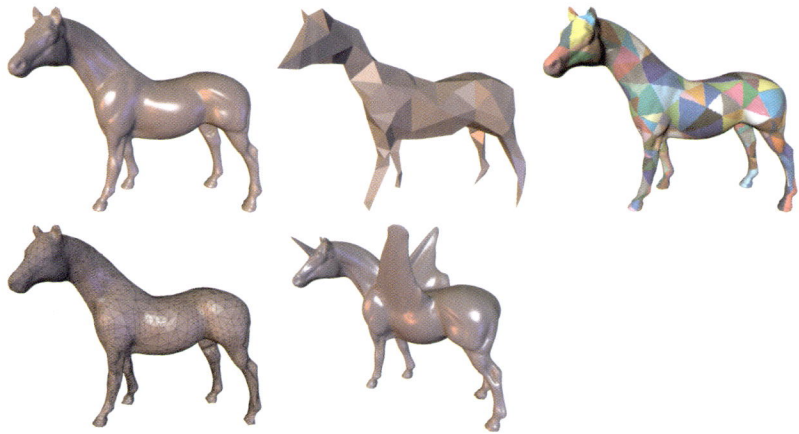

Fig. 8.15. Overview of the remeshing algorithm. Top left: a scanned input mesh (courtesy Cyberware). Next the parameter or base domain, obtained through mesh simplification. Top right: regions of the original mesh colored according to their assigned base domain triangle. Bottom left: adaptive remeshing with subdivision connectivity ($\epsilon = 1\%$). Bottom right: multiresolution edit.

in practice: a mesh which consists of hundreds of thousands of triangles is trivially parameterized over its combinatorial complex using the geometric realization φ. But this object is unwieldy and not smooth (piecewise linear only). What we seek instead is a parameterization of the original mesh over a homeomorphic complex with very few triangles. Once the original mesh, or a close approximation to it, is realized in this fashion many numerical modeling tasks are greatly facilitated. In particular it is then an easy matter to resample the original mesh onto a semi-regular mesh as it appears in traditional subdivision, allowing us to apply the algorithms described in the previous section.

A procedure for establishing such a parameterization is the subject of this section. Figure 8.15 shows the outline of the procedure: beginning with an irregular input mesh (top left), in a first step a base domain is established through mesh simplification (top middle). Concurrent with simplification, a mapping is constructed which assigns every vertex from the original mesh to a base triangle (top right). Using this mapping an adaptive remesh with subdivision connectivity can be built (bottom left) which is now suitable for such applications as multiresolution editing (bottom middle).

There are many possible ways of approaching this problem and before describing the details of one such algorithm we first consider some other possibilities. These fall into two broad categories: those that are geared towards approximating an initial unstructured mesh with smooth patches; and those which explicitly aim at constructing parameterizations for remeshing.

8.5.1 Approximation of a given set of samples

Hoppe and co-workers (Hoppe et al. (1994)) describe a fully automatic algorithm for approximating a given polyhedral mesh with (modified) Loop subdivision patches (Loop (1987)) respecting features such as edges and corners. Their algorithm uses a nonlinear optimization procedure taking into account approximation error and the number of triangles of the base domain. The result is a smooth parameterization of the original polyhedral mesh with a (hopefully) small number of patches. Since the approach only uses subdivision, small features in the original mesh can only be resolved accurately by increasing the number of patches accordingly. From the point of view of constructing parameterizations, the main drawback of algorithms in this class is that the number of triangles in the base domain depends heavily on the *geometric* complexity of the original surface.

This latter problem was addressed in work of Krishnamurthy and Levoy (1996). They approximate densely-sampled geometry with bi-cubic spline patches and displacement maps. Arguing that a fully automatic system cannot put isoparameter lines where a skilled modeler or animator would want them, they require the user to lay out the entire network of top level spline patch boundaries. A coarse-to-fine matching procedure with relaxation is used to arrive at a high quality patch-based surface whose base domain need not mimic small scale geometric features. Any fine level features are captured by allowing the addition of a displacement map on top of the smooth spline patches. The principal drawback of their procedure is that the user is required to define the *entire* base domain rather then only selected features. Additionally, given that the procedure works from coarse to fine, it is possible for the procedure to "latch" onto the wrong part of the surface in regions of high curvature (Krishnamurthy and Levoy (1996), Figure 7).

8.5.2 Remeshing

Lounsbery and co-workers (Lounsbery (1994); Lounsbery et al. (1997)) were the first to propose algorithms that extend classical multiresolution analysis to arbitrary topology surfaces. Because of its connection to the mathematical foundations of wavelets, this approach has proven very attractive

(Schröder and Sweldens (1995); Eck et al. (1995); Zorin et al. (1996); Eck and Hoppe (1996); Certain et al. (1996); Zorin et al. (1997)). The algorithm described in the first half of this chapter belongs to this class. Since the central requirement of these methods is that the input mesh have subdivision connectivity-remeshing is generally required to apply these approaches. Eck and co-workers (Eck et al. (1995)) were the first to develop such a remeshing algorithm, computing a smooth parameterization of high resolution triangle meshes over a low face-count base domain, followed by a semi-regular resampling step. After this conversion step, adaptive simplification, compression, progressive transmission, rendering, and editing become simple and efficient operations (Certain et al. (1996); Eck and Hoppe (1996); Zorin et al. (1997)).

Eck et al. arrive at the base domain through a Voronoi tiling of the original mesh. Using a sequence of local harmonic maps, a parameterization which is smooth over each triangle in the base domain and which meets with C^0 continuity at base domain edges is constructed (Eck et al. (1995), Plate 1(f)). Runtimes for the algorithm can be long because of the many harmonic map computations. This problem was recently addressed by Duchamp and co-workers (Duchamp et al. (1997)), who reduced the harmonic map computations from their initial $O(N^2)$ complexity to $O(N \log N)$ through hierarchical preconditioning.† The initial Voronoi tile construction relies on a number of heuristics which render the overall algorithm fragile (for an improved version see Klein et al. (1997)). Moreover, there is no explicit control over the number of triangles in the base domain or the placement of patch boundaries. The latter is critical in applications as one wants to avoid unnecessary subdivision near sharp features.

8.5.3 An alternative parameterization algorithm

We describe here an algorithm which was designed to overcome the drawbacks of previous work as well as to introduce new features. A fast coarsification strategy is used to define the base domain, avoiding the potential difficulties of finding Voronoi tiles (Eck et al. (1995); Klein et al. (1997)). Since the algorithm proceeds from fine to coarse, correspondence problems found in coarse-to-fine strategies (Krishnamurthy and Levoy (1996)) are avoided, and all features are correctly resolved. Piecewise linear approximations to conformal maps are used during coarsification to produce a global parameterization of the original mesh. This map is then further improved through the use of a hierarchical Loop smoothing procedure obviating the

† The hierarchy construction they employed for use in a multigrid solver is closely related to the hierarchy construction described below.

need for iterative numerical solvers (Eck et al. (1995)). Since the procedure is performed globally, derivative discontinuities at the edges of the base domain are avoided (Eck et al. (1995)). In contrast to fully automatic methods (Eck et al. (1995)), the algorithm supports vertex and edge tags (Hoppe et al. (1994)) to constrain the final parameterization to align with selected features; however, the user is not required to specify the entire patch network (Krishnamurthy and Levoy (1996)). During remeshing we take advantage of the original fine-to-coarse simplification hierarchy to output a sparse, adaptive, semi-regular mesh directly without resorting to a depth-first oracle (Schröder and Sweldens (1995)) or the need to produce a uniform subdivision connectivity mesh at exponential cost followed by wavelet thresholding (Certain et al. (1996)).

8.6 Hierarchical surface representation

An important part of the algorithm is the construction of a mesh hierarchy. The original mesh $(\mathcal{P},\mathcal{K}) = (\mathcal{P}^L,\mathcal{K}^L)$ is successively simplified into a series of homeomorphic meshes $(\mathcal{P}^l,\mathcal{K}^l)$ with $0 \leq l < L$, where $(\mathcal{P}^0,\mathcal{K}^0)$ is the coarsest or base mesh (see Figure 8.16). Several approaches for such a mesh simplification have been proposed, most notably progressive meshes (PM) (Hoppe (1996)). In PM the basic operation is the "edge collapse." A sequence of such atomic operations is prioritized based on approximation error. The linear sequence of edge collapses can be partially ordered based on topological dependence (Xia and Varshney (1996); Hoppe (1997)), which defines levels in a hierarchy. The depth of these hierarchies appears "reasonable" in practice, though can vary considerably for the same dataset (Hoppe (1997)).

Our approach is similar in spirit, but inspired by the hierarchy proposed by Dobkin and Kirkpatrick (1985) (DK), which guarantees that the number of levels L is $O(\log N)$. While the original DK hierarchy is built for convex meshes, we show how the idea behind DK can be used for general ones. The DK atomic simplification step is a *vertex remove*, followed by a retriangulation of the hole. The two basic operations "vertex remove" and "edge collapse" are related since an edge collapse into one of its endpoints corresponds to a vertex removal with a particular retriangulation of the resulting hole (see Figure 8.17). The main reason for choosing an algorithm based on the ideas of the DK hierarchy is that it guarantees a logarithmic bound on the number of levels.

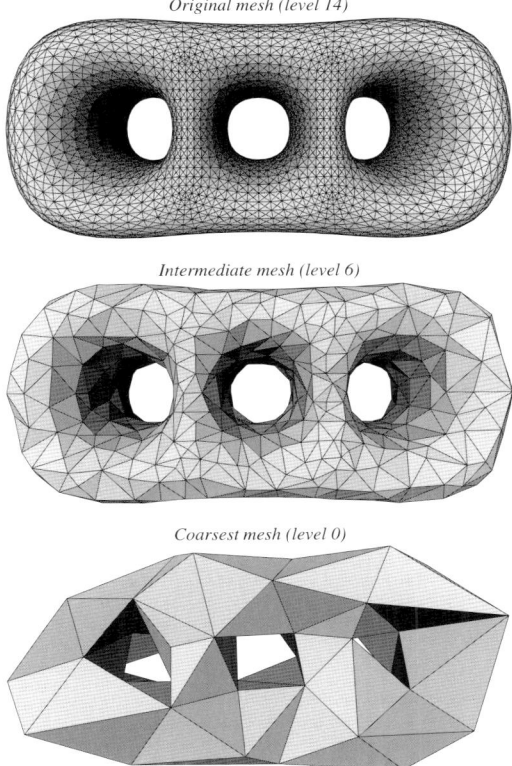

Fig. 8.16. Example of a modified DK mesh hierarchy. At the top the finest (original) mesh $\varphi(|\mathcal{K}^L|)$ followed by an intermediate mesh, and the coarsest (base) mesh $\varphi(|\mathcal{K}^0|)$ at the bottom (original dataset courtesy University of Washington).

8.6.1 Vertex removal

One DK simplification step $\mathcal{K}^l \to \mathcal{K}^{l-1}$ consists of removing a maximally independent set of vertices with low outdegree (see Figure 8.18). Recall that an independent set of vertices has the property that no two vertices in the set are connected by an edge. To find such a set, the original DK algorithm used a greedy approach based only on *topological* information. Instead it is desirable to take geometric information into account as well. This can be done by selecting candidates for the independent set based on a priority queue.

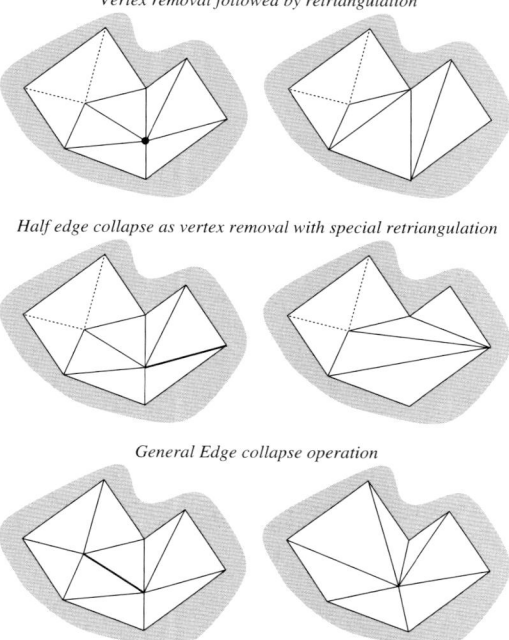

Fig. 8.17. Examples of different atomic mesh simplification steps. At the top vertex removal, in the middle half-edge collapse, and edge collapse at the bottom.

At the start of each level of the DK algorithm, none of the vertices are marked and the set to be removed is empty. The algorithm randomly selects a non-marked vertex of outdegree less than 12, removes it and its star from $\mathcal{K}^l$, marks its neighbors as unremovable and iterates this until no further vertices can be removed. In a triangulated surface the average outdegree of a vertex is 6. Consequently, no more than half of the vertices can be of outdegree 12 or more. Thus it is guaranteed that at least 1/24 of the vertices will be removed at each level (Dobkin and Kirkpatrick (1985)). In practice, it turns out one can remove roughly 1/4 of the vertices reflecting the fact that the graph is four-colorable. In any case, given that a constant fraction can be removed on each level, the number of levels behaves as $O(\log N)$. The entire hierarchy can thus be constructed in linear time. A better approach is to stay in the DK framework, but to replace the random selection of vertices by a priority queue based on geometric information.

Chapter 8 Scalable algorithms in computer graphics   237

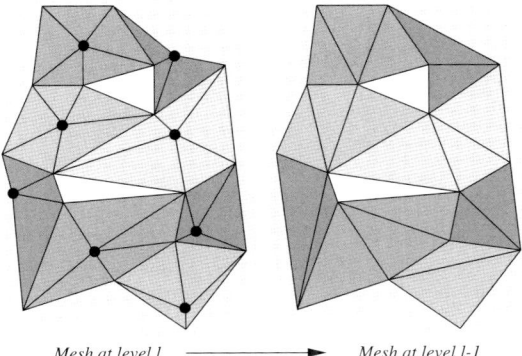

Mesh at level l ⟶ Mesh at level l-1

Fig. 8.18. On the left a mesh with a maximally independent set of vertices marked by heavy dots. Each vertex in the independent set has its respective star highlighted. Note that the stars of the independent set do not tile the mesh (two triangles are left white). The right side gives the retriangulation after vertex removal.

Roughly speaking, vertices with small and flat 1-ring neighborhoods should be chosen first. Note that the complexity of the overall construction grows to $O(N \log N)$ because of the priority queue.

Figure 8.16 shows three stages (original, intermediary, coarsest) of such a DK hierarchy. The main observation is that the coarsest mesh can be used as the domain of a parameterization of the original mesh since all meshes in the DK hierarchy are homeomorphic. The construction of these homeomorphisms is the subject of the next section.

8.6.2 Flattening and retriangulation

To find $\mathcal{K}^{l-1}$, we need to retriangulate the holes left by removing the independent set. One possibility is to find a plane into which to project the 1-ring neighborhood $\varphi(|\text{star}(i)|)$ of a removed vertex $\varphi(|i|)$ without overlapping triangles and then retriangulate the hole in that plane. However, finding such a plane, which may not even exist, can be expensive and involves linear programming (Cohen et al. (1997)).

Instead, it is much easier and more robust to use the conformal map z^α (Duchamp et al. (1997)) which minimizes metric distortion when mapping the neighborhood of a removed vertex into the plane. Let $\{a\}$ be a vertex to be removed. Enumerate cyclically the K_a vertices in the 1-ring $\mathcal{N}(i) = \{b_k \mid 1 \leq k \leq K_a\}$ such that $\{b_{k-1}, a, b_k\} \in \mathcal{K}^l$ with $b_0 = b_{K_a}$. A piecewise linear approximation of z^α, which we denote by μ_a, is defined by its values for the center point and 1-ring neighbors; namely, $\mu_a(p_a) = 0$ and $\mu_a(p_{b_k}) = r_k^\alpha \exp(i\theta_k \alpha)$, where $r_k = \|p_a - p_{b_k}\|$,

$$\theta_k = \sum_{l=1}^{K_a} \angle(p_{b_{l-1}}, p_a, p_{b_l}),$$

and $\alpha = 2\pi/\theta_{K_a}$. The advantages of the conformal map are numerous: it always exists, it is easy to compute, it minimizes metric distortion, and it is a bijection and thus never maps two triangles on top of each other. Once the 1-ring is flattened, we can retriangulate the hole using a constrained Delaunay triangulation (CDT), for example (see Figure 8.19). This tells us how to build $\mathcal{K}^{l-1}$.

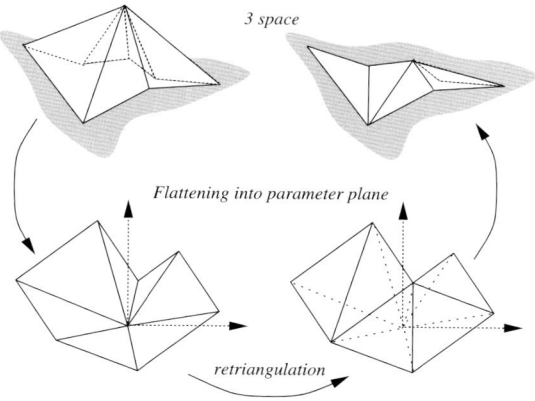

Fig. 8.19. In order to remove a vertex p_a, its star (a) is mapped from 3-space to a plane using the map z^α. In the plane the central vertex is removed and the resulting hole retriangulated (bottom right).

When the vertex to be removed is a boundary vertex, we map to a half disk by setting $\alpha = \pi/\theta_{K_a}$ (assuming b_1 and b_{K_a} are boundary vertices and setting $\theta_1 = 0$). Retriangulation is again performed with a CDT.

8.7 Initial parameterization

To find a parameterization, we begin by constructing a bijection Π from $\varphi(|\mathcal{K}^L|)$ to $\varphi(|\mathcal{K}^0|)$. The parameterization of the original mesh over the base domain follows from $\Pi^{-1}(\varphi(|\mathcal{K}^0|))$. In other words, the mapping of a point

$p \in \varphi(|\mathcal{K}^L|)$ through Π is a point $p^0 = \Pi(a) \in \varphi(|\mathcal{K}^0|)$, which can be written as

$$p^0 = \alpha\, p_a + \beta\, p_b + \gamma\, p_c,$$

where $\{a, b, c\} \in \mathcal{K}^0$ is a face of the base domain and α, β and γ are barycentric coordinates, i.e., $\alpha + \beta + \gamma = 1$.

The mapping can be computed concurrently with the hierarchy construction. The basic idea is to successively compute piecewise linear bijections Π^l between $\varphi(|\mathcal{K}^L|)$ and $\varphi(|\mathcal{K}^l|)$ starting with Π^L, which is the identity, and ending with $\Pi^0 = \Pi$.

Notice that we only need to compute the value of Π^l at the vertices of $\mathcal{K}^L$. At any other point it follows from piecewise linearity.† Assume we are given Π^l and want to compute Π^{l-1}. Each vertex $\{a\} \in \mathcal{K}^L$ falls into one of the following categories:

(i) $\{a\} \in \mathcal{K}^{l-1}$: The vertex is not removed on level l and survives on level $l-1$. In this case nothing needs to be done. $\Pi^{l-1}(p_a) = \Pi^l(p_a) = p_a$;

(ii) $\{a\} \in \mathcal{K}^l \setminus \mathcal{K}^{l-1}$: The vertex gets removed when going from l to $l-1$. Consider the flattening of the 1-ring around p_a (see Figure 8.19). After retriangulation, the origin lies in a triangle which corresponds to some face $t = \{d, e, f\} \in \mathcal{K}^{l-1}$ and has barycentric coordinates (α, β, γ) with respect to the vertices of that face, i.e., $\alpha\,\mu_a(p_d) + \beta\,\mu_a(p_e) + \gamma\,\mu_a(p_f)$ (see Figure 8.20). In that case, let $\Pi^{l-1}(p_a) = \alpha\, p_d + \beta\, p_e + \gamma\, p_f$;

(iii) $\{a\} \in \mathcal{K}^L \setminus \mathcal{K}^l$: The vertex was removed earlier, thus $\Pi^l(p_a) = \alpha'\, p_{a'} + \beta'\, p_{b'} + \gamma'\, p_{c'}$ for some triangle $t' = \{a', b', c'\} \in \mathcal{K}^l$. If $t' \in \mathcal{K}^{l-1}$, nothing needs to be done; otherwise, the independent set property guarantees that exactly one vertex of t' is removed, say $\{a'\}$. Consider the conformal map $\mu_{a'}$ (Figure 8.20). After retriangulation, the $\mu_{a'}(p_a)$ lies in a triangle which corresponds to some face $t = \{d, e, f\} \in \mathcal{K}^{l-1}$ with barycentric coordinates (α, β, γ) (black dots within highlighted face in Figure 8.20). In that case, let $\Pi^{l-1}(p_a) = \alpha\, p_d + \beta\, p_e + \gamma\, p_f$ (i.e., all vertices in Figure 8.20 are reparameterized in this way).

Note that on every level, the algorithm requires a sweep through all the vertices of the finest level resulting in an overall complexity of $O(N \log N)$. Figure 8.21 visualizes the mapping we just computed.

† In the vicinity of vertices in $\mathcal{K}^l$ a triangle $\{a, b, c\} \in \mathcal{K}^L$ can straddle multiple triangles in $\mathcal{K}^l$. In this case the map depends on the flattening strategy used (see Section 8.6.2).

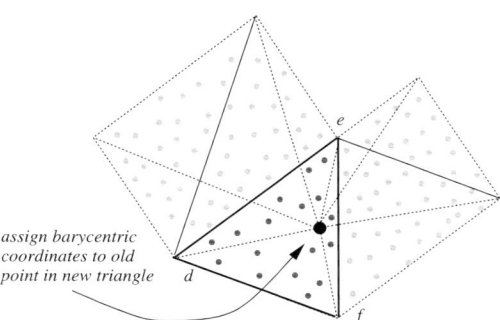

Fig. 8.20. After retriangulation of a hole in the plane (see Figure 8.19), the just removed vertex gets assigned barycentric coordinates with respect to the containing triangle on the coarser level. Similarly, all the finest level vertices that were mapped to a triangle of the hole now need to be reassigned to a triangle of the coarser level.

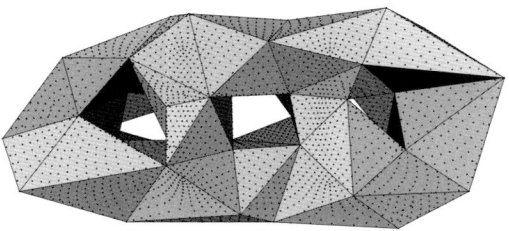

Fig. 8.21. Base domain $\varphi(|\mathcal{K}^0|)$. For each point p_a from the original mesh, its mapping $\Pi(p_a)$ is shown with a dot on the base domain.

8.7.1 Tagging and feature lines

In the algorithm described so far, there is no *a priori* control over which vertices end up in the base domain or how they will be connected. However, often there are features such as sharp creases in the input mesh. It is desirable to ensure that these features align with particular isoparameter lines in the parameterization. Such features could be detected automatically or specified by the user. We consider two types of features on the finest mesh: vertices and paths of edges. Guaranteeing that a certain vertex of the original mesh ends up in the base domain is straightforward. Simply mark that vertex as unremovable throughout the DK hierarchy.

We now describe an algorithm for guaranteeing that a certain path of edges on the finest mesh gets mapped to an edge of the base domain. Let

$\{v_a \mid 1 \leq a \leq I\} \subset \mathcal{K}^L$ be a set of vertices on the finest level which form a path, i.e., $\{v_a, v_{a+1}\}$ is an edge. Tag all the edges in the path as feature edges. First tag v_1 and v_I, so-called *dart points* (Hoppe et al. (1994)), as unremovable so they are guaranteed to end up in the base domain. Let v_a be the first vertex on the interior of the path which gets marked for removal in the DK hierarchy, say, when going from level l to $l-1$. Because of the independent set property, v_{a-1} and v_{a+1} cannot be removed and therefore must belong to $\mathcal{K}^{l-1}$. When flattening the hole around v_a, tagged edges are treated the same as boundary edges. We first straighten out the edges $\{v_{a-1}, v_a\}$ and $\{v_a, v_{a+1}\}$ along the x-axis, and use two boundary-type conformal maps to the half disk above and below (see the last paragraph of Section 8.6.2). When retriangulating the hole around v_a, we put the edge $\{v_{a-1}, v_{a+1}\}$ in $\mathcal{K}^{l-1}$, tag it as a feature edge, and compute a CDT on the upper and lower parts (see Figure 8.22). If we apply similar procedures on coarser levels, we ensure that v_1 and v_I remain connected by a path (potentially a single edge) in the base domain. This guarantees that Π maps the curved feature path onto the coarsest level edge(s) between v_1 and v_I.

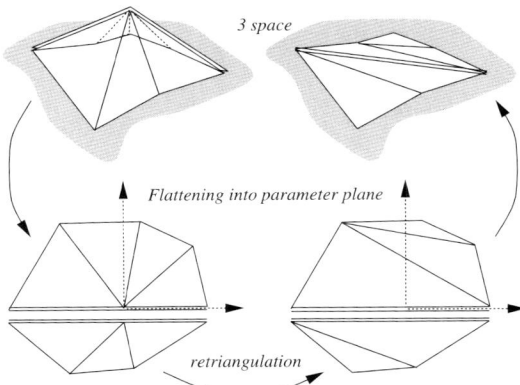

Fig. 8.22. When a vertex with two incident feature edges is removed, we want to ensure that the subsequent retriangulation adds a new feature edge to replace the two old ones.

In general, there will be multiple feature paths which may be closed or cross each other. As usual, a vertex with more than two incident feature edges is considered a corner, and marked as unremovable.

The feature vertices and paths can be provided by the user or detected automatically. As an example of the latter case, we consider every edge whose dihedral angle is below a certain threshold to be a feature edge, and every vertex whose curvature is above a certain threshold to be a feature vertex. An example of this strategy is illustrated in Figure 8.25.

8.8 Remeshing

In this section, we consider remeshing using semi-regular triangulations. In the process, we compute a smoothed version of our initial parameterization. We also show how to efficiently construct an adaptive remesh with guaranteed error bounds.

8.8.1 Uniform remeshing

Since Π is a bijection, we can use Π^{-1} to map the base domain to the original mesh. We follow the strategy used in Eck et al. (1995), quadrisecting the base domain and using the inverse map to obtain a semi-regular connectivity remeshing. This introduces a hierarchy of semi-regular meshes $(\tilde{\mathcal{P}}^m, \tilde{\mathcal{K}}^m)$ ($\tilde{\mathcal{P}}$ is the point set and $\tilde{\mathcal{K}}$ is the complex) obtained from m-fold midpoint subdivision of the base domain $(\mathcal{P}^0, \mathcal{K}^0) = (\tilde{\mathcal{P}}^0, \tilde{\mathcal{K}}^0)$. Midpoint subdivision implies that all new domain points lie *in* the base domain, $\tilde{\mathcal{P}}^m \subset \varphi(|\tilde{\mathcal{K}}^0|)$ and $|\tilde{\mathcal{K}}^m| = |\tilde{\mathcal{K}}^0|$. All vertices of $\tilde{\mathcal{K}}^m \setminus \tilde{\mathcal{K}}^0$ have outdegree 6. The uniform remeshing of the original mesh on level m is given by $(\Pi^{-1}(\tilde{\mathcal{P}}^m), \tilde{\mathcal{K}}^m)$.

We thus need to compute $\Pi^{-1}(q)$ where q is a point in the base domain with dyadic barycentric coordinates. In particular, we need to compute which triangle of $\varphi(|\mathcal{K}^L|)$ contains $\Pi^{-1}(q)$, or, equivalently, which triangle of $\Pi(\varphi(\mathcal{K}^L|))$ contains q. This is a standard *point location* problem in an irregular triangulation. This problem can be solved with the point location algorithm of Brown and Faigle (1997) which avoids looping that can occur with non-Delaunay meshes (Guibas and Stolfi (1985); Garland and Heckbert (1995)) Once we have found the triangle $\{a,b,c\}$ which contains q, we can write q as

$$q = \alpha\,\Pi(p_a) + \beta\,\Pi(p_b) + \gamma\,\Pi(p_c),$$

and thus

$$\Pi^{-1}(q) = \alpha\,p_a + \beta\,p_b + \gamma\,p_c \in \varphi(|\mathcal{K}^L|).$$

Figure 8.23 shows the result of this procedure: a level 3 uniform remeshing of a 3-holed torus using the Π^{-1} map.

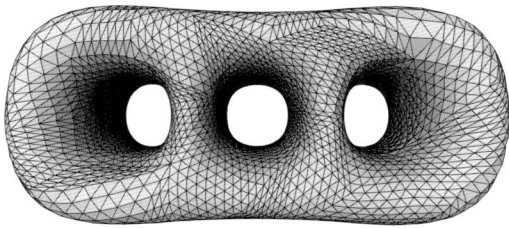

Fig. 8.23. Remeshing of 3 holed torus using midpoint subdivision. The parameterization is smooth within each base domain triangle, but clearly not across base domain triangles.

A note on complexity The point location algorithm is essentially a walk on the finest level mesh with complexity $O(\sqrt{N})$. Hierarchical point location algorithms, which have asymptotic complexity $O(\log N)$, exist (Kirkpatrick (1983)) but have a much larger constant. Given that we schedule the queries in a systematic order, we almost always have an excellent starting guess and observe a constant number of steps. In practice, the finest level "walking" algorithm beats the hierarchical point location algorithms for all meshes we encountered (up to $100K$ faces).

8.8.2 Smoothing the parameterization

It is clear from Figure 8.23 that the mapping we used is not smooth across global edges. One way of obtaining global smoothness is to consider a map that minimizes a global smoothness functional and goes from $\varphi(|\mathcal{K}^L|)$ to $|\mathcal{K}^0|$ rather than to $\varphi(|\mathcal{K}^0|)$. This would require an iterative PDE solver. However, computation of mappings to topological realizations that live in a high-dimensional space are needlessly cumbersome and a much simpler smoothing procedure suffices.

The main idea is to compute Π^{-1} at a smoothed version of the dyadic points, rather than at the dyadic points themselves. This can equivalently be viewed as changing the parameterization. To that end, define a map $\mathcal{L}$ from the base domain to itself by the following modification of classic Loop subdivision.

- If all the points of the stencil needed for computing either a new point or smoothing an old point are inside the same triangle of the base domain, we can simply apply the Loop weights and the new points will be in that same face.

- If the stencil stretches across two faces of the base domain, we flatten them out using a "hinge" map at their common edge. We then compute the point's position in this flattened domain and extract the triangle in which the point lies together with its barycentric coordinates.
- If the stencil stretches across multiple faces, we use the conformal flattening strategy discussed earlier.

Note that the modifications to Loop force $\mathcal{L}$ to map the base domain onto the base domain. We emphasize that we do *not* apply the classic Loop scheme (which would produce a "blobby" version of the base domain). Nor are the surface approximations that we later produce Loop surfaces.

The composite map $\Pi^{-1} \circ \mathcal{L}$ is our *smoothed parameterization* that maps the base domain onto the original surface. The mth level of uniform remeshing with the smoothed parameterization is $(\Pi^{-1} \circ \mathcal{L}(\tilde{\mathcal{P}}^m), \tilde{\mathcal{K}}^m)$, where $\tilde{\mathcal{P}}^m$, as before, are the dyadic points on the base domain. Figure 8.24 shows the result of this procedure: a level-3 uniform remeshing of a 3-holed torus using the smoothed parameterization.

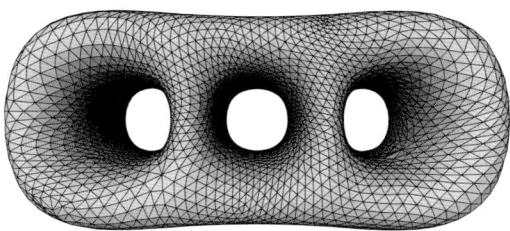

Fig. 8.24. The same remeshing of the 3-holed torus as in Figure 8.23, but this time with respect to a Loop smoothed parameterization. **Note** Because the Loop scheme only enters in smoothing the *parameterization*, the surface shown is still a sampling of the original mesh, *not* a Loop surface approximation of the original.

When the mesh is tagged, we cannot apply smoothing across the tagged edges since this would break the alignment with the features. Therefore, we use a modified versions of Loop which can deal with corners, dart points and feature edges (Hoppe et al. (1994); Schweitzer (1996); Zorin (1997)) (see Figure 8.25).

8.8.3 Adaptive remeshing

One of the advantages of meshes with subdivision connectivity is that classical multiresolution and wavelet algorithms can be employed. The standard wavelet algorithms used, e.g., in image compression, start from the finest level, compute the wavelet transform, and then obtain an efficient representation by discarding small wavelet coefficients. Eck et al. (1995); Eck and Hoppe (1996) as well as Certain et al. (1996) follow a similar approach. They remesh using a uniformly subdivided grid followed by decimation through wavelet thresholding. This has the drawback that in order to resolve a small local feature on the original mesh, one may need to subdivide to a very fine level. Each extra level quadruples the number of triangles, most of which will later be decimated using the wavelet procedure. Imagine, for example, a plane which is coarsely triangulated except for a narrow spike. Making the spike width sufficiently small, the number of levels needed to resolve it can be made arbitrarily high.

Instead of first building a uniform remesh and then pruning it, we can immediately build the adaptive mesh with a guaranteed conservative error bound. This is possible because the DK hierarchy contains the information on how much subdivision is needed in any given area. Essentially, one can let the irregular DK hierarchy "drive" the adaptive construction of the semi-regular pyramid.

First compute for each triangle $t \in \mathcal{K}^0$ the following error quantity:

$$E(t) = \max_{p_a \in \mathcal{P}^L \text{ and } \Pi(p_a) \in \varphi(|t|)} \text{dist}(p_a, \varphi(|t|)).$$

This measures the distance between one triangle in the base domain and the vertices of the finest level mapped to that triangle.

The adaptive algorithm is now straightforward. Set a certain relative error threshold ϵ. Compute $E(t)$ for all triangles of the base domain. If $E(t)/B$, where B is the largest side of the bounding box, is larger than ϵ, subdivide the domain triangle using the Loop procedure above. Next, we need to reassign vertices to the triangles of level $m = 1$. This is done as follows: for each point $p_a \in \mathcal{P}^L$ consider the triangle t of $\mathcal{K}^0$ to which it is currently assigned. Next consider the four children of t on level 1, t_j with $j = 0, 1, 2, 3$ and compute the distance between p_a and each of the $\varphi(|t_j|)$. Assign p_a to the closest child. Once the finest level vertices have been reassigned to level 1 triangles, the errors for those triangles can be computed. Now iterate this procedure until all triangles have an error below the threshold. Because all errors are computed from the finest level, we are guaranteed to resolve all features within the error bound. Note that we are not computing the true

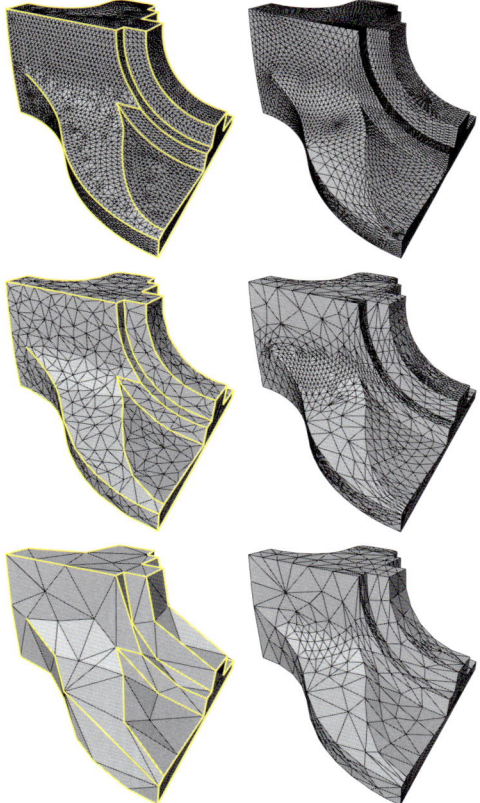

Fig. 8.25. Left (top to bottom): three levels in the DK pyramid, finest ($L = 15$) with 12946, intermediate ($l = 8$) with 1530, and coarsest ($l = 0$) with 168 triangles. Feature edges, dart and corner vertices survive on the base domain. Right (bottom to top): adaptive mesh with $\epsilon = 5\%$ and 1120 triangles (bottom), $\epsilon = 1\%$ and 3430 triangles (middle), and uniform level 3 (top). (Original dataset courtesy Hugues Hoppe.)

distance between the original vertices and a given approximation, but rather an easy-to-compute upper bound for it.

In order to be able to compute the Loop smoothing map $\mathcal{L}$ on an adaptively subdivided grid, the grid needs to satisfy a *vertex restriction criterion*, i.e., if a vertex has a triangle incident to it with depth i, then it must have a

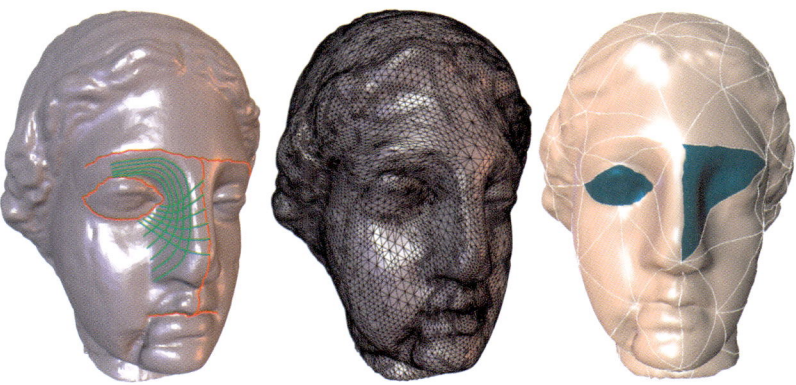

Fig. 8.26. Example of a constrained parameterization based on user input. Left: original input mesh (100,000 triangles) with edge tags superimposed in red. Green lines show some smooth isoparameter lines of our parameterization. The middle shows an adaptive subdivision connectivity remesh. The right one patches corresponding to the eye regions (right eye was constrained, left eye was not) are highlighted to indicate the resulting alignment of top level patches with the feature lines. (Dataset courtesy Cyberware.)

complete 1-ring at level $i-1$ (Zorin et al. (1997)). This restriction may necessitate subdividing some triangles even if they are below the error threshold. Examples of adaptive remeshing can be seen in Figure 8.15 (lower left) and Figure 8.25.

Figure 8.26 shows an example of a constrained parameterization and subsequent adaptive remeshing. The original dataset of 100,000 triangles is shown on the left. The red lines indicate user-supplied feature constraints which may facilitate subsequent animation. The green lines show some representative isoparameter lines of our parameterization subject to the red feature constraints. The middle image shows an adaptive subdivision connectivity remesh with 74698 triangles ($\epsilon = 0.5\%$). On the right we have highlighted a group of patches, two over the right (constrained) eye and one over the left (unconstrained) eye. This indicates how user-supplied constraints force domain patches to align with desired features. Other enforced patch boundaries are the eyebrows, center of the nose, and middle of lips (see red lines in left image).

Finally, the example shown in Figure 8.15 starts with an original mesh (96966 triangles) on the top left and results in an adaptive, subdivision connectivity remesh on the bottom left. This remesh was subsequently edited in the interactive multiresolution editing system described earlier (bottom middle).

8.9 Summary

Exploiting multiresolution is now widely considered an important ingredient in the construction of scalable algorithms for many computer graphics applications. Such constructions fall into two basic categories, those generalizing classical notions of multiresolution which are intricately connected with subdivision and wavelets, and those based on mesh simplification. The former can leverage the considerable mathematical machinery developed for the classical setting. The latter are still very new and little is known so far about the construction of smooth functions over the associated mesh hierarchies (for first steps in this direction see Daubechies et al. (1999); Daubechies et al. (1998); Guskov et al. (1999)).

In this chapter we primarily considered subdivision for surfaces and its multiresolution extension through the introduction of detail vectors at successive levels of subdivision. The resulting representations are very flexible since they can represent arbitrary 2-manifold surfaces (with boundary). Because of the hierarchical structure many algorithms which are geared towards delivering interactive performance on small computers are immediately applicable For example, adaptive approximation based on local criteria is easy to include.

Unfortunately many examples of complex geometry, in particular those coming from 3D scanning sources are not imbued with the particular semi-regular mesh structure required by subdivision-based algorithms. One way of addressing this challenge is through remeshing, i.e., resampling of the original geometry onto a semi-regular mesh. Once this step is performed all the advantages of the semi-regular setting apply. We described one such algorithm in detail. It exploits mesh simplification to discover a low-complexity base complex which can be employed as the domain for the coordinate functions describing the original geometry. It is then an easy matter to resample the original geometry with guaranteed error bounds.

8.10 Outlook

The demands of real applications are often not satisfied by the traditional constructions arising from the "mathematical clean room." Examples include irregular samples, non-trivial measures, arbitrary topology 2-manifolds, small computational budgets, and very large datasets. Because of this we have witnessed over the last few years increasing attention being devoted to the generalization of multiresolution from the infinite, regular, tensor product setting, to the semi-regular, and, much more recently, to the irregular setting. In many of these cases algorithms are running ahead of the associated theory, which often requires completely new tools for analyzing settings such as irregular samples, or manifolds with non-trivial structure.

The interchange between applications in computer graphics and, more generally, computer modeling, simulation, and design, and the classical mathematical treatment of multiresolution has led to significant advances in both theory and applications and we expect this to continue into the future. While processing power keeps increasing there appears no end in sight for our ability to absorb all available cycles and still be left wanting more. Similar observations apply to available network bandwidth. Multiresolution in all its forms will continue to play a central role and the field is wide open and in need of further, significant advances.

Acknowledgments

The author was supported in part by NSF (ACI-9624957, ACI-9721349, DMS-9874082, DMS-9872890), DOE (W-7405-ENG-48), and the NSF STC for Computer Graphics and Scientific Visualization. Other support was provided by Alias|wavefront and a Packard Fellowship.

None of this work could have happened without my students and collaborators and in particular Wim Sweldens with whom I have been developing these ideas for the last 5 years.

References

Brown, P.J.C. and Faigle, C.T. (1997). A robust efficient algorithm for point location in triangulations. Tech. Rep., Cambridge University.

Burt, P.J. and Adelson, E.H. (1983). Laplacian pyramid as a compact image code. *IEEE Trans. Commun.*, **31(4)**, 532–540.

Catmull, E. and Clark, J. (1978). Recursively generated B-spline surfaces on arbitrary topological meshes. *Comput. Aided Design*, **10(6)**, 350–355.

Certain, A., Popović, J., DeRose, T., Duchamp, T., Salesin, D. and Stuetzle, W. (1996). Interactive multiresolution surface viewing. In *Computer Graphics (SIGGRAPH '96 Proceedings)*, ed. H. Rushmeier, pp. 91–98. Addison Wesley.

Cohen, J., Manocha, D. and Olano, M. (1997). Simplifying polygonal models using successive mappings. In *Proceedings IEEE Visualization 97*, pp. 395–402.

Curless, B. and Levoy, M. (1996). A volumetric method for building complex models from range images. In *Computer Graphics (SIGGRAPH '96 Proceedings)*, ed. H. Rushmeier, pp. 303–312. Addison Wesley.

Daubechies, I., Guskov, I. and Sweldens, W. (1998). Commutation for irregular subdivision. Tech. Rep., Bell Laboratories, Lucent Technologies.

Daubechies, I., Guskov, I. and Sweldens, W. (1999). Regularity of irregular subdivision. *Const. Approx.*, **15**, 381–426.

DeRose, T., Kass, M. and Truong, T. (1998). Subdivision surfaces in character animation. In *Computer Graphics (SIGGRAPH '98 Proceedings)*, pp. 85–94.

Dobkin, D. and Kirkpatrick, D. (1985). A linear algorithm for determining the separation of convex polyhedra. *J. Algorithms*, **6**, 381–392.

Doo, D. and Sabin, M. (1978). Analysis of the behaviour of recursive division surfaces near extraordinary points. *Comput. Aided Design*, **10(6)**, 356–360.

Duchamp, T., Certain, A., DeRose, T. and Stuetzle, W. (1997). Hierarchical computation of PL harmonic embeddings. Tech. Rep., University of Washington.

Dyn, N., Levin, D. and Gregory, J.A. (1990). A butterfly subdivision scheme for surface interpolation with tension control. *ACM Trans. Graphics*, **9(2)**, 160–169.

Eck, M., DeRose, T., Duchamp, T., Hoppe, H., Lounsbery, M. and Stuetzle, W. (1995). Multiresolution analysis of arbitrary meshes. In *Computer Graphics (SIGGRAPH '95 Proceedings)*, ed. R. Cook, pp. 173–182.

Eck, M. and Hoppe, H. (1996). Automatic reconstruction of B-spline surfaces of arbitrary topological type. In *Computer Graphics (SIGGRAPH '96 Proceedings)*, H. Rushmeier, pp. 325–334. Addison Wesley.

Finkelstein, A. and Salesin, D.H. (1994). Multiresolution curves. In *Computer Graphics Proceedings*, Annual Conference Series, pp. 261–268.

Forsey, D. and Wong, D. (1995). Multiresolution surface reconstruction for hierarchical B-splines. Tech. Rep., University of British Columbia.

Forsey, D.R. and Bartels, R.H. (1988). Hierarchical B-spline refinement. *Computer Graphics (SIGGRAPH '88 Proceedings)*, **22(4)**, 205–212.

Garland, M. and Heckbert, P.S. (1995). Fast polygonal approximation of terrains and height fields. Tech. Rep. CMU-CS-95-181, Carnegie Mellon University.

Gortler, S.J. and Cohen, M.F. (1995). Hierarchical and variational geometric modeling with wavelets. In *Proceedings Symposium on Interactive 3D Graphics*.

Guibas, L. and Stolfi, J. (1985). Primitives for the manipulation of general subdivisions and the computation of Voronoi diagrams. *ACM Trans. Graphics*, **4(2)**, 74–123.

Guskov, I., Sweldens, W. and Schröder, P. (1999). Multiresolution signal processing for meshes. Tech. Rep. PACM TR 99-01, Princeton University, submitted.

Heckbert, P.S. and Garland, M. (1997). Survey of polygonal surface simplification algorithms. Tech. Rep., Carnegie Mellon University.

Hoppe, H. (1996). Progressive meshes. In *Computer Graphics (SIGGRAPH '96 Proceedings)*, ed. H. Rushmeier, pp. 99–108. Addison Wesley.

Hoppe, H. (1997). View-dependent refinement of progressive meshes. In *Computer Graphics (SIGGRAPH '97 Proceedings)*, pp. 189–198.

Hoppe, H., DeRose, T., Duchamp, T., Halstead, M., Jin, H., McDonald, J., Schweitzer, J. and Stuetzle, W. (1994). Piecewise smooth surface reconstruction. In *Computer Graphics (SIGGRAPH '94 Proceedings)*, pp. 295–302.

Kirkpatrick, D. (1983). Optimal search in planar subdivisions. *SIAM J. Comput.*, **12**, 28–35.

Klein, A., Certain, A., DeRose, T., Duchamp, T. and Stuetzle, W. (1997).

Vertex-based Delaunay triangulation of meshes of arbitrary topological type. Tech. Rep., University of Washington.

Kobbelt, L. (1996a). Interpolatory subdivision on open quadrilateral nets with arbitrary topology. In *Proceedings of Eurographics '96*, pp. 409–420. Computer Graphics Forum.

Kobbelt, L. (1996b). A variational approach to subdivision. *Comput. Aided Geom. Des.*, **13**, 743–761.

Krishnamurthy, V. and Levoy, M. (1996). Fitting smooth surfaces to dense polygon meshes. In *Computer Graphics (SIGGRAPH '96 Proceedings)*, ed. H. Rushmeier, pp. 313–324. Addison Wesley.

Levin, A. (1999). Combined subdivision schemes for the design of surfaces satisfying boundary conditions. *Comput. Aided Geom. Des.*, to appear.

Loop, C. (1987). *Smooth subdivision surfaces based on triangles*. Master's thesis, University of Utah, Department of Mathematics.

Lounsbery, M. (1994). *Multiresolution analysis for surfaces of arbitrary topological type*. PhD thesis, Department of Computer Science, University of Washington.

Lounsbery, M., DeRose, T. and Warren, J. (1997). Multiresolution analysis for surfaces of arbitrary topological type. *ACM Trans. Graphics*, **16(1)**, 34–73.

Peters, J. and Reif, U. (1997). The simplest subdivision scheme for smoothing polyhedra. *ACM Trans. Graphics*, **16(4)**.

Prautzsch, H. and Umlauf, G. (1998). Improved triangular subdivision schemes. In *Proceedings of the CGI '98*.

Schröder, P., and Sweldens, W. (1995). Spherical wavelets: Efficiently representing functions on the sphere. In *Computer Graphics (SIGGRAPH '95 Proceedings)*, ed. R. Cook.

Schweitzer, J.E. (1996). *Analysis and application of subdivision surfaces*. PhD thesis, University of Washington, 1996.

Stam, J. (1998). Exact evaluation of Catmull–Clark subdivision surfaces at arbitrary parameter values. In *Computer Graphics (SIGGRAPH '98 Proceedings)*, pp. 395–404.

Sweldens, W. (1996). The lifting scheme: A custom-design construction of biorthogonal wavelets. *Appl. Comput. Harmon. Anal.*, **3(2)**, 186–200.

Sweldens, W. and Schröder, P. (1996). Building your own wavelets at home. In *Wavelets in Computer Graphics*, pp. 15–87. ACM SIGGRAPH Course Notes.

Taubin, G. (1995). A signal processing approach to fair surface design. In *Computer Graphics (SIGGRAPH '95 Proceedings)*, ed. R. Cook, pp. 351–358. Annual Conference Series.

Xia, J.C. and Varshney, A. (1996). Dynamic view-dependent simplification for polygonal models. In *Proc. 'Visualization 96*, pp. 327–334.

Zorin, D. (1997). *Stationary subdivision and multiresolution surface representations*. PhD thesis, California Institute of Technology.

Zorin, D., Schröder, P. and Sweldens, W. (1996). Interpolating subdivision for meshes with arbitrary topology. In *Computer Graphics (SIGGRAPH '96 Proceedings)*, ed. H. Rushmeier, pp. 189–192. Addison Wesley.

Zorin, D., Schröder, P. and Sweldens, W. (1997). Interactive multiresolution mesh editing. In *Computer Graphics (SIGGRAPH '97 Proceedings)*, pp. 259–268.

9
Mathematical methods in reverse engineering
J. HOSCHEK

Abstract

In many areas of industrial applications it is desirable to create a computer model of existing objects for which no such model is available. This process is called *reverse engineering*. In this chapter we will be concerned with reverse engineering of the shape of surfaces of objects. We will develop computer models which provide gains in efficiency of design, modification and manufacture. Reverse engineering typically starts with digitising an existing object. These discrete data must then be converted into smooth surface models.

9.1 Introduction

In the computer-assisted manufacturing process there often remain objects which are not originally described in a CAD-system. The reasons for this are:

- modifications (grinding down or putting on of material) of existing parts, which are required in order to improve the quality of the product;
- copying a part, when no original drawings are available;
- real-scale clay or wood models which are needed as stylists and management often prefer real 3D objects for evaluation, rather than projections of objects on 2D screens at reduced scale.

These physical objects are measured by mechanical digitising machines or laser scanners or Moiré-based optical sensors. The resulting dataset may be partially ordered (digitising machines) or unorganised (laser scanners), and has to be transformed into smooth surfaces. As is mentioned in Varady et al. (1997) the main purpose of reverse engineering is to convert discrete datasets into piecewise smooth, continuous models.

We will not discuss technical aspects of rapid prototyping and of data acquisition (Varady et al. (1997)). Rather, we assume that the data acquiring is finished and we have a suitable organisation of data: regular data with boundary curves, scattered data with boundary curves or unorganised clouds of points. We will be concerned with the two most critical parts of reverse engineering: segmentation and surface fitting.

We mainly consider approximations by free-form surfaces with rectangular structure of boundary curves. The most important methods in this field are the feature-line-oriented and the region-growing approach. The natural division lines (feature lines) of a physical object are determined by computer-vision techniques (especially sharp edges), by numerical or geometric criteria or by the experience of the user. Each feature line L of the physical object corresponds to a polygonal or to a piecewise polynomial curve. The set of all curves may be denoted by $C = C(L)$. The user has to subdivide manually or with the help of suitable algorithms, the set $C(L)$ into a finite number of (rectangular) regions R_i. Afterwards for each region a surface $S(R_i)$ is generated by least squares fitting with respect to a prescribed error tolerance and satisfying the continuity requirements imposed by the user.

Region-growing methods start at a seed point or a (small) seed region (Sapidis and Besl (1995)). The seed region R is approximated in the least squares sense by a polynomial surface S, followed by an iterative process, where first R is growing iteratively to include all points in the neighbourhood of the seed region which hold the prescribed error tolerance between the approximation surface S and the given points. Thus, algorithms are developed for maximising the amount of data represented by a polynomial surface which guarantees that the distance error between the given points and the surface is smaller than a prescribed tolerance.

These two strategies are called *bottom-up methods* (Varady et al. (1997)). On the other hand *top-down methods* are developed starting with the premise that all points belong to a single surface. If a prescribed error tolerance is exceeded by a single approximating surface the points in the critical areas are subdivided by the user into a suitable number of new sets, which are further approximated. Although users would like to have a fully automatic system which can detect and classify feature lines without any user interaction, with the present state of the art all CAD-systems need more or less a priori information or assumptions on the object, connected with manual interactions. Fig. 9.1 shows a real example: a dataset of 400,000 points from a joystick and a manual segmentation in 48 B-spline patches. The engineer needed 5 hours whilst using the CAD-system SURFACER from Imageware,

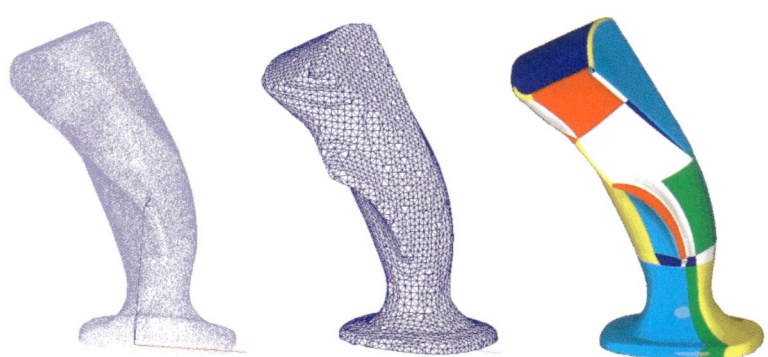

Fig. 9.1. Manual segmentation of a joystick: on the left-hand side a point-cloud, in the middle a triangulation, on the right-hand side one can see a patch structure manually developed by the user (courtesy of W. Wilke, Daimler–Chrysler)

the obtained maximum error is 1.7% (with respect to the length l of the diagonal of the bounding box, $l = 190$mm).

In what follows we will first describe automatic or semi-automatic algorithms for feature line detections. We will subsequently discuss surface approximation methods.

9.2 Triangulation and segmentation

We start with a large point-cloud and would like to partition or segment the data points into subsets, to each of which localised surface fitting can be applied. A good segmentation should represent the underlying structure of the object in a natural manner. Therefore the segmentation lines should be natural division lines (feature lines). Feature lines can be sharp edges or smooth edges. In order to get automatic or semi-automatic algorithms, suitable numerical estimates must be used. Sharp edges can be detected, where surface normals estimated from the point data change direction suddenly, while smooth edges are determined by places where surface normals estimates change rapidly or where estimates of surface curvatures have extreme values.

In Milroy et al. (1997) orthogonal cross-sections of the physical model are used to get estimates for the normal vector, for the normal curvature and for the principal curvature. To determine these estimates an osculating paraboloid is introduced. For the identification of a single edgepoint **P**, the principal curvature directions are used. The surface curvature on either side of the point **P** in a principal direction $\mathbf{m}_1$ is estimated by neighbouring osculating paraboloids. If the principal curvature κ_1 is a local maximum, the point is flagged as an edge point. Similarly the principal direction $\mathbf{m}_2$ is used to decide whether the principal curvature κ_2 is a maximum. The set of edge points is connected by energy-minimising snake splines (Kass et al. (1987)).

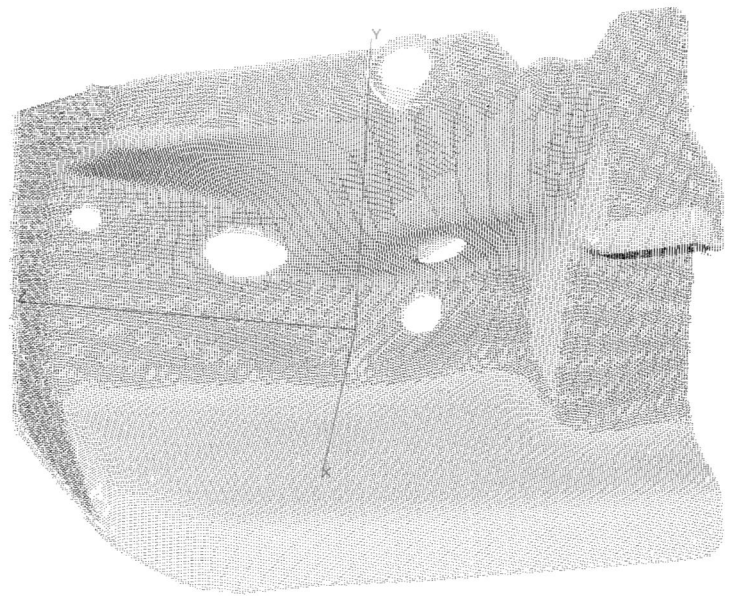

Fig. 9.2. Cloud of 95,436 disorganized points digitised from a mechanical model (courtesy of W. Wilke, Daimler–Chrysler)

In Lukács and Andor (1998), a differential geometric approach is developed based on triangulated point data with normal vector and curvature estimations at these points. On the triangulation a set of points is determined so that the principal curvature is stationary (i.e., the principal curvature has an extremal value) in the corresponding principal direction. The curves connecting such stationary points are called *ridges* (local positive maximum of principal curvature) or *ravines* (local negative minimum of principal curvature). Ridges and ravines are natural dividing lines between regions of monotone principal curvature.

We will now describe in detail a fast bottom-up algorithm which determines feature lines using an edge-based technique combined with region-based properties (Hoschek et al. (1998), Wilke (2000)). As an example we will use a cloud of disorganised points scattered on the surface of a body having some holes (see Fig. 9.2). This cloud contains 95,436 points and was generated by an optical sensor with 2-range images.

To determine the feature lines, we will triangulate the given cloud of points. In the past many algorithms were developed for 3D-triangulations of disorganised data. Well-known examples are: 3D-Delaunay triangulations or marching-cube based algorithms.

For surface reconstruction with a large dataset (more than 100,000 points), these algorithms are mostly too slow and often lead only to the convex hull of the given dataset. The fast algorithm using a multistep procedure developed in Wilke (2000) is as follows.

- The dataset is first subdivided into regular solids (voxels) of a basis size chosen by the user. The points in a voxel are then projected on its least squares plane. If this projection is not unique, the voxels are refined. With the help of an octree data structure the topology of the object can be reconstructed, since the octree data structure gives implicit neighbourhood information between the voxels. We remark that the given data cloud may be related to an open or a closed surface.
- In each least squares plane the points are triangulated by a planar Delaunay triangulation and then lifted to a 3D-triangulation of the points in the voxel (see Fig. 9.3).
- The edges of each voxel triangulation are now connected with neighbouring voxels using a stitching algorithm introduced in Shewchuck (1995). Due to measurement errors some gaps may remain. These must be closed interactively.

On a Pentium II 266 MHz this multistep algorithm requires four seconds for the planar Delaunay triangulations of 100,000 points (Fig. 9.3), and five seconds for closing the gaps.

For point-clouds with unknown topology a region growing method† should be used as follows:

- split the point-cloud into voxels, and compute the voxel centre points;

† This algorithm may fail for nonsensical data like a flat triangulation connected with high curvature values. In that case no starting point can be found automatically.

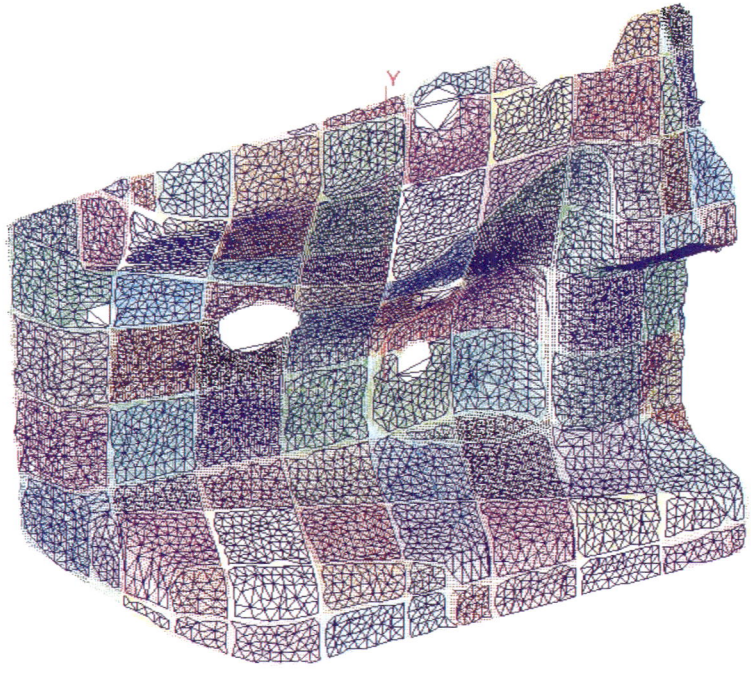

Fig. 9.3. The set of points in Fig. 9.2 is automatically subdivided into 114 voxels and triangulated into 35,200 triangles larger than 1mm (4870 open edges),(courtesy of W. Wilke, Daimler–Chrysler)

- find a starting point and create a ring of triangles with centres of neighbouring voxels;
- append new triangles only to the actual boundaries of the triangulation (the neighbouring information for the voxels is given by the octree data structure);
- append stepwise the voxel centres until all centres are included in the triangulation.

We assume that feature lines appear where the oriented normal vector field of the required surface changes rapidly. To estimate the normal vector at a point $\mathbf{P}_k$, we consider the normal vectors $\mathbf{D}_i$ (with outwards orientation) of the neighbouring triangles $\{T_i\}$, and introduce as normal vector $\mathbf{N}_k$ at

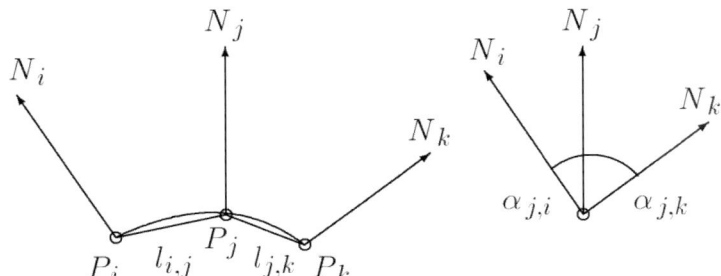

Fig. 9.4. Estimation of angular variation: normal vectors and their variation at a point P_j (courtesy of W. Wilke, Daimler–Chrysler)

$\mathbf{P}_k$ the weighted average

$$\mathbf{N}_k = \frac{\sum w_i \mathbf{D}_i}{|\sum w_i \mathbf{D}_i|}, \qquad i \in \{T_i\} \tag{9.1}$$

in which w_i are suitable weights, like the inverse area of each triangle T_i

$$w_i = \frac{2}{|(\mathbf{P}_i - \mathbf{P}_k) \times (\mathbf{P}_{i+1} - \mathbf{P}_k)|}.$$

To estimate the variation of the normal vectors at the points $\{\mathbf{P}\}_k$, we use the angles (see Fig. 9.4)

$$\alpha_{j,k} := \arccos(\mathbf{N}_j \cdot \mathbf{N}_k) \tag{9.2}$$

between $\mathbf{N}_k$ at $\mathbf{P}_k$ and a neighbouring normal vector $\mathbf{N}_j$ at $\mathbf{P}_j$ and the Euclidean distance

$$l_{j,k} := |\mathbf{P}_j - \mathbf{P}_k|.$$

We introduce as an estimate of the *angular variation* of the normal vector $\mathbf{N}_k$ at the point $\mathbf{P}_k$

$$\beta_k = \sum_j \frac{\alpha_{j,k}}{l_{j,k}} \Big/ \sum_j \frac{1}{l_{j,k}}, \tag{9.3}$$

where the sum is over indices j of points $\mathbf{P}_j$ in the vicinity of $\mathbf{P}_k$. If all $l_{j,k}$ are equal, then β_k is the arithmetic mean of the $\alpha_{j,k}$'s.

For this special case we can give a geometric interpretation of the angular variation. Consider the normal curvature κ_n at the point $\mathbf{P}_k$. By the well-known Euler formula (Do Carmo (1976)) the normal curvature equals

$$\kappa_n = \kappa_1 \cos^2 \varphi + \kappa_2 \sin^2 \varphi , \qquad (9.4)$$

with κ_i as principal curvatures and φ as the angle formed with the principal curvature direction corresponding to κ_i. Equation (9.4) leads to an integral representation of the mean curvature H_k at the point $\mathbf{P}_k$

$$\begin{aligned} \frac{1}{2\pi} \int_0^{2\pi} \kappa_n(\mathbf{P}_k) \, d\varphi &= \frac{1}{2\pi} \int_0^{2\pi} (\kappa_1(\mathbf{P}_k) \cos^2 \varphi + \kappa_2(\mathbf{P}_k) \sin^2 \varphi) \, d\varphi \\ &= \frac{\kappa_1(\mathbf{P}_k) + \kappa_2(\mathbf{P}_k)}{2} = H_k. \end{aligned}$$

If we discretize the continuous function κ_n in n_k directions φ_j ($j = 1, ..., n_k$) with normal curvatures κ_n^j we obtain

$$H_k \sim \frac{1}{n_k} \sum_{j=1}^{n_k} \kappa_n^j(\mathbf{P}_k).$$

We can estimate the normal curvatures κ_n^j by the angle α_{jk} between the normal vector $\mathbf{N}_k$ at $\mathbf{P}_k$ and the normal vector $\mathbf{N}_j$ at the points $\mathbf{P}_j$ of the neighbourhood of $\mathbf{P}_k$. Thus the last equation can be written as

$$H_k \sim \beta_k = \frac{1}{n_k} \sum_j \alpha_{kj}.$$

The angular variations can be used to recognise flat regions in the sense of a region-based technique. Regions with very small or no changes of β could be expected to be flat. The corresponding triangles are cancelled and the arising gaps are suitably closed. Moreover, this step reduces the data volume.

Before starting the segmentation procedure it may be necessary to smooth out parts of the cloud of points in order to reduce the effect of measurement errors during the laser-scanning process. To decide whether a data point $\mathbf{P}_0$ of the cloud should be corrected, we consider the least squares plane ε_0 determined by the points $\{\mathbf{P}_j\}$ ($j = 1, ..., n$) in the vicinity of $\mathbf{P}_0$. The plane ε_0 may have normal vector $\mathbf{N}$ passing through the centre of gravity $\mathbf{S}$ of $\{\mathbf{P}_j\}$. To avoid rounding of sharp edges we only smooth those points whose angular variation β of their normal vector is less than a given threshold angle σ. On the other hand points must be excluded where large changes of curvature can be expected. Therefore a maximal deviation $d_{\max}$ of the

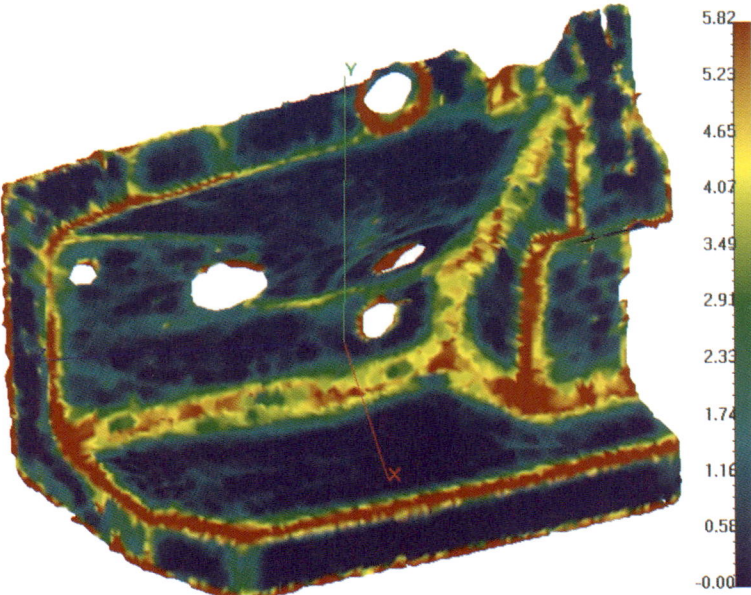

Fig. 9.5. Changes of angular variation are the loci of the expected feature lines. They separate planar and nonplanar parts of the object (courtesy of W. Wilke, Daimler–Chrysler)

points is prescribed, which will be based on the noise of the measurement (e.g. $d_{\max} = 0.1\text{mm}$). The smoothed point will assume the position

$$\widehat{\mathbf{P}}_0 = \mathbf{P}_0 - \mathbf{N}d \quad \text{with} \quad d = \min(\mathbf{N} \cdot (\mathbf{P}_0 - \mathbf{S}), d_{\max}).$$

Feature lines can be expected in regions in which the angular variation changes rapidly. To avoid indication errors, the data are smoothed with the help of an erosion operator† (Haberäcker (1991)) before feature line detection. In Sarkar and Menq (1991) a Laplace operator was used but from our experience the erosion operator leads to much better results.

To indicate the changes of the angular variation, we use a colour map for the angular variation β. For each triangle the values β_i at the various colours

† Erosion in a point $\mathbf{P}_0$ is defined by $\min(\kappa_0, \kappa_1, \ldots \kappa_k)$ with κ_j as curvature values of the neighbouring points $\{\mathbf{P}_j\}$ of $\mathbf{P}_0$.

are interpolated linearly: *very small (less* $1°$) = *blue, small = cyanine, medium = green, medium large= yellow and large = red.* The absolute values depend on the maximal values of the angular variation of a given example. Regions with feature lines are visible by jumps in the coloured map; the feature lines themselves are the boundaries of these regions (in Fig. 9.5, colours are transformed into grey values).

To obtain points on the feature lines automatically, we use an algorithm based on the principle of region growing. In order to start the algorithm we choose an initial triangle in the neighbourhood of a potential feature line and prescribe a threshold value σ. An estimate of σ can be extracted from the angular variation data depicted in Fig. 9.5. We choose the value at which the colour changes from dark grey to medium grey (in Fig. 9.6 we used $\sigma = 4°$). The algorithm connects neighbouring triangles satisfying the following criteria:

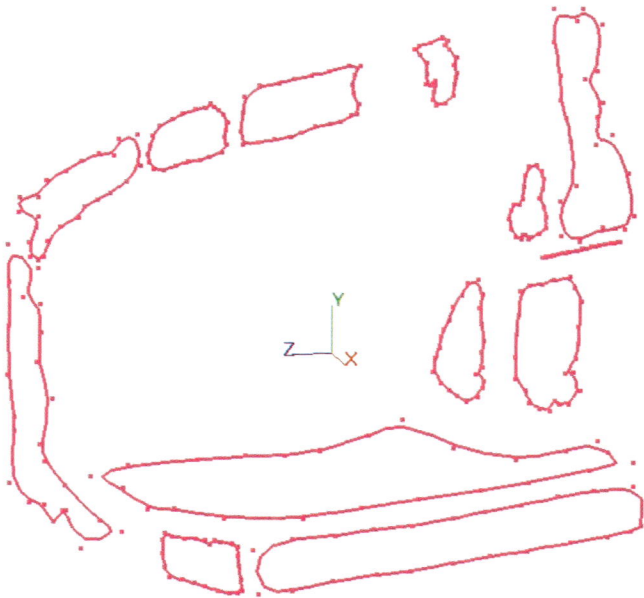

Fig. 9.6. Feature lines (with control points) of the model in Fig. 9.2 as boundaries of regions with low curvature (courtesy of W. Wilke, Daimler–Chrysler)

- each triangle is inspected only once.
- the minimal angular variations at the vertices of the neighbouring triangles are less than σ. The triangles which fulfil this criteria are also the candidates for the initial triangles of the first step in the region growing process.
- the changes between minimal angular variation at the vertices of one triangle and the maximal angular variation at the vertices of the neighbouring triangles are less than σ.

As a result we obtain triangulated feature regions and polygonal lines as the boundaries. These polygonal lines are approximated by cubic B-spline curves with the use of energy constraints (minimal length of the control polygon) leading to the required B-spline representation of the feature lines (see Fig. 9.6).

After determining feature lines we have various alternatives on how to proceed.

- Detect quadrilateral faces or loops in the network of feature lines or introduce manually (or automatically) connecting curves to get a quadrilateral network covering the whole cloud of points. Afterwards all these quadrilateral regions have to be approximated by B-spline surfaces with a prescribed error tolerance and continuity requirements imposed by the user (*bottom-up strategies*);
- Approximate the whole cloud of points with a single overall B-spline surface using free boundaries and energy constraints. If the given error tolerance is exceeded by the single B-spline surface in some parts of the surface, improve the quality of approximation in these bad regions with the help of the local feature lines in the vicinity of the bad region. These local feature lines can be boundary curves or trimming curves of a new local approximation surface approximating the bad regions of the single overall approximation surface with respect to the given error tolerance. Thus we get a patchwork of approximation surfaces containing the overall B-spline approximation surface and the local B-spline approximation surfaces improving the approximation quality of the bad regions of the overall approximation. At the trimming or boundary curves of the new surface patches continuity requirements must be fulfilled (*top-down strategy*). Hoschek et al. (1998) call this a geometric concept of reverse engineering (see Fig. 9.7).

Chapter 9   Reverse engineering   263

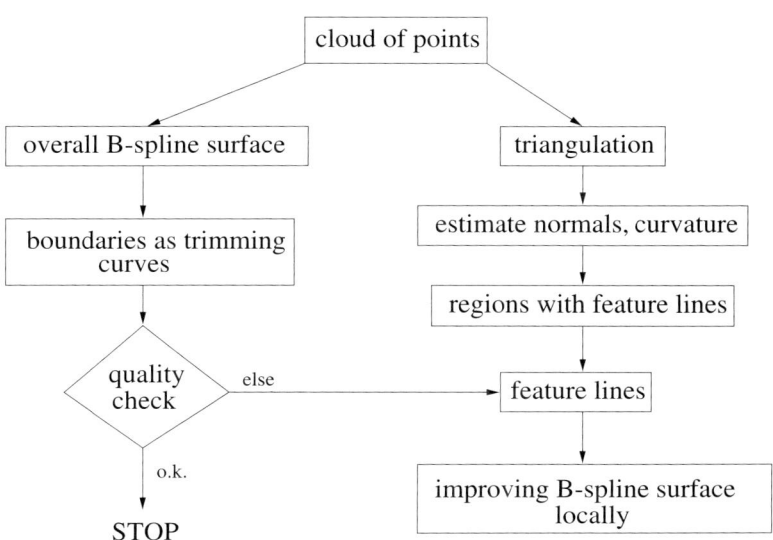

Fig. 9.7. Geometric concept of reverse engineering

9.3 Approximation

Following the first alternative (bottom-up) we assume that the boundaries of the B-spline surface patches (faces) have a rectangular structure. In this case we first approximate the four boundary curves by four (open) B-spline curves of given order k and a given knot-sequence by the least squares method. We start with a suitable initial parametrisation (chordal, centripetal) (Hoschek and Lasser (1993)) and improve the parametrisation by local reparametrisation (Hoschek and Lasser (1993)) or global reparametrisation (Speer et al. (1998)).

The four boundary curves determine a reference surface for parametrisation of the interior points $\mathbf{P}_i$ ($i = 1, ..., r$) of the face. As reference surface a bilinear Coons patch or tensor-product B-spline surface C with a small number of knots can be chosen. Now we project (perpendicular to C) the interior points $\mathbf{P}_i$ of the given cloud on C. The closest point $\overline{\mathbf{P}}_i$ of C determines an initial guess of parameter values (u_i, v_i) of $\mathbf{P}_i$ in the corresponding parameter domain. With this initial parametrisation of $\mathbf{P}_i$ the given set of points is approximated by one tensor-product B-spline surface with fixed order k in the parametric representation (Farin (1996),

Hoschek and Lasser (1993))

$$\mathbf{X}(u,v) = \sum_{j=0}^{n}\sum_{l=0}^{m} \mathbf{d}_{jl}\, N_{lk}(u) N_{lk}(v), \tag{9.5}$$

where the parameter values are defined with respect to suitably chosen knot sequences U, V, while the $\mathbf{d}_{ij}$ are the unknown control points.

The objective function of the approximation problem is the reduction of the distances between the given points $\mathbf{P}_i$ and the required surface. Thus we would like to minimise

$$\mathbf{Q}_{\text{dist}} = \sum_{i=1}^{r}\left(\mathbf{P}_i - \sum_{j=0}^{n}\sum_{l=0}^{m} \mathbf{d}_{jl} N_{jk}(u_i) N_{lk}(v_i)\right)^2. \tag{9.6}$$

The minimisation of (9.6) is obtained by well-known least squares techniques, in which the boundary control points $\mathbf{d}_{0l}, \mathbf{d}_{nl}, \mathbf{d}_{j0}, \mathbf{d}_{jm}$ are known from the approximation of the boundary curves.

For the second alternative (overall approximation) the use of the usual least squares surface approximation is limited in several ways. To obtain a unique approximation surface the Schoenberg–Whitney conditions must be satisfied. This requires well-distributed point-sets without holes. But an overall approximation must take into account holes in the dataset and non-rectangular domains. Therefore, we should use an approximation with free boundaries. Free boundaries of an approximation surface can be obtained as the boundaries of a suitable rectangle in a least squares plane of the given set of points, where the boundaries of the given set of points lie totally in the interior of the rectangle. The planar rectangle is an initial guess of the required approximation surface. If the point-set is more complicated, more sophisticated surface types must be chosen. The only condition which must be fulfilled by the surface initially guessed is that the perpendicular projection of the given points onto the surface be bijective. This initial guess of an approximation surface will be deformed towards the given point-set during the approximation process.

To fulfill the Schoenberg–Whitney condition for the holes and the domain without points between the given boundaries of the dataset and the rectangular free boundaries of the approximation surface, we introduce a new objective combined with a fairness measure $\mathbf{Q}_{\text{fair}}$, and we minimise a new objective functional

$$\mathbf{Q} = \mathbf{Q}_{\text{dist}} + \lambda\, \mathbf{Q}_{\text{fair}}. \tag{9.7}$$

Here λ is a penalty factor controlling the influence of the fairness functional.

It can be determined automatically during an iterative algorithm (Dietz (1996)). Using this new objective functional we can handle point-sets of varying density, point-sets with holes and with nonrectangular boundary curves, and point-sets with measurement errors. A number of suggestions for such fairness functionals can be found in Bloor et al. (1995), Greiner (1994), Moreton and Séquin (1992), Rando and Roulier (1991). Applications are described in Dietz (1996), Hadenfeld (1995), Hagen and Schulze (1987), Sinha and Schunck (1992), Welch and Witkin (1992). Many of these functionals are related to physical energies. For example, $\int \left(\kappa_1^2 + \kappa_2^2 \right) dS$ is the bending energy of a thin plate with small deflections (with κ_1, κ_2 principal curvatures). But this energy, like many others, depends highly nonlinearly on the unknown control points. If a fairness functional is quadratic in the unknowns, the resulting system of equations is linear. The following functionals are some examples of simplified measures for the area, bending energy and change of curvature of a surface $\mathbf{X}$. They are all quadratic in the control points,

$$\mathbf{Q}_1 = \iint \left(\mathbf{X}_u^2 + \mathbf{X}_v^2 \right) du\, dv, \tag{9.8}$$

$$\mathbf{Q}_2 = \iint \mathbf{X}_{uu}^2 + 2\mathbf{X}_{uv}^2 + \mathbf{X}_{vv}^2 du\, dv, \tag{9.9}$$

$$\mathbf{Q}_3 = \iint \left(\mathbf{X}_{uuu} + \mathbf{X}_{uvv} \right)^2 + \left(\mathbf{X}_{uuv} + \mathbf{X}_{vvv} \right)^2 du\, dv. \tag{9.10}$$

These quantities may be combined in a convex manner to define a fairness functional

$$\mathbf{Q}_{\text{fair}} = \sum_{p=1}^{3} \alpha_p \mathbf{Q}_p, \quad \text{with} \quad \sum_{p=1}^{3} \alpha_p = 1 .$$

The factors α_p control the influence of the individual functionals and hence the shape of the resulting surface. These so-called design factors α_p are to be determined by the user.

A necessary condition for the functional $\mathbf{Q}$ to be minimal is

$$\nabla \mathbf{Q} \stackrel{!}{=} 0 .$$

This leads to the linear system

$$\mathbf{A}\mathbf{d} = \mathbf{b} \quad \text{with} \quad \mathbf{A} = \mathbf{A}_{\text{dist}} + \lambda \mathbf{A}_{\text{fair}} \quad \text{and} \quad \mathbf{A}_{\text{fair}} = \sum_p \alpha_p \mathbf{A}_p . \tag{9.11}$$

Here $\mathbf{A}_{\text{dist}}$ is the matrix of the usual normal equations and $\mathbf{b}$ is the corresponding right-hand side. The vector $\mathbf{d}$ contains the unknown control points. Equation (9.11) is actually a system of equations with three right-hand sides (one for each coordinate). The matrices $\mathbf{A}_p$ correspond to the energies $\mathbf{Q}_v$ and do not depend on the point parametrisation. The matrix $\mathbf{A}_{\text{fair}}$ can therefore be precalculated. Only the matrix $\mathbf{A}_{\text{dist}}$ changes after the reparametrisation described below.

By using B-splines, the matrices $\mathbf{A}_p, \mathbf{A}_{\text{fair}}$ are symmetric, positive semi-definite and banded with the same bandwidth as the matrix $\mathbf{A}_{\text{dist}}$ for the usual least squares fit. The elements of $\mathbf{A}_p$ are determined by integrals over products of B-spline basis functions. These integrals can be calculated exactly as the integrals are piecewise polynomials (Dietz (1996)). Subject to some weak conditions on the dataset, the approximation surface is uniquely determined (Dietz (1996)).

After solving (9.11) we update the initial parametrisation to improve the quality of the approximation surface. Both steps, approximation and reparametrisation, are repeated in a loop until an end condition is satisfied (e.g., the approximation error is sufficiently small). For reparametrisation we can use a local approach (Hoschek and Lasser (1993), Hoschek and Dietz (1996)). For each point $\mathbf{P}_i$ the corresponding parameter values (u_i, v_i) are replaced by the parameter values $(\tilde{u}_i = u_i + \Delta u, \tilde{v}_i = v_i + \Delta v)$ of the closest point to $\mathbf{P}_i$ on the surface $\mathbf{X}$. This is done by Newton's method which is locally convergent of order 2. The correction terms are given by

$$\begin{pmatrix} \Delta u \\ \Delta v \end{pmatrix} = \begin{pmatrix} \mathbf{X}_u^2 + (\mathbf{X}-\mathbf{P})\mathbf{X}_{uu} & \mathbf{X}_u\mathbf{X}_v + (\mathbf{X}-\mathbf{P})\mathbf{X}_{uv} \\ \mathbf{X}_u\mathbf{X}_v + (\mathbf{X}-\mathbf{P})\mathbf{X}_{uv} & \mathbf{X}_v^2 + (\mathbf{X}-\mathbf{P})\mathbf{X}_{vv} \end{pmatrix}^{-1} \begin{pmatrix} (\mathbf{X}-\mathbf{P})\mathbf{X}_u \\ (\mathbf{X}-\mathbf{P})\mathbf{X}_v \end{pmatrix}. \quad (9.12)$$

In general, we obtain sufficient accuracy after 1 to 3 iterations beginning with $P = P_i$. The reparametrisation together with the approximation, including the fairing functional, is very robust and the approximation surfaces are both smooth and stiff (Dietz (1996)).

Our overall approximation may be insufficient on domains with a rapid change of curvature. An additional local improvement is therefore introduced. As a local energy constraint we again use the thin-plate energy

$$\mathbf{Q}_{\text{fair}} = \iint \lambda(u,v) \left(\mathbf{X}_{uu}^2 + 2\mathbf{X}_{uv}^2 + \mathbf{X}_{vv}^2 \right) du\, dv$$

weighted by the real-valued B-spline function

$$\lambda(u,v) = \sum_{i=0}^{n} \sum_{j=0}^{m} w_{ij} N_{ir}(u) N_{js}(v) \geq 0 \qquad (w_{ij} \in \mathbf{R}).$$

This functional surface λ allows us to locally control the influence of the energy term. In flat regions λ can be chosen large with a strong smoothing effect. Near the edges λ has to be small in order to allow an appropriate approximation result. For a fixed $\lambda(u,v)$, Q_{fair} is obviously quadratic in the control points $\mathbf{d}_{ij}$.

Now we combine all these functionals and set up the objective functional

$$\mathbf{Q} = \mathbf{Q}_{\text{dist}} + \mathbf{Q}_{\text{fair}} + \gamma \mathbf{Q}_{\text{cdev}} \qquad (9.13)$$

of our approximation step. The last expression

$$\mathbf{Q}_{\text{cdev}} = \iint (\lambda_0 - \lambda(u,v))^2 \, du \, dv$$

in (9.13) is necessary to avoid the trivial solution $\lambda(u,v) \equiv 0$, which means no fairing is applied at all. The variable λ_0 is a positive constant (Dietz (1998)); γ is a positive scalar penalty factor controlling the influence of the corresponding terms. Each of the introduced functionals is quadratic in the unknown control points $\mathbf{d}_{ij}$.

The optimal function $\lambda(u,v)$ can be determined by minimising $\mathbf{Q}$ with respect to the scalar control points w_{ij} while fixing the control points $\mathbf{d}_{ij}$ and the scalars γ and λ_0. We call this step a *weight correction*. It is a constrained optimisation problem due to $\lambda(u,v) > 0$, but it can be simply solved by considering the unconstrained problem and then choosing the penalty factor γ in such a way that $\min_{u,v} \lambda(u,v) = \sigma > 0$. The necessary conditions of the unconstrained problem yield a linear system of equations for the unknown parameters w_{ij} (Dietz (1998)).

In order to satisfy continuity requirements, we use the normal vector field of the neighbouring patches of a patch C and we improve the approximation of C with additional restrictions on the normal vectors. The deviation of the prescribed normal vectors $\mathbf{N}_i$ ($i = 1, \ldots, M$) forming the surface normals of the approximation surface can be approximately measured by the quadratic functional (Ye et al. (1996))

$$\mathbf{Q}_{\text{norm}} = \sum_{i=1}^{M} (\mathbf{N}_i \cdot \mathbf{X}_{u_i})^2 + (\mathbf{N}_i \cdot \mathbf{X}_{v_i})^2 \ .$$

Unfortunately this deviation measure depends on the parametrisation. Therefore in Dietz (1998) a nearly parametric independent quadratic functional is proposed, namely,

$$\mathbf{Q}_{\text{norm}} = \sum_{i=1}^{M} (\mathbf{N}_i \cdot \mathbf{X}_{r_i})^2 + (\mathbf{N}_i \cdot \mathbf{X}_{s_i})^2 \qquad (9.14)$$

with

$$\mathbf{X}_{r_i}|\mathbf{X}_{s_i} = \left(\frac{\mathbf{X}_{u_i}}{|\widehat{\mathbf{X}}_{u_i}|} \pm \frac{\mathbf{X}_{v_i}}{|\widehat{\mathbf{X}}_{v_i}|}\right) \Big/ \left|\frac{\widehat{\mathbf{X}}_{u_i}}{|\widehat{\mathbf{X}}_{u_i}|} \pm \frac{\widehat{\mathbf{X}}_{v_i}}{|\widehat{\mathbf{X}}_{v_i}|}\right|,$$

where $\widehat{\mathbf{X}}_{u_i}, \widehat{\mathbf{X}}_{v_i}$ denote quantities of the surface from the previous iteration step. It follows that

$$|\mathbf{X}_{r_i}| \approx |\mathbf{X}_{s_i}| \approx 1, \ \mathbf{X}_{r_i} \stackrel{\perp}{\sim} \mathbf{X}_{s_i}, \ \mathbf{N}_i \cdot \mathbf{X}_{r_i} = \mathbf{N}_i \cdot \mathbf{X}_{s_i} = 0\ .$$

The functional $\mathbf{Q}_{\text{norm}}$ is a good approximation of the exact functional $\sum \alpha_i^2$ for small angles α_i, since $\alpha_i^2 \approx \sin^2 \alpha_i$ and the sum and the difference of two unit vectors are orthogonal.

We combine this functional with (9.13) and get a new objective function

$$\mathbf{Q} = \mathbf{Q}_{\text{dist}} + \mathbf{Q}_{\text{fair}} + \gamma \mathbf{Q}_{\text{cdev}} + \mu \mathbf{Q}_{\text{norm}}\ .$$

with μ as a new design factor. The necessary conditions for a minimum of $\mathbf{Q}$ with respect to the $\mathbf{d}_{ij}$s yield the linear equation system

$$(\mathbf{A}_{\text{dist}} + \mathbf{A}_{\text{fair}} + \mu \mathbf{A}_{\text{norm}})\,\mathbf{d} = \mathbf{B}\ , \qquad (9.15)$$

where the $3(n+1)(m+1) \times 1$-vector $\mathbf{d}$ contains all components of the control points $\mathbf{d}_{ij}$. The $3(n+1)(m+1) \times 3(n+1)(m+1)$-matrices $\mathbf{A}_{\text{dist}}$, $\mathbf{A}_{\text{norm}}$ and $\mathbf{A}_{\text{fair}}$ are symmetric, positive semi-definite, and due to the local support of the B-spline basis functions are sparse and banded. Due to the dot products in $\mathbf{Q}_{\text{norm}}$, the system (9.15) has triple size and bandwidth of that of the usual least squares fit working only with components.

We start our iterations with a least squares fitting plane as a surface of lowest bending energy. An initial (and for the plane, optimal) parametrisation can be simply obtained by orthogonal projection of the points onto this plane. A sequence of surfaces is then successively computed (see Fig. 9.8), minimising thin-plate energy whilst deforming towards the points, which serve as a mould. The parametrisation is adapted after each iteration step according to the actual surface with the help of the parameter correction (9.12). After each deformation of the approximation surface an adaptive parametrisation (orthogonal distance vectors) is restored by minimising the total error sum with respect to the new parametrisation.

For the final (overall) approximation surface the boundary curves will, in general, not coincide with the boundaries of the point-cloud. The surface regions outside the point-set should therefore be trimmed away by B-spline curves defined in the parametric domain of the surface. This can be done as follows.

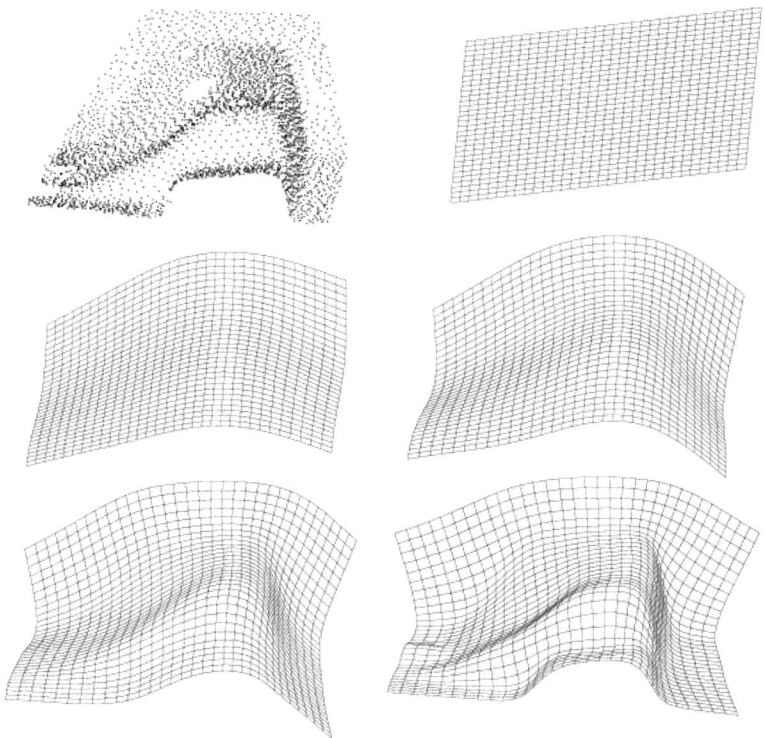

Fig. 9.8. 3234 data points with surfaces after 1, 7, 10, 14, 18 iteration steps (courtesy of U. Dietz)

The boundary points $\{\mathbf{R}\}_i$ are projected perpendicularly onto the surface and then for each projected point the preimage (u_i, v_i) in the parametric domain is calculated.

These points (u_i, v_i) are approximated by a 2D B-spline curve (Hoschek and Schneider (1992))

$$\mathbf{C}(t) = \bigl(u(t),\, v(t)\bigr) = \sum_j \mathbf{d}_j\, N_{jk(t)} \qquad (9.16)$$

using least squares techniques and parameter correction. The error distance is measured in $\mathbf{R}^3$ whereas the approximation works in the parametric domain. The final trimming curve is obtained by inserting $\mathbf{C}(t)$ into the surface

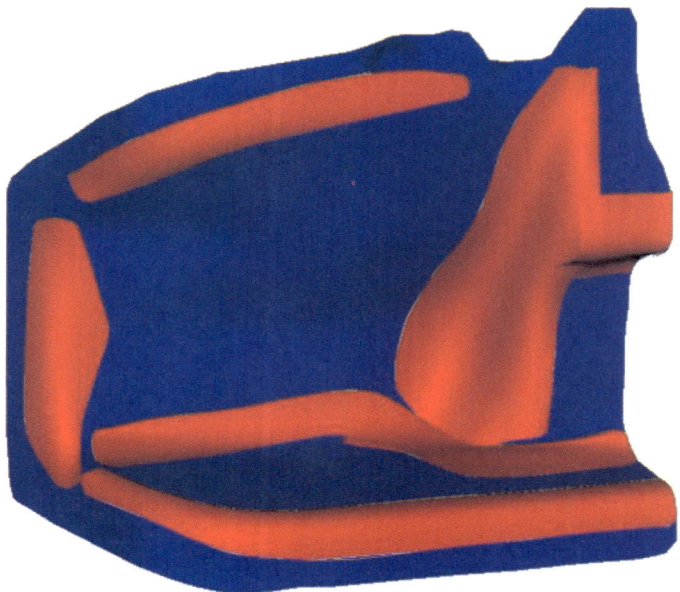

Fig. 9.9. An overall approximation of the cloud of points in Fig. 9.2 with trimmed boundary curves as natural boundaries of the given cloud of points and blending patches determined by feature lines (courtesy of W. Wilke, Daimler–Chrysler)

representation of $\mathbf{X}$, i.e., $\mathbf{X}(u(t), v(t))$. In a similar way holes can be performed in the interior of the surface if the point-cloud requires it.

Fig. 9.9 contains an overall approximation of the given cloud of points of Fig. 9.2 with trimming curves as natural boundaries of the given set of points with a maximal approximation error of 1.1%. To improve the approximation surface, six trimmed (blending) surface patches (bicubic B-spline surfaces) are constructed on the basis of the corresponding feature lines (see Fig. 9.6) for the domains with large deviations from the given set of points. Thus, the maximal error is reduced to 0.6% (with respect to the length of the diagonal of the bounding box, 228mm).

To demonstrate the effectiveness of the developed methods we again consider the joystick from Fig. 9.1. Because of the topology of the object, the cloud of points was manually split into three parts to get appropriate approximation surfaces by an overall approximation. For both sides of the object, 30×30 bicubic B-spline surfaces were used, while the top was app-

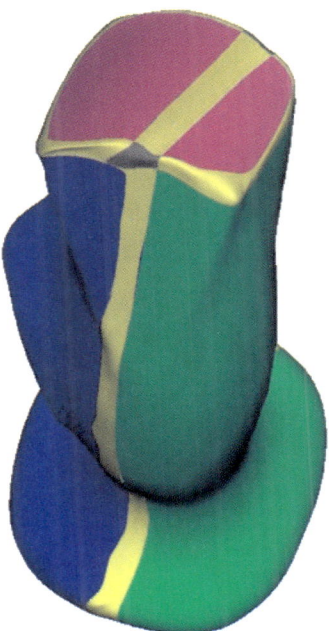

Fig. 9.10. Semi-automatic approximation of the joystick from Fig. 9.1 (compare with Fig. 9.1) (courtesy of W. Wilke, Daimler–Chrysler)

roximated by 10×10 bicubic B-spline patches. Afterwards, domains with large deviations from the given set of points were blended with the help of the feature lines. For the semi-automatic solution, 11 patches were used with a maximum error of 0.9% (see Fig. 9.10). For the semi-automatic segmentation fewer than 2 hours were needed, while for the manual solution 5 hours were used. Also, the maximum error with the manned solution was 1.7% and the data volume was much larger (see Section 9.1).

9.4 Topology-preserving methods

The algorithms developed in Sections 9.2 and 9.3 can only be used if the topology of the required surface is known. For surfaces of arbitrary topology, an automatic working algorithm for generating a patch structure was proposed in Eck and Hoppe (1996). One starts with a basic triangulation M_0 of the given points with respect to a given error tolerance δ, by using the triangulation algorithm introduced by Hoppe et al. (1992). The given points are projected onto the corresponding faces of the triangulation in order to obtain an initial parametrisation and a parametric domain of the same topology as the required surface. This triangulation M_0 consists of many faces. These faces are reparametrised in order to get a simple base complex $K_\triangle$ and a continuous map $\rho_\triangle : K_\triangle \longrightarrow M_0$. In order to obtain a low metric torsion $\rho_\triangle$ is constructed with the help of harmonic maps (Eck et al. (1995)). By merging pairs of triangles of $K_\triangle$, a new base complex $K_\square$ is constructed, the faces of which consist of quadrilaterals. One starts with a user-chosen seed point in order to automatically generate a quadrilateral patch structure. In addition the parametrisation of $K_\triangle$ is transformed to $K_\square$. Over $K_\square$ a control mesh M_x is constructed using Peters' scheme (Peters (1995)), which collectively defines an initial control network of G^1-continuous tensor product B-spline patches of low degree. This network is improved upon by least squares with energy constraints (see Section 9.3) in order to obtain surface patches which are as close as possible to the corresponding given points. If these approximating patches differ more than the user's specified distance error ϵ, $K_\square$ will be subdivided into smaller quadrilateral surfaces. This method leads to a refined domain complex and a new B-spline control mesh which covers the given set of points completely and also attains the required error tolerance.

In Guo (1997) a similar approach was developed. As an initial guess of the required surface a simplicial surface M (a polygonal mesh with triangular faces) capturing the topology structure of the set of points is constructed by interactive use of α-shapes (Edelsbrunner and Mücke (1994)). Based on this structure a curvature continuous surface $\widetilde{M}$ is built using least square methods and generalised B-splines (Loop and DeRose (1990)).

These approaches are very useful for visualisation of objects with unknown topology. For practical applications in mechanical engineering these techniques have one disadvantage. The triangulation depends on the choice of some parameters and cannot divide a surface into series of natural division lines like sharp edges and feature lines.

9.5 Convexity-preserving methods

In this section we present methods for fitting convex (or concave) tensor-product spline surfaces. There are two types of methods available in the literature. These are the use of Powell–Sabin spline functions over special triangulations in $\mathbf{R}^2$ and the use of B-spline functions over a rectangular grid. Unfortunately, convexity-preserving algorithms for parametric surfaces are presently not available.

A Powell–Sabin spline based least squares algorithm for fitting convex surfaces has been developed in Willemans and Dierckx (1994). The algorithm uses the convexity criteria in Chang and Feng (1984) which leads to a set of linear and quadratic inequalities necessary and sufficient for convexity of the Powell–Sabin splines. The coefficients of the surface representation are found by minimising the squared error sum with linear and quadratic inequality constraints. The resulting nonlinear optimisation problem is time consuming and is solved with the help of a suitable function from the NAG-library.

The B-spline approach as developed in Jüttler (1997a), Jüttler (1997b), Jüttler (1998) uses a suitable linearisation of convexity conditions and leads to a quadratic programming problem (minimisation of the quadratic objective function with linear constraints). The given knot sequences U, V define a partition of the parametric domain of the spline function $X(u, v)$ with scalar control factors d_{ij} according to the B-spline representation (9.5) into cells

$$C_{ij} = [u_i, u_{j+1}] \times [v_j, v_{j+1}] \quad \begin{matrix} i = k - 1, \ldots, n \\ j = k - 1, \ldots, m \end{matrix}. \quad (9.17)$$

The restriction $X(u,v)|_{C_{ij}}$ of the spline function (9.5) to each cell is a bivariate tensor-product polynomial of degree $(k-1, k-1)$. The user has to decide whether to fit a convex or a concave spline segment or an unconstrained segment. In Hoschek and Jüttler (1999) algorithms are proposed for this decision. Each cell C_{ij} is labelled by a value $\gamma_{ij} \in \{-1, +1\}$ in order to specify whether $X(u,v)|_{C_{ij}}$ should be convex or concave. No constraints are imposed if $\gamma_{ij} = 0$. A surface cell is convex (concave) if and only if the second derivatives in all directions $\mathbf{r} = (r_1, r_2)^T$

$$\begin{aligned} \frac{\partial^2}{\partial \mathbf{r}^2} X(u,v) &= \left. \frac{d^2}{dt^2} X(u + tr_1, v + tr_2) \right|_{t=0} \\ &= (r_1, r_2) \underbrace{\begin{pmatrix} X_{uu}(u,v) & X_{uv}(u,v) \\ X_{uv}(u,v) & X_{vv}(u,v) \end{pmatrix}}_{=H(u,v)} \begin{pmatrix} r_1 \\ r_2 \end{pmatrix} \end{aligned} \quad (9.18)$$

are nonnegative (resp. nonpositive) for all $(u,v) \in C_{ij}$. That is, the Hessian matrix $H(u,v)$ of the spline function is positive (resp. negative) semi-definite. By using knot insertion at the knots of the knot sequences U, V we get a Bézier representation of the Hessian

$$H(u,v) = \sum_{r=0}^{d}\sum_{s=0}^{d} B_r^d(u, u_i, u_{i+1})\, B_s^d(v, v_j, v_{j+1})\, B_{r,s} \qquad (9.19)$$

over the cell C_{ij}. Here we use the Bernstein polynomials

$$B_k^d(t, a, b) = \binom{d}{k} \frac{(t-a)^k (b-t)^{d-k}}{(b-a)^d} \qquad (9.20)$$

of degree $d = k - 1$ with respect to the interval $[a, b]$. The second partial derivatives X_{uu}, X_{uv} and X_{vv} are bivariate tensor-product polynomials of degree $(d-2, d)$, $(d-1, d-1)$, and $(d, d-2)$, respectively. Thus, the Hessian is a matrix-valued polynomial of degree (d, d). The components of the $(d+1)^2$ symmetric coefficient matrices $B_{r,s} \in \mathbf{R}^{2 \times 2}$ are certain constant linear combinations of the B-spline control points $d_{i,j}$ the coefficients of which depend upon the given knot sequences. In the following, in order to simplify notation, we will not refer to the indices i, j.

To guarantee the desired shape features of the spline surface (i.e., convexity and/or concavity of the segments) one has to determine conditions which imply that the Hessian matrix is nonnegative (resp. nonpositive) definite. For $(u, v) \in C_{i,j}$, the Hessian matrix is a nonnegative linear combination of the coefficient matrices $B_{r,s}$. Hence it is sufficient to find such conditions for the individual coefficient matrices. A symmetric 2×2 matrix $B = B_{r,s} = (b_{k,l})_{k,l=1,2}$ is positive (resp. negative) semi-definite if and only if the inequalities

$$b_{1,1} \geq 0,\ b_{2,2} \geq 0 \ \text{(resp. } b_{1,1} \leq 0,\ b_{2,2} \leq 0)\ \text{ and }\ b_{1,1} b_{2,2} - b_{1,2}^2 \geq 0 \qquad (9.21)$$

hold. These conditions, however, produce quadratic inequalities for the control points $d_{i,j}$, as the components of the matrices $B_{r,s}$ are linear combinations of them. It is advantageous to use a stronger set of sufficient linear conditions, as the resulting optimisation problem is easier to deal with. Linear conditions can be generated with the help of the following result (see Hoschek and Jüttler (1999) Jüttler (1997b), Jüttler (1998)).

Let $N \geq 2$ be an integer. Consider the $2N$ linear inequalities

$$\begin{aligned}(1 - \tau_{k-1})(1 - \tau_k) b_{1,1} + (\tau_{k-1} + \tau_k - 2\tau_{k-1}\tau_k) b_{1,2} + \tau_{k-1}\tau_k b_{2,2} \geq 0 \\ (1 - \tau_{k-1})(1 - \tau_k) b_{1,1} - (\tau_{k-1} + \tau_k - 2\tau_{k-1}\tau_k) b_{1,2} + \tau_{k-1}\tau_k b_{2,2} \geq 0\end{aligned} \qquad (9.22)$$

with $\tau_k = k/N$, $k = 1, \ldots, N$.

(∗) If the components of the symmetric 2×2 matrix B satisfy the linear inequalities (9.22), then B is positive semi-definite.

(∗∗) Consider finitely many positive definite symmetric 2×2 matrices. If N is sufficiently big, then the inequalities (9.22) are satisfied for all matrices.

On the basis of these results, we can formulate the following shape-preserving surface fitting procedure.

(i) For each individual cell choose the desired shape $\gamma_{i,j} \in \{-1, 0, +1\}$ of the approximating spline function.
(ii) For all cells with $\gamma_{i,j} = 1$ (resp. $= -1$) generate linear sufficient convexity (resp. concavity) conditions, by applying the construction of (∗), (∗∗) (with appropriate refinement level, e.g., $N = 4$) to the $(d+1)^2$ coefficient matrices $B_{r,s}$ (resp. $-B_{r,s}$) from (9.19). This leads to a system of linear inequalities for the unknown control points $d_{i,j}$.
(iii) Choose the quadratic objective function Q according to (9.7).
(iv) The scalar control coefficients $d_{i,j}$ in the spline surface (9.5) are found by minimising the above quadratic objective function subject to the linear inequality constraints obtained from step 2. This is a so-called *quadratic programming problem*, see, e.g., Fletcher (1990). A number of powerful algorithms for quadratic programming are available (see Vanderbei (1992)).

As an example we consider a cloud of 200 points. Fig. 9.11 contains the knot lines of the required approximation surface. Grey cells correspond to concave regions of the surface while white cells are unconstrained.

Fig. 9.12 contains corresponding (functional) approximation surfaces. The left-hand side figure is obtained with the help of the above described shape constrained approximation. The right-hand side is the result of unconstrained approximation. One can clearly see that the constrained approximation gives the desired shape, while the unconstrained one has a lot of undesired oscillations. The B-spline function used is bicubic. The corresponding knot vector has 4×4 segments. The maximal error for the constrained case is 3.39% and for the unconstrained case 2.92% (with respect to the height $z_0 = 6$ of the bounding box).

9.6 Simple surfaces

Engineers often prefer simple surfaces like planes, spheres, quadrics, cones and cylinders of revolution, tori or surfaces revolution, helical surfaces, pipe

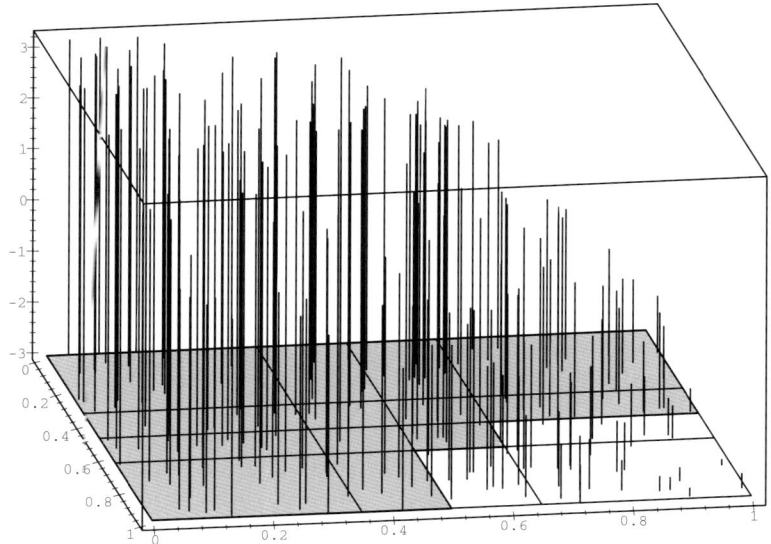

Fig. 9.11. Set of points and associated shape indicators $\gamma_{i,j}$ (courtesy of B. Jüttler)

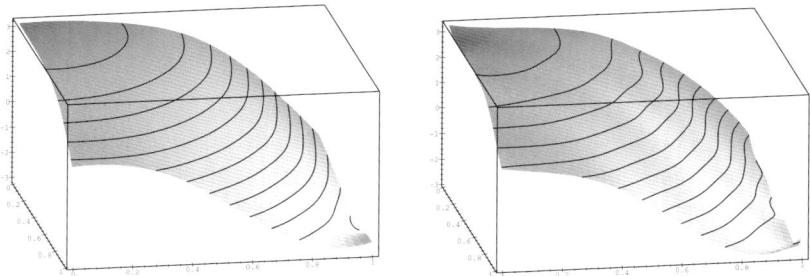

Fig. 9.12. Approximation of the point-set from Fig. 9.11 with convexity constraints (left) and unconstrained (courtesy of B. Jüttler)

surfaces, developable surfaces and ruled surfaces rather than the most general free-form surfaces. The surfaces of revolution, pipe surfaces and developable surfaces can be interpreted as profile surfaces. These are special sweeping surfaces traced out by a planar curve if the plane which contains the curve is rolling on a developable surface.

For representation in CAD-systems, as well as for the manufacturing of a part using geometric properties of the surface, it is often helpful to recognise the type of the surface and fit the given cloud of points by surfaces of the determined type. For spheres, cones, cylinders of revolution and tori, a nonlinear least squares approach was developed in Lukács et al. (1998).

Approximations of cloud of points by surfaces of revolution were developed in Elsässer and Hoschek (1996). A surface of revolution is obtained by rotating a planar curve (meridian) about an axis of revolution A that lies in the plane of the curve and is rigidly connected to the plane. In Elsässer and Hoschek (1996) two different cases were discussed. If the axis of revolution is known, the given points are rotated around the axis into a suitable plane through the given axis. These rotated points determine the meridian. They are first approximated by a rational B-spline curve with a linear scheme for the control points and a Newton method for the weights of the rational B-spline curve. This procedure is iteratively repeated using an additional reparametrisation. In the second case an algorithm was presented in order to determine the unknown axis of revolution, with the help of a suitable analytic description of a motion in $\mathbf{R}^3$.

In Pottmann and Randrup (1998) an approach was presented that allows for the determination of surfaces of revolution and helical surfaces with similar algorithms. A helical surface is determined by rotating a planar generator curve around an axis of revolution A and a proportional translation parallel to A. The algorithms used in Pottmann and Randrup (1998) apply methods of line geometry to the set of surface normals, whereas the normals of a surface lie in a linear complex if and only if the surface is part of a cylinder of revolution, a surface of revolution or a helical surface. First, the generating motion M is determined with the help of an eigenvector problem. The given data points are projected into an auxiliary plane π. For a rotation, π passes through the axis. For a translation, π is perpendicular to its direction. For a helical motion, the plane π is determined approximately orthogonal to the paths of the given points under the motion M. In π the projection of the cloud of points has to be fitted by a curve C, the generator curve of the required surface. The generator curve is determined by a combination of thinning binary images and least squares fits as proposed in Randrup (1998). This approach was extended in Pottman et al. (1998a) to profile surfaces.

Other special classes of simple surfaces are ruled surfaces and developable surfaces. A ruled surface is one which has the property that through every point of the surface there passes a straight line which lies entirely on the surface. Thus, the ruled surface is covered by a one-parameter set of straight lines, which are called *rulings* or *generators*. Each ruling lies in a tangent plane at every point of the ruling. If the tangent plane varies along a ruling from one point to another, the ruled surface is called *nondevelopable* (general ruled surfaces); the ruled surface is *developable* if and only if the tangent planes at every point of a ruling coincide. Thus, the tangent plane of a developable surface form a family which depends on one parameter, whilst the tangent planes of a general ruled surface form a family depending on two parameters.

One can express ruled surfaces as tensor-product surfaces of degree $(1, n)$. In this case the developable surfaces are then determined by nonlinear side conditions expressing the developability (Do Carmo (1976), Lang and Röschel (1992)). In order to avoid such nonlinear side conditions we will use dual B-splines (Hoschek (1983)), where dual in the sense of projective geometry means that the control points of a rational tensor-product B-spline representation are exchanged by control planes. A dual $(1, n)$ rational tensor-product B-spline surface describes a ruled surface (Hoschek and Schwanecke (1998)), whilst a dual B-spline curve (or a $(0, n)$ rational tensor-product B-spline surface) determines a one-parameter set of planes, i.e., a developable surface.

For reverse engineering with ruled or developable surfaces the following problem has to be solved. Given a set of scattered data points $\{\mathbf{P}\}_\mathbf{j}$, it is required to find a ruled or a developable surface of which the generators g_j are as close as possible to the given points. The critical point of approximation with these classes of surfaces is the choice of an appropriate error measurement. In Pottman et al. (1998b) Pluecker coordinates were used for an error measurement concerning developable surfaces, whereas in Hoschek and Schneider (1997) the distances from a point to a line were used, which unfortunately leads to a nonlinear optimisation problem. In Pottman and Wallner (1999) and Hoschek and Schwanecke (1998) an error measurement between planes was independently introduced, which leads to linear algorithms. For this measurement the assumption that each tangent plane of the required surface intersect the z-axis of a suitable coordinate system must hold. Geometrically the error measurement used minimises the z-distances between the tangent planes of the given and of the approximation surface. In order to obtain an initial guess of tangent planes for a set of scattered data an approximate triangulation using the methods developed in Hoppe et al. (1992). Based on this triangulation, the approximations of the tangent planes were determined in suitable strips by least squares fit. Special attention was paid to the developable surfaces in order to control the regression curve (Pottman and Wallner (1999), Schneider (1998)).

Using these methods of reverse engineering one cannot expect that a given cloud of points can be approximated by a single surface of the type discussed. In general, one has to construct a combination of surfaces connected with imposed continuity conditions. Such a segmentation strategy may be based on a region growing algorithm (Pottman and Wallner (1999)). Starting from a seed region chosen by the user, the data has to be checked for their type and approximated by the corresponding surface type.

9.7 Subdivision surfaces

Until now we have used, in general, free-form surfaces for geometric shape description. For numerical computations on these surfaces the geometry is usually sampled at discrete locations and converted into a piecewise linear approximation, i.e., into a polygonal mesh. Between the continuous representation of the geometric shapes by free-form surfaces and the discrete representation of polygonal meshes, there is a compromise emerging from the theory of subdivision surfaces (Dyn (1991)). These surfaces are defined by a base mesh (triangular, quadrilateral) roughly describing their shape, and a refinement rule that allows one to split the edges and the faces on the mesh in order to obtain a finer and smoother version of the mesh (Kobbelt (1998)).

For the use of subdivision surfaces for reverse engineering the topology of the given cloud of points $\{\mathbf{P}\}_i$ has first to be reconstructed by determining the neighbourhood of the points $\{\mathbf{P}\}_i$. After recovering the topology the patch layout (Kobbelt (1999)) has to be defined. The point-cloud is divided into disjoint sub-clouds with triangular or quadrilateral faces. The corners of these sub-regions uniquely define a polygonal mesh. The point-cloud can be parametrised according to the methods proposed in Sections 9.3 and 9.4.

Depending upon the coarseness of the patch layout, different refining strategies can be used to reconstruct the required surfaces. If the patch layout overly ignores the relevant shape, then the relevant faces have to be subdivided.

The classical subdivision technique uses fixed refinement rules of local support which only depend on the vertices and the edges of the initial coarse mesh. In Kobbelt (1996a), Kobbelt (1996b), Kobbelt (1997), Kobbelt (1998) the ideas of energy-oriented surface fairing (see Section 9.3) and subdivision are combined in order to exploit the flexibility of subdivision schemes and to improve the quality of the required surface. In this approach, for each subdivision step the new vertices are not placed according to some fixed refinement rules but the refinement parameters are solutions of a constrained optimisation problem. If the approximation error of the refined mesh (with respect to the given cloud of points $\{\mathbf{P}\}_i$) is incorporated into a fairness functional (see Section 9.3), modified subdivision schemes are determined which generate smooth approximating surfaces. Figs. 9.13 and 9.14 contain a comparison of the subdivision technique and the approximation by B-spline surfaces using feature lines.

As an example we consider a point-set of a model (see Fig. 9.13 left-hand side) with more than 4000 points. The right-hand side of Fig. 9.13 gives an overall approximation with improvement of the planar areas with the help of feature lines as trimming curves of planar B-spline surfaces (maximal error 0.63%, length of the diagonal of the bounding box 98mm).

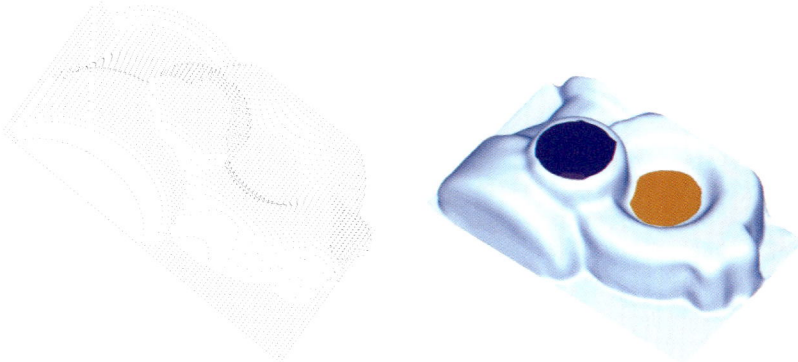

Fig. 9.13. Overall approximation of a cloud of points with local improvements in B-spline representation (20 × 20 bicubic patches)(courtesy of W. Wilke, Daimler–Chrysler)

Fig. 9.14 contains the same surface in a subdivision approximation. In the left-hand side figure one can see a coarse mesh. In the right-hand side a refined mesh. The maximal distance error is 0.36%, the CPU-time 13.5sec on a SGI with R10K processor at 250 MHz.

When comparing the free-form and the subdivision techniques one has to consider the results of both methods. The free-form technique leads to a B-spline surface representation connected with trimming curves. The data volume is very small. In CAD-systems the corresponding evaluation algorithms are available. Unfortunately, for the explicit representation of surfaces by triangular faces all data must be stored (for Fig. 9.14, 127,000 triangles), which leads to a large data volume. If the user considers adaptive

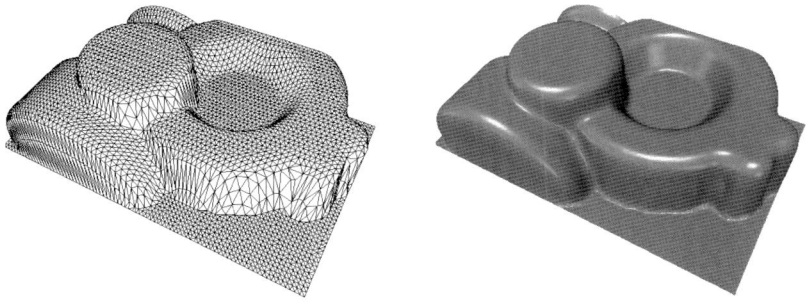

Fig. 9.14. Subdivision approximation of the cloud of points in Fig. 9.13 by a coarse and a refined mesh (courtesy of L. Kobbelt)

refinements the corresponding refinement rules must be available on the CAD-system.

References

Bajaj, Ch.L., Bernardini, F. and Xu, G. (1995). Automatic reconstruction of surfaces and scalar fields from 3D scans. *Computer Graphics (SIGGRAPH '95 Proceedings)*, **29**, 109–118.

Besl, P.J. and Jain, R.C. (1988). Segmentation through variable-order surface fitting. *IEEE PAMI*, **10**, 167–192.

Bloor, M.I.G., Wilson, M.J. and Hagen, H. (1995). The smoothing properties of variational schemes for surface design. *Computer Aided Geometric Design*, **12**, 381–394.

Chang, G.-Z. and Feng, Y.-Y. (1984). An improved condition for the convexity of Bernstein–Bézier surfaces over triangles. *Computer Aided Geometric Design*, **1**, 279–283.

Dietz U. (1996). B-Spline Approximation with Energy Constraints. In *Advanced Course on Fairshape*, ed. J. Hoschek and P. Kaklis, pp. 229–240. Teubner Stuttgart.

Dietz, U. (1998). Fair surface reconstruction from point clouds. In *Mathematical Methods for Curves and Surfaces II*, ed. M. Dæhlen, T. Lyche and L.L. Schumaker, pp. 79–86. Vanderbilt University Press Nashville.

Do Carmo, M.P. (1976). *Differential Geometry of Curves and Surfaces*. Prentice Hall Englewood Cliffs.

Dong, J. and Vijayan, S. (1997). Manufacturing feature determination and extraction — Part I; Part II. *Computer-Aided Design*, **29**, 427–440; 475–484.

Dyn, N. (1991). Subdivision schemes in computer aided geometric design. In *Advances in Numerical Analysis II, Wavelets, Subdivisions and Radial Functions*, ed. W.A. Light, pp. 36–104. Oxford University Press, Oxford.

Eck, M. and Hoppe, H. (1996). Automatic reconstruction of B-spline surface of arbitrary topological type. *Computer Graphics (SIGGRAPH '96 Proceedings)*, **30**, 325–334.

Eck, M., DeRose, T., Duchamp, T., Hoppe, H., Lounsbery, M. and Stuetzle, W. (1995). Multiresolution analysis on arbitrary meshes. *Computer Graphics (SIGGRAPH '95 Proceedings)*, **29**, 173–182.

Edelsbrunner, H. and Mücke, E. (1994). Three-dimensional alpha shapes. *ACM Trans. Graphics*, **13**, 43–72.

Elsässer, B. and Hoschek, J. (1996). Approximation of digitized points by surfaces of revolution. *Comput. & Graphics* **20**, 85–94.

Farin, G. (1996). *Curves and Surfaces for CAGD – A Practical Guide*, 4th ed. Academic Press, New York.

Fletcher, R. (1990). *Practical Methods of Optimisation*. Wiley-Interscience, Chichester.

Greiner, G. (1994). Variational design and fairing of spline surfaces. *Computer Graphics Forum*, **13**, 143–154.

Guo, B. (1997). Surface reconstruction: from points to splines. *Computer-Aided Design*, **29**, 269–277.

Haberäcker, P. (1991). *Digitale Bildverarbeitung*. Hansa Verlag, Hamburg.

Hadenfeld J. (1995). Local energy fairing of B-spline surfaces. In *Mathematical Methods for Curves and Surfaces*, ed. M. Dæhlen, T. Lyche and L.L. Schumaker, pp. 203–212. Vanderbilt University Press, Nashville.

Hagen, H. and Schulze, G. (1987). Automatic smoothing with geometric surface patches. *Computer Aided Geometric Design*, **4**, 231–235.

Hoppe, H., DeRose, T., Duchamp, T., McDonald, J. and Stuetzle, W. (1992). Surface reconstruction from unorganized points. *Computer Graphics (SIGGRAPH '92 Proceedings)*, **26**, 71–78.

Hoschek, J. (1983). Dual Bézier curves and surfaces. In *Surfaces in Computer Aided Geometric Design*, ed. R.E. Barnhill and W. Boehm, pp. 147–156. North Holland, Amsterdam.

Hoschek, J. and Dietz, U. (1996). Smooth B-spline surface approximation to scattered data. In *Reverse Engineering*, ed. J. Hoschek and W. Dankwort, pp. 143–152. Teubner, Stuttgart.

Hoschek, J. Dietz, U. and Wilke, W. (1998). A geometric concept of reverse engineering of shape: Approximation and feature lines. In *Mathematical Methods for Curves and Surfaces II*, ed. M. Dæhlen, T. Lyche and L.L. Schumaker, pp. 253–262. Vanderbilt University Press, Nashville.

Hoschek, J. and Jüttler, B. (1999). Techniques for a fair and shape-preserving surface fitting with tensor-product B-splines. In *Shape Preserving Representations in Computer Aided Design*, ed. J.M. Pena. Nova Science Publishers, New York.

Hoschek, J. and Lasser, D. (1993). *Fundamentals of Computer Aided Geometric Design*. AK Peters, Wellesley.

Hoschek, J. and Schneider, F.-J. (1992). Approximate spline conversion for integral and rational Bézier and B-spline surfaces. In *Geometry Processing for Design and Manufacturing*, ed. R.E. Barnhill, pp. 45–86. SIAM, Philadelphia.

Hoschek, J. and Schneider, M. (1997). Interpolation and approximation with developable surfaces. In *Curves and Surfaces with Applications in CAGD*, ed. A. Le Méhauté, C. Rabut and L.L. Schumaker, pp. 185–203. Vanderbilt University Press, Nashville.

Hoschek, J. and Schwanecke U. (1998). Interpolation and approximation with ruled surfaces. In *The Mathematics of Surfaces VIII*, ed. R. Cripps, pp. 213–232. Information Geometers, Winchester.

Jüttler, B. (1997a). Linear convexity conditions for parametric tensor-product Bézier surface patches. In *The Mathematics of Surfaces VII*, ed. T.N.T. Goodman and R. Martin, pp. 189–208. Information Geometers, Winchester.

Jüttler, B. (1997b). Surface fitting using convex tensor-product splines. *J. Comput. App. Math.*, **84**, 23–44.

Jüttler, B. (1998). Convex surface fitting with tensor-product Bézier surfaces. In *Mathematical Methods for Curves and Surfaces II*, ed. M. Dæhlen, T. Lyche and L.L. Schumaker, pp. 263–270. Vanderbilt University Press, Nashville.

Kass, M., Witkin, A. and Terzopuolos, D. (1987). Snakes: active contour models. *Proceedings of the First International Conference on Computer Vision*, **1**, 259–269.

Kobbelt, L. (1996a). A variational approach to subdivision. *Computer Aided Geometric Design*, **13**, 743–761.

Kobbelt, L. (1996b) Interpolatory subdivision on open quadrilateral nets with arbitrary topology. *Computer Graphics Forum*, **15**, Eurographics '96 Conference Issue, 409–420.

Kobbelt, L. (1997). Discrete fairing. In *The Mathematics of Surfaces VII*, ed. T.N.T Goodman and R. Martin, pp. 101–131. Information Geometers, Winchester.

Kobbelt, L. (1998). Variational design with parametric meshes of arbitrary topology. In *Creating fair and shape preserving curves and surfaces*, ed. H. Nowacki and P. Kaklis, pp. 189–198. Teubner, Stuttgart.

Kobbelt, L. (1999) *Reverse engineering based on subdivision surfaces*. Private communication.

Lang, J. and Röschel, O. (1992). Developable $(1,n)$ Bézier surfaces. *Computer Aided Geometric Design*, **9**, 291–298.

Loop, C. and DeRose, T. (1990). Generalised B-spline surfaces of arbitrary topology. *Computer Graphics (SIGGRAPH '90 Proceedings)* **24**, 347–357.

Lukács, G. and Andor, L. (1998). Computing natural division lines on free-form surfaces based on measured data. In *Mathematical Methods for Curves and Surfaces II*, ed. M. Dæhlen, T. Lyche and L.L. Schumaker, pp. 319–326. Vanderbilt University Press, Nashville.

Lukács, G., Marshall, D. and Martin, R. (1998). Faithful least-squares fitting of spheres, cylinders, cones and tori. In *Computer Vision–ECCV*, ed. H. Burkhardt and B. Neumann, pp. 671–686. Lecture Notes in Computer Science, **1406**, Springer–Verlag, Heidelberg.

Milroy, M.J., Bradley, C. and Vickers, G.W. (1997). Segmentation of a wrap-around model using an active contour. *Computer-Aided Design*, **29**, 299–320.

Moreton, H.P. and Séquin, C.H. (1992). Functional optimisation for fair surface design. *Computer Graphics (SIGGRAPH '92 Proceedings)*, **26**, 167–176.

Peters, J. (1995) Biquartic C^1-surface splines over irregular meshes. *Computer-Aided Design*, **27**, 895–903.

Pottmann, H., Chen, H.-Y. and Lee, I.-K. (1998a). Approximation by profile surfaces. In *The Mathematics of Surfaces VIII*, ed. R. Cripps, pp. 17–36. Information Geometers, Winchester.

Pottmann, H., Peternell, M. and Ravani, B. (1998b). Approximation in line space.

In *Advances in Robot Kinematics: Analysis and Control*, ed. J. Lenarcic and M. Husty, pp. 403–412. Kluwer, Dordrecht.

Pottmann, H. and Randrup, T. (1998). Rotational and helical surface approximation for reverse engineering. *Computing*, **60**, 307–322.

Pottmann, H. and Wallner, J. (1999). Approximation algorithms for developable surfaces. *Computer Aided Geometric Design*. To be published in 1999.

Rando, T. and Roulier, J.A. (1991). Designing faired parametric surfaces. *Computer-Aided Design*, **23**, 492–497.

Randrup, T. (1998). *Reverse engineering methodology to recognize and enhance manufacturability of surfaces — with examples from shipbuilding*. PhD Thesis, Technical University of Denmark, Lyngby.

Sapidis, N.S. and Besl, P.J. (1995). Direct construction of polynomial surfaces from dense range images through region growing. *ACM Trans. Graphics*, **14**, 171–200.

Sarkar, B. and Menq, C.H. (1991). Smooth-surface approximation and reverse engineering. *Computer-Aided Design*, **23**, 623–628.

Schneider, M. (1998). Interpolation with developable strip-surfaces consisting of cylinders and cones. In *Mathematical Methods for Curves and Surfaces II*, ed. M. Dæhlen, T. Lyche and L.L. Schumaker, pp. 437–444. Vanderbilt University Press, Nashville.

Shewchuck, J.R. (1995). Triangle – a two-dimensional quality mesh generator and Delaunay Triangulator. Tech. Report Carnegie Mellon University, Pittsburgh.

Sinha, S.S. and Schunck, B.G. (1992). A two-stage algorithm for discontinuity-preserving surface reconstruction. *IEEE Trans. on Pattern Analysis and Machine Intelligence*, **14**, 36–55.

Speer, T., Kuppe, M., and Hoschek, J. (1998). Global reparametrization of curve approximation. *Computer Aided Geometric Design*, **15**, 869–878.

Vanderbei, R. (1992). LOQO Version 1.08, available via anonymous ftp from `elib.zib-berlin.de` at `/pub/opt-net/software/loqo/1.08`.

Varady, T., Martin, R.R. and Cox, J. (1997). Reverse engineering of geometric models – an introduction. *Computer-Aided Design*, **29**, 255–268.

Welch, W. and Witkin, A. (1992). Variational surface modeling. *Computer Graphics (SIGGRAPH '92 Proceedings)*, **26**, 157–166.

Wilke, W. (2000). *Segmentierung und approximation großer punktwolken*. PhD Thesis, Darmstadt University of Technology.

Willemans, K. and Dierckx, P. (1994). Surface fitting using convex Powell–Sabin splines. *J. Comput. Appl. Math.*, **56**, 263–282.

Ye, X., Jackson, T.R. and Patrikalakis, N.M. (1996). Geometric design of functional surfaces. *Computer-Aided Design*, **28**, 741–752.

Index

adaptive
  methods, 59, 190
  partitions, 192, 193, 210
  splitting, 192
analysis operator, 115–117, 227
angular variation, 258–261
approximation
  N-term, 189, 190, 196, 198, 210
  order, 38, 44, 58, 73, 74, 81, 82, 85, 86, 88–90, 98, 104, 105

B-spline, 85, 153, 157–160, 162, 163, 165, 171–175, 182, 185, 214, 219, 266, 272
  Gram matrices, 159, 185
  surfaces, 262, 280
B-wavelet, 153, 154, 164, 166, 168–171, 174, 182–184
Besov spaces, 195, 196, 212
Bessel
  function, 7–9, 13, 14, 19
  systems, 129–131, 144
bottom-up method, 253
bracket product, 75, 76, 83, 86, 127, 128, 132

Cesàro means, 47
collocation matrix, 31, 158, 159
compact Riemannian manifold, 50, 51
completely monotone, 3, 5, 6
conditionally positive definite, 26, 30, 38, 40, 52
continuous wavelet transform (CWT), 45, 59, 60

data compression, 188, 190, 203, 211
developable surfaces, 276–278
dilation matrix, 87, 88, 103–106
discrete wavelet transform (DWT), 45, 59, 60
dual basis, 117, 120, 121, 123, 135, 143

energy constraints, 262, 272
erosion operator, 260

error
  indicators, 207
  tolerance, 253, 262, 272

fairness functional, 264, 265, 280
fast multipole methods, 33, 36
fiberization, 115, 132, 142, 143
Fourier transform, 5–9, 18, 19, 21, 30, 32, 45, 64, 74, 76, 112, 113, 117, 119, 124, 134, 142, 143, 203
frames, 114, 115, 122, 129, 132, 142, 143, 213, 217–219, 229
free-form surface, 253, 276, 279
free-knot splines, 198
FSI space, 73, 75, 76, 81, 82, 85, 104, 106, 113, 114, 132, 142, 143

Gelfand transform, 124
Gram matrices, 159, 176, 178, 185
Gramian analysis, 143

Hölder spaces, 192
hierarchy, 218, 225, 234, 236, 242
homogeneous harmonic polynomial, 46

interpolation, 25, 39, 44, 45, 49, 52, 53, 56, 133
  scattered data, 53

knot
  insertion, 159, 160, 162, 198, 273
  multiplicity of, 164, 166, 182

Laplace–Beltrami operator, 46, 51, 61, 62
Legendre polynomials, 47, 49, 53
linear independence
  local, 115, 125, 145, 146

Marcinkiewicz–Zygmund inequality, 57, 67, 68
matrix
  collocation, 158, 159
  dilation, 87, 88, 103–106

285

Index

mesh
  refinement, 207
  semi-regular, 230, 231, 234, 242, 248
minimally supported wavelets, 162, 163, 169
minimax rate, 204
multiplicity of knots, 164, 166
multiquadrics 2–5, 7, 13, 14
multiresolution, 74, 85, 103, 155, 214, 215, 217, 248, 249

N-term approximation, 189, 190, 196, 198, 210
native spaces 28, 38
normal vector field, 257, 267

oracle estimator, 205, 206
orthogonal projection, 47, 48, 82, 83, 128, 136, 169, 180, 191, 209, 268
Oslo algorithm, 172

parameterizations, 213, 216, 218–220, 231–234, 237, 238, 240, 242–244, 247, 263, 266–268, 272
positive definite function, 1–3, 26, 38, 48, 49
PSI space, 73, 75, 80, 81, 86, 113–115, 122, 125, 129, 132, 138

quadrature, 44, 63, 66
quadrilateral network, 262
quasi-interpolation, 45
Quillen–Suslin Theorem, 140

radial basis function (RBF), 1–3, 6, 9, 13, 19, 22, 25–32, 35–37, 39, 56, 62
refinable function, 86, 87, 103, 115, 145, 148
refinement
  equation 87–89, 92, 93, 103, 104, 174
    mask symbol, 88, 94
    space, 207
  strategies, 190, 207
remeshing, 216, 232, 234, 242–244, 248
Riemannian manifold
  compact, 50, 51
Riesz basis, 77, 79, 155, 156, 162, 174, 194, 200
ruled surface, 276–278

scattered data interpolation, 1, 53
Schwartz space, 9
segmentation, 253, 254, 259
semi-regular meshes, 230, 231, 234, 242, 248
shift-invariant space (SI), 73–75, 85, 86, 112–114, 132, 136, 137, 141–143, 145
  band limited, 113
  finitely generated, 73, 75, 107, 113, 138

local, 112, 138, 142
principal, 73, 75, 112, 138
Sobolev spaces, 26, 29, 50–52, 63, 74, 195, 204
space refinement, 207
spectral analysis, 117
spectral synthesis, 117
spherical basis functions (SBF), 45, 50, 56, 60
spherical harmonics, 38, 45–48, 52–55, 60, 63–66
spline
  free-knot, 198
  surface, 6, 274, 275
  thin-plate, 6, 25–27, 29, 33, 35
  wavelet, 152–154, 159, 162, 171, 174, 178, 182
stability, 9, 38, 83, 87, 90, 106, 122, 127, 129, 132, 134, 174–176
statistical estimation, 188, 190, 211
Strang–Fix conditions, 87, 89–91, 93
subdivision, 104, 207, 215–217, 219, 221, 224, 225, 227, 232, 233, 245, 248, 279, 280
superfunction, 73, 82, 93, 94
surface
  B-spline, 262, 280
  developable, 276–278
  free-form, 253, 276, 279
  ruled, 276–278
synthesis operator, 114, 115, 119, 122–125, 129

tensor product spline wavelets, 176
thin-plate energy, 266, 268
thin-plate spline, 1, 6
top-down method, 253
transfer operator, 73, 88, 96, 98, 105, 106
tree structure, 198, 202
triangulation, 178, 179, 181, 254, 255, 257, 272, 273, 278
trimming curve, 220, 262, 269, 270, 280

uncertainty principle, 45, 60–62

wavelet, 45, 59, 60, 74, 104, 112, 144, 152–155, 157, 163, 164, 171, 177, 178, 180, 181, 194, 195, 198, 208–211, 216, 218–220, 248
  continuous transform (CWT), 45, 59, 60
  discrete transform (DWT), 45, 59, 60
  minimally supported, 162, 163, 169
  spline, 152–154, 159, 162, 171, 174, 178, 182
  thresholding, 193, 204, 211, 234, 245
  transform, 59, 154, 156, 170, 171, 228, 245
Weyl–Heisenberg system, 114, 143

zonal functions, 59